Fall '87

3/84

Darren Payne

SE2 75 3677

Technical
Writing

Technical Writing

FIFTH EDITION

Gordon H. Mills

John A. Walter
The University of Texas at Austin

With the assistance of Vody Mills
and Marion K. Smith, Brigham Young University

Holt Rinehart and Winston
New York Chicago San Francisco Philadelphia Montreal Toronto London
Sydney Tokyo Mexico City Rio de Janeiro Madrid

Acquisitions Editor	Charlyce Jones Owen
Special Projects Editor	Jeanette Ninas Johnson
Production Manager	Robin B. Besofsky
Design Supervisor	Gloria Gentile
Interior Design	Caliber Design Planning, Inc.
Cover Photo	Marc David Cohen

Library of Congress Cataloging-in-Publication Data
Mills, Gordon H., 1914–
  Technical writing.
  Bibliography: p.
  Includes index.
  1. Technical writing.   I. Walter, John A., 1914–
II. Title.
T11.M53   1986   808'.066'6021   85-24914
ISBN 0-03-062019-8

CBS COLLEGE PUBLISHING
Holt, Rinehart and Winston
The Dryden Press
Saunders College Publishing

Preface to the Fifth Edition

In the preface to the fourth edition we wrote, ''It is both a pleasure and an obligation to begin by remarking that in its successive editions the content of this book has more and more come to represent the experience, research, and generous counsel of many people. Although quite unanticipated, this support for our own efforts is something we have valued deeply. We are grateful for an opportunity to express publicly our awareness of the considerable degree to which, over the years, the character of the book has been formed by the cumulative advice and assistance provided by our co-workers. We can make such a statement about the sources of the content of the book without pretending to derive from these sources any claims concerning the quality of the book. Because all the help offered to the book had to be funneled through us, the book's faults — whatever they may be — are our faults.''

The words just quoted are as relevant to the present edition as they were to the last. Before undertaking this fifth edition, we had the benefit of a dozen quite detailed critiques of the fourth, all from experienced teachers of technical writing who had used the book in their classes. Where there has been general agreement about changes that should be made, we have tried to carry out the suggestions of these teachers. Additional changes have been made as a result of our own critical examination of the text in the light of our own teaching experience.

The changes we have made, then, reflect the following aims: to improve the text stylistically wherever possible; to expand discussion of some topics so as to provide a more complete treatment of them; to delete some portions of the text which have ceased to be useful; to make some organizational changes for a more logical sequence; and to update illustrative materials.

In attempting to improve the book stylistically, we were reminded of the words of Jacques Barzun and Henry Graff in *The Modern Researcher:* "To the general public 'revise' is a noble word and 'tinker' is a trivial one, but to the writer the difference between them is only the difference between the details of the hard work and the effect it achieves. The successful revision of a book in manuscript is made up of an appalling number of small, local alterations. Rewriting is nothing but able tinkering." We have tinkered with the text throughout in an effort to make it clearer, simpler, and more concise. We hope we have succeeded.

We have added a discussion of the writing process to the first chapter, extended our discussion of reader adaptation, often called "reader analysis," and added a discussion of making familiar comparisons, both in the chapter on style. A note on giving instructions or directions has been added to the discussion of describing a process. There is a new appendix on the use of library materials in relation to the writing of a library research report. Other additions have been designed to clarify the text. Finally, the selective bibliography which appears as Appendix A has been updated.

Apart from tightening up the style throughout in the interest of conciseness, the principal deletions have been the appendices on organizing the research report and on the Galt Manuscripts. It appears that most teachers of technical writing have not found time for much use of these materials.

Organizational changes include altering the order of the chapters in Section Three, placing the chapter on introductions first since that seems more logical, putting the chapter on proposals before the one on the progress report in Section Four, and combining the chapters on the forms of report organization and on the format of reports for the same reason. We continue to believe that the organization of the text is basically sound as it stands, although individual teachers may, of course, alter the sequence of assignments to suit the particular needs of their students and of the course they are offering.

In selecting new illustrative material for this edition, we have been guided by two considerations: first, the new material should be of contemporary interest to students in a wide range of disciplines, and, second, new selections should suitably exemplify the type of expository writing discussed. Most of the new illustrative material appears in Section Two, dealing with special techniques of technical writing. Those selections from earlier editions which serve a useful illustrative purpose and which have not become "dated" have been retained.

There remains the pleasant task of expressing our gratitude to those people who have been kind enough to assist us in the preparation of this edition. Two individuals in particular deserve special thanks: Marion Smith of Brigham Young University and Vody Mills.

Professor Smith has assisted throughout the preparation of this edition, and his assistance has been both stylistic and substantive. His stylistic suggestions are too numerous to mention here, but his substantive contributions deserve special mention. He has, for example, made useful contributions to the chapter on style in the discussion of reader adaptation and the making of familiar comparisons, to the chapter on describing a process in the section on giving directions, in the chapter on proposals in the discussion of small-scale proposals, to many of the chapters through supplying new illustrative material, and to the bibliography by suggesting the addition of new titles. He also made helpful recommendations about the organization of the book, in particular the combining of the chapters on the forms of report organization and on the format of reports. For all of his assistance, he deserves high praise.

Vody Mills has worked diligently and devotedly from the very beginning in the preparation of this edition. She carefully collated and evaluated the suggestions made by reviewers, meticulously went through the text searching for (and finding) inconsistencies of style, redundancies, and typographical errors, edited and correlated the efforts of Professor Smith with those made here, handled necessary correspondence, and prepared the final manuscript for submission to the publishers. Her help has been essential in making this edition possible.

Many additional obligations are expressed in the text, but we want to acknowledge our special gratitude to Paul Anderson of Miami University, Herman Estrin of the New Jersey Institute of Technology, John Harris of Brigham Young University, Donald Cunningham of Texas Tech University, Maxine Hairston of the University of Texas at Austin, Stanley Higgins of Westinghouse, Frank Smith of McDonnell Douglas Corporation, and our friends and associates in the Association of Teachers of Technical Writing and the Society for Technical Communication. We would particularly like to thank Kay Nichols of the University of Texas library staff for her help in revising the chapter on finding published information. And, of course, we remain grateful to those who have helped us with earlier editions.

We also wish to express our thanks to the editors we have worked with at Holt, Rinehart and Winston: Susan Katz, Nedah Abbott, Charlyce Jones Owen, and Jeanette Ninas Johnson, as well as to colleagues who provided us with detailed and constructive reviews of the fourth edition and the fifth edition manuscript: Mary M. Engesser, Oregon State University; Herman A. Estrin, New Jersey Institute of Technology; Ruth E. Falor, The Ohio State University; David F. Goslee, University of Tennessee — Knoxville; Mitchell H. Jarosz, Delta College; John W. Johnston, Tyler Junior College; Michael Keene, University of Tennessee — Knoxville; Rodney D. Keller, Ricks College; Judith Killen, University of Louisville; Katherine E. Staples, Austin Community College; Thomas Tryzna, Seattle Pacific University; and M. Anne Yeoman, Virginia Polytechnic Institute.

Finally, the pronoun "we" is used in this preface even though Gordon Mills suffered a fatal heart attack after the publication of the fourth edition. The use of the plural pronoun is appropriate here, and in the text itself, since almost all of the book remains the result of our long collaboration. The present book reflects a number of discussions after the appearance of the fourth edition on plans for any future edition. Gordon's thoughts and ideas have been borne in mind throughout the preparation of this present edition.

John A. Walter

Austin, Texas
November 1985

Preface to
the First Edition

This book had its inception in our need for a logical bridge between the professional writing of scientists and engineers and the content of a course for students of technical writing. Certain widespread practices had developed in such courses, as we knew both from personal experience and from such published studies as A. M. Fountain's *A Study of Courses in Technical Writing* (1938), the American Society for Engineering Education's report on *Instruction in English in Engineering Colleges* (1940), and M. L. Rider's *Journal of Engineering Education* article, "Some Practices in Teaching Advanced Composition for Engineers" (1950). We felt that many of these practices were unquestionably proving their value, but about others we weren't sure, and there seemed to be no clearly established basis upon which to decide about them. The difficulty was partly that the limits of the subject were uncertain; apparently nobody had ever seriously explored the concept of technical writing with the purpose of trying to say precisely what technical writing is. There were, of course, numerous systems of classification of articles and reports; but, unfortunately, these systems were dissimilar at many points and were often more puzzling than helpful in relation to our question of what materials and instruction were most needed by our students.

In an effort to find practical solutions to the problems just noted, as well as to others not mentioned, we undertook three investigations. We began by seeking examples of reports and articles, and expressions of opinion about important problems; altogether we incurred an indebtedness to over three hundred industrial and research organizations in making our survey. We also worked out, in writing, a theory of what technical writing is (later published as Circular No. 22 of The University of Texas, Bureau of Engineering Research, under the title, *The Theory of Technical Writing*). Thirdly, we studied the content and organization of college courses in the subject. The content of this book rests primarily upon these investigations, together with numerous other studies of a more limited scope. Perhaps it is proper to say here that these investigations did not constitute our introduction to the subject, since we had both had considerable experience in the field, in the capacity of teachers and editors. On the other hand, we did try hard to avoid letting the particularities of our personal experience affect the conclusions we drew from these systematic studies. We realize, of course, that the nature of our own experience, both academic and nonacademic, has no doubt been reflected in our text; and if in spite of the good counsel and abundant materials furnished to us we have fallen into error, the fault is entirely our own. We do believe, however, that our methods have been sound; we hope that our book is sound too.

Perhaps we should add, about ourselves, that our collaboration has extended to all parts of the text. Almost every page of it represents a joint effort.

A few comments on the text itself need to be made here. As we said, the organization of the book was determined by a study of the needs and practices of courses in technical writing, as well as by the internal logic of the subject matter. One problem, however, resisted solution: we could not find any clear grounds on which to decide when to introduce certain elements of our subject that would not themselves usually be the basis of writing assignments. Section Three (Transitions, Introductions, and Conclusions) and Section Five (Report Layout) are chiefly involved, although the same difficulty exists with Chapter 3 (Style in Technical Writing). We have no pat answers as to how these elements should be introduced into a course. On the contrary, we believe that a suitable decision can be made only by the instructor.

We should also like to remark that we are aware we have sometimes been blunt in criticizing quoted materials. We hope all readers of the book will understand that these materials were not prepared especially for our use. They are, instead, routine products, and many of them were doubtless written under great pressure. We have been critical in order to help students learn, not because of any fancied superiority to the writers whose work we criticize.

We regret that a complete list of those organizations and persons who have helped us is too long to present here. We are deeply grateful to all of them, and we have acknowledged our specific indebtedness to many in the text. A few have requested anonymity. Our greatest single debt is to John Galt, Manager, Phenolic Products Plant, Chemical Materials Department, General Electric Company, Pittsfield, Massachusetts. Mr. Galt permitted us to quote the extremely interesting manuscripts in Appendix B. We should like to mention also The Civil Aeronautics Administration, Technical Development and Evaluation Center; and the Research Laboratories Division, General Motors Corporation. Dr. W. E. Kuhn, Manager of the Technical and Research Division, The Texas Company, deserves special thanks for repeated favors.

Austin, Texas　　　　　　　　　　　　　　　　　　　　　　　　　　　　G.M.
January 1954　　　　　　　　　　　　　　　　　　　　　　　　　　　　J.W

Contents

SECTION I   Preliminary Considerations   1

APPENDIXES

Technical Writing

SECTION I

Preliminary Considerations

Some topics are more obviously preliminary to a study of technical writing than others. Most obvious of all is the question of what the term "technical writing" means. In Chapter 1, therefore, technical writing is defined, its major subdivisions are designated, and some evidence is examined as to its role in science and engineering.

Closely related to the need for defining and subdividing the subject of technical writing is that of identifying what might be called its basic concepts. That is, can the practice of technical writing be reduced to a few general propositions? Probably not; on the other hand, five concepts about technical writing are so fundamental that recognition of them does help give perspective and meaning to later study of the many different aspects of the subject. These five fundamental concepts are presented in Chapter 2, which is limited to one short page of text.

The third chapter in this section is concerned with style. Of course the style of any piece of writing is an integral part of that writing and should not be thought of as preliminary to it. Nevertheless, style also reflects responses to a broad array of questions that present themselves the instant the writer picks up a pen. One of these questions, for example, is whether to seek a style permitting what is usually called self-expression, as opposed to concentrating attention on the needs of the reader. The importance in technical writing of a number of such issues draws the subject of style into the area of preliminary consideration. Style is also given repeated attention in subsequent sections of this book, as occasion requires.

The fourth and last chapter in Section I is a review of the logic of organization, in the form of outlining and abstracting. This subject of the elements of organization is preliminary to a study of technical writing only in the practical sense that it is common to almost any study of writing and that in some degree it will already be familiar to any reader of this book.

1

Introduction

The purpose of this book is to discuss the principles and practice of the kind of writing required of engineers and scientists as part of their professional work. The reader to whom the book is directed is primarily the technical student who has had enough training in the fundamentals of composition to be ready for consideration of some of the more specialized problems of technical writing.

In this chapter we will first explain what technical writing is, and then go on to discuss the importance of writing as compared with other elements of technical work, what kinds of writing engineers and scientists are expected to do, what aspects of writing they particularly need to study, and the writing process. At the end of the chapter will be found some related comments by experienced scientists and engineers.

What Technical Writing Is

Although one of the obvious characteristics of technical writing is its technical subject matter, it would be very difficult to say precisely what a technical subject is. For our purposes, however, it will be sufficient to say merely that a technical subject is one that falls within the general field of science and engineering. Subjects in this general field are usually recognized by a strong emphasis on quantitative measurement, such as cubic yards, metric tons, light years, foot/pounds, nanoseconds.

Technical writing has other characteristics besides its subject matter, of course. One of these characteristics is its "formal" aspect — a term hard to define but easy to illustrate. There are, for example, certain forms of reports, like progress reports, that are used in technical writing. There are also certain forms of style and diction used, and certain forms of graphic aids such as sketches, graphs, and flowsheets.

Another characteristic of technical writing is its scientific point of view. Ideally, technical writing is impartial and objective, clear and accurate in the presentation of facts, concise and unemotional. In practice, naturally, some of these qualities are often lacking, particularly clarity and conciseness. An additional fact about point of view is that technical writing is usually designed for a specific reader or group of readers, perhaps the staff of a certain research group, rather than for a great mass of readers, as is newspaper writing.

A fourth characteristic of technical writing relates to the way information is extracted by the reader. The reader of a novel, a short story, a poem, a biography, or, say, a foreign policy statement normally begins with the first line of the document and reads continuously through to the last line. Thus, literary works, biographies, news reports, and the like, are read sequentially, and the writer assumes that at any given point the reader has read all the preceding material. No such assumption can be made, however, for encyclopedias, handbooks, parts catalogues, specifications, or instruction manuals. Automotive mechanics, technicians, and quality control specialists, for instance, often flip to specific pages for very small pieces of information. Such people want to find information quickly and usually have neither the time nor the interest required for reading everything in a document. In the light of such use, large quantities of technical writing are prepared to allow random access to contents. Liberal use is made of headings and subheadings, tables of contents, chapter outlines, tables and graphs, color coding of sections, and such devices as finger tabs. Coordinate and subordinate blocks of information may be read as self-contained units.

The last major characteristic can be called the special techniques of technical writing. What this means can easily be explained by an analogy. A person who decides to learn how to write fiction usually finds it necessary to study the writing of dialogue. No one questions the logic of this. But the writer of fiction is not the only one who uses dialogue: probably we all occasionally write down some conversation, in a letter or elsewhere. Nevertheless, the writer of fiction uses dialogue more often than most people, and it is more important in fiction than in most other types of writing. Similarly, there are certain techniques that the technical writer uses particularly often. These techniques also appear in other kinds of writing, but not so frequently, and not often as important parts of the whole. Consequently, the technical writer should learn everything possible about them. The most important are definitions, descriptions of mechanisms, descriptions of processes, classification, and interpretation. Each one of these writing problems is complex enough to need careful attention, and each one of them appears frequently in technical writing.

It should be clearly understood that these special techniques are not types of technical reports. Several of them may appear in a single report; but for an entire report to be nothing more than, say, the description of a mechanism would be unusual. Again, it is like dialogue in a short story, which may take an important part, but is seldom the whole story.

In summary, then, we can define technical writing as follows:

1. Technical writing is exposition about scientific subjects and about various technical subjects associated with the sciences.[1]
2. Technical writing is characterized by certain formal elements, such as its scientific and technical vocabulary, its use of graphic aids, and its use of conventional report forms.
3. Technical writing is ideally characterized by the maintenance of an attitude of impartiality and objectivity, by extreme care to convey information accurately and concisely, and by the absence of any attempt to arouse emotion.
4. Much technical writing is often read in non-linear sequence and is thus designed for random access to the information presented.
5. Technical writing is writing in which there is a relatively high concentration of certain complex and important writing techniques — in particular definition, description of mechanisms, description of a process, classification, and interpretation.

What People in the Technical Professions Are Required to Write

People in the technical professions are called upon for a considerable variety of writing: reports of many kinds, memoranda, technical notebooks, proposals, professional papers and magazine articles, patent disclosures, letters, promotional brochures, specifications, technical bulletins, instruction manuals, handbooks, and sometimes even books. On the other hand, as you would expect, recent college graduates are usually asked to do only rather routine writing, not greatly different from reports written in school. What importance, then, does writing have in the work of someone just starting out in a profession?

Let's consider a reasonably representative situation. Suppose a beginning engineer who has been assigned to check some aspect of the quality control in the production of a certain type of computer discovers a problem that needs to be called to the attention of the assistant manager of the plant. A report must be written. Who will write it — the novice engineer, or the immediate production supervisor? Obviously, it would be to the engineer's advantage to be able to write the report, and to write it well. What is fairly likely to happen in such circumstances is that the beginner is asked to write the report. If it is good, it is forwarded; if it is not good, the supervisor rewrites it and sends it on over his or her own name.

As beginners in the technical professions acquire experience, more and more often they may be asked to make recommendations about decisions, or to make the decisions themselves. Customers of the company write letters asking for advice

---

[1] But see John A. Walter's "Technical Writing: Species or Genus?" in *Technical Communication, 24* (Second Quarter 1977), 6–9, in which it is argued that the subject of technical writing may embrace almost any field of knowledge when the objective of the writing is to convey information in an effective way.

about their technical problems, and these letters must be answered. Letters and reports must be written to other people within the organization, and both informal and formal oral discussions of joint problems must be conducted. In one small company we know, each member of the engineering staff is required once a year to make an oral report concerning his or her year's work to the assembled officials of the company. Other common writing assignments are the submittal of progress reports at regular intervals and long reports at important stages of a project. People who are ambitious to establish a reputation often submit articles to professional journals — and many companies encourage their staff members to do so. Of course circumstances differ, but among all these kinds of writing the demand for decisions about action to be taken may well prove most significant, as suggested in the statement from the Ethyl Corporation on p. 11.

In any event, the bulk of the writing you will be asked to do will probably be in the form of reports and other routine documents such as memoranda. What is a report? There is little point in attempting an exact definition. Perhaps as good a definition as any is that a report is a piece of technical writing designed to meet a specific need. In the introduction to Section IV you will find a list of 30 "types" of reports. Many of these types differ from each other only in minor details, however, and in some cases probably in none at all. What happens is that a group of people decide they need to have information about certain types of projects written up in a certain form, and perhaps at certain stages of progress. They make up some rules and give this "type" of report a name — perhaps "preliminary," or "partial," or "shop," or "test." That is exactly what they should do. If the form of report they devise serves their purposes, no one can ask for more.

There are, nevertheless, a few types of reports that are well standardized. Three that deserve mention are the progress report, the recommendation report, and the proposal. These three will be discussed in detail in Section IV along with several other reports distinguished primarily by form.

Basic Aspects of Technical Writing

In the most elementary terms, technical writing can be broken down into two parts, or aspects: (1) the "end products" (like reports and letters), the concrete "package" that you deliver; (2) the skills that enter into the preparation of the end product. This distinction is useful in pointing out specific aspects of writing that are of particular importance and that we will accordingly be concerned with in this book.

The important "end products" of technical writing are these:

1. Business letters
2. Various kinds of reports — both written and oral
3. Articles for technical journals — and possibly books
4. Abstracts
5. Graphic aids
6. Instruction manuals
7. Handbooks

8. Brochures
9. Proposals
10. Memoranda
11. Specifications

It is quite possible that you may never be interested in writing for technical journals, but the other items in the list above are all routine work. Oral reports seem less tangible than the others, perhaps more like a skill than a "product," but the spoken word has as real an existence as the written. And we should add that "oral reports" refers not only to formal speechmaking, but also to informal discussions of technical problems. The heading "graphic aids," by which we mean graphs, drawings, and other nonverbal supplements, also looks a little odd in this list. Are graphic aids a skill, an end product, or neither? Whatever they are, it is good to know about them. It doesn't particularly matter what we call them.

The skills that deserve particular attention are the following:

1. Special techniques of technical writing
2. Style
3. Introductions, transitions, and conclusions
4. Outlines (or organization)
5. The layout, or format, of reports

The special techniques of technical writing have already been commented on. The other items in the list need brief explanation.

The word "style" usually suggests an aesthetic quality of prose, a quality determined by the relative smoothness or awkwardness with which sentences are put together. Many eminent scientists and engineers have developed a splendid prose style. Naturally, we'd like to encourage you to develop a good style; but above all else we will emphasize clarity. Since technical writing is by definition a method of communicating facts, it is absolutely imperative that it be clear. At the same time, the nature and complexity of the subject matter of technical writing often involve the writer in particularly difficult stylistic problems. These problems will be discussed in Chapter 3. One other important aspect of style in technical writing is point of view. In brief, the point of view should be scientific: objective, impartial, and unemotional.

The third item is introductions, transitions, and conclusions. The technique here is to learn to tell your readers what you're going to tell them, then to tell them, and then to tell them what you've told them. This skill is one of the most important a technical writer, or any writer, can possess.

The fourth element is outlines. A more accurate phrase might be "the theory of organizing your writing" because that is what we are really interested in.

And the fifth and last element, the layout of reports, has to do with such matters as margins, spacing, subheads, the title page, and the like.

Doubtless you have noticed that grammar and punctuation were not included in either of the two lists above. The reason for the omission of these fundamentals is that they are not properly part of the formal subject of technical writing. Constant attention to these fundamentals is nevertheless a necessity. We suggest that you get

ot already have one, and — if you have not already done so — develop the habit of using it. Professional writers do no less.

Altogether, the topics which have been listed are those that are most important to the beginner in technical writing. It should be understood, however, that they are not the only aspects of technical writing that deserve attention. Others, for instance the handling of footnotes and bibliography, and the use of the library, will be discussed in the appropriate place.

Our purpose has been to state in the simplest terms what the study of technical writing involves. Setting up a practical course of study naturally requires rearrangement and regrouping of the topics listed. Reference to the table of contents will indicate how that has been done in this text.

The Writing Process

In the foregoing discussion of the basic aspects of technical writing, we listed a number of the "end products" you will likely have to produce in your professional life. Now we would like to emphasize that whatever label may be attached to the end result of your writing, whether it be a letter, a report, a proposal, an article, or some other product, you will find yourself involved in a writing process made up of the following steps or stages: (1) accepting the writing task or recognizing the need to write, (2) preparing to write, (3) writing a rough draft or drafts, and (4) polishing and revising your final draft. Let's consider each of these steps or stages briefly.

In your school work, the first step of the writing process has probably been, simply, accepting a teacher's assignment to write on a given topic or on one of your own choice. In the world of work, this first step is usually taken as a result of the work you have been doing. You may accept the fact, for instance, that you need to report on the progress you have made on a project, or if the project has been completed you recognize that you have to prepare a final report to be submitted to your superiors or to your client. Or you may realize that you need to write a letter or memorandum to report information or to request information. Whatever the end product of your work-related writing may be, this first step of the writing process leaves you little choice: you write because you must. An exception, of course, is writing you may choose to do because of your desire to enhance your professional status or because you wish to share information with others in your field. Such writing includes articles for professional journals and books.

The second step or stage, preparing to write, involves a number of substeps: review and study of the information accumulated and, perhaps, deciding what additional work or research needs to be done; identifying the purpose your writing will serve, particularly with respect to your readers; careful analysis of your readers — their needs and desires, their knowledge of the subject matter to be communicated, their probable attitudes toward the conclusions and recommendations you may wish to make, and so on; and, finally, deciding what must be included in and excluded from your presentation and deciding, at least tentatively, on the most effective order of presentation. This last substep, of course, results in your organization of your material, perhaps in the form of an outline.

With your information in mind or at hand, with an understanding of your purpose and knowledge of your readers, and with an organizational plan, you are now ready to undertake the third step in the process, the writing of an initial draft. Most writers find that it is best at this stage to concentrate on getting down on paper all the information they should communicate, without worrying very much about details of grammar, usage, and style. After you have completed an initial draft, you will probably find it necessary to write an additional draft, or drafts, during the writing of which you may need to clarify ideas, delete irrelevant material and repetitions, discover better word choices, and so on. In short, in preparing your final draft, you want to ensure that you have communicated the information you wish to convey and that you have done so in a manner that will make it clear and interesting to your readers. You want to be able to say to yourself, ''This is what I want and need to say, and this is the way I want to say it.''

The final step or stage of the process is polishing and revising your manuscript, preferably after you have put it aside for a time so you can approach the work with a fresh and critical eye. This last stage involves a number of things: checking for errors in grammar, spelling, punctuation, and usage; making sure that the words you have used will be clear to your readers (and defining those you think may be unclear); making sure that any illustrations you have used serve their purpose and are properly placed; and checking the physical appearance of your presentation — the major headings, the subheadings, the margins, captions for figures — making certain, in short, that the layout and format are exactly the way you want them to be.

Two final observations need to be made about the writing process. First, the process is not necessarily linear or strictly time sequential. You may find, for example, that in the process of producing a first draft that you want to make an organizational change; you may run across additional information you want to include; you may want to rework your introduction so that it will do a better job of introducing your subject and preparing your readers for your main presentation. You may, in short, find yourself doing a good deal of backtracking or leapfrogging forward to work on another part of your writing. The second observation we want to make is this: in many if not most of the remaining portions of this book you will find the writing process dealt with, either explicitly or implicitly.

The Place of Writing in Technical Work

Finally, you have probably been wondering what place technical writing might take in the special area you plan to work in.

The particular circumstances of your own job will determine how much writing you will have to do, of course, as well as how much importance writing will have in your career, but there are some facts available about what you are likely to encounter. For example, one large corporation has prepared the following account of how its junior technical employees spend their time:

Type of Work	Percentage of Time
1. Collection and correlation of data	26
2. Calculations	34

3. Writing reports and letters   20
4. Selling results of their work   12
5. Other (literature reviews, attend-
   ance at meetings, consulting with
   others)   8

According to these figures, college graduates entering this particular corporation can, on the average, expect to spend a fifth of their working time in writing, or the equivalent of at least one whole day each week.[2] As a matter of fact, if we include the spoken word as well as the written word, they will evidently spend a great deal more than a fifth of their time communicating ideas in words, since items 4 and 5 clearly require talking.

These figures represent only the averages reported by one corporation. Let us approach the problem in another way. Suppose we ask whether the ability to write well is likely to play an important part in your professional career.

A significant answer to this question can be found in the results of a study carried out by Paul V. Anderson entitled *Research into the Amount, Importance, and Kinds of Writing Performed on the Job by Graduates of Seven University Departments That Send Students to Technical Writing Courses.* With a 37 percent response to the 2,335 questionnaires Anderson sent to Miami University (Oxford, Ohio) graduates of departments of chemistry, home economics, manufacturing engineering, office administration, paper science and engineering, systems analysis, and zoology, he found that 93 percent reported that the ability to write well was of at least "some importance" in the performance of their jobs, 67 percent that writing was of "great importance," and 16 percent that it was of "critical importance." Anderson's study is reproduced at the end of the chapter on interpretation, pp. 190–212.

Another survey, conducted by Exxon Research and Engineering Company, revealed that their junior technical employees believed English to be the most important of 15 nontechnical subjects they took in college; they placed public speaking and report writing second and third, respectively.

We hope these few figures will encourage you to think about the part that writing is likely to play in your own future work. The best thing you can do will be to take every opportunity to ask questions on this subject when talking with people who are actively engaged in your professional field. The illustrative material with which this chapter is concluded will provide a further insight.

Illustrative Material

Many examples of technical writing are presented in this book, both within the text of the discussion and in special sections at the end of chapters. These special sections

---

[2] See also Lester Faigley and Thomas P. Miller's article "What We Learn from Writing on the Job" in *College English, 44* (October 1982), 557–569. From interviews of 200 college-trained people in a wide variety of occupations (professional and technical, management and administration, sales, clerical, crafts, and blue-collar and service work) in agriculture, mining and construction, manufacturing, transportation, wholesale and retail trades, finance and insurance, service, and government, they found that these people spent an average of 23.1 percent of their time writing, or more than one day of a five-day week.

are labeled, as this one is, "Illustrative Material." Some of these examples were written by students; more of them were written by scientists and engineers as part of their regular work. Examination of the latter will help you acquire an understanding of what you might expect in your own career.

For the pages that close this introductory chapter, we have chosen a few items which give direct expression to attitudes toward writing held by people in science and industry. As you read these materials, we suggest you ponder a sentence from a pamphlet entitled *Stromberg-Carlson "Accucode" Supervisory Control Systems:*[3]

> Stromberg-Carlson is staffed to undertake the "end-to-end" (engineer, furnish and install) system responsibilities to solve any telecommunication or supervisory control problem.

The emphasis placed on "staff" in this statement underscores the fact that any technical group is a complex team. And, of course, the effectiveness of the team is to a considerable degree dependent upon the ability of its members to communicate clearly with one another, and with "outsiders," concerning the technical problems the team undertakes to solve. Sometimes it is a temptation to decide that this necessary communication can be accomplished entirely through the language of mathematics, without having to depend on words. While contemplating this idea, however, we remembered a highly amusing essay entitled "Mathmanship," which is famous in the world of science and engineering. It is the last of the four items reprinted below.

The first item, "Training the Professional-Technical Employee," was presented as a paper at a meeting of a division of the American Petroleum Institute, and subsequently printed and circulated by the Ethyl Corporation. Only a small portion is presented here.

Training the Professional-Technical Employee[4]
T. C. Carron

CLEAR-WRITING PROGRAM

A research laboratory is unique in that its sole output is ideas. And its only tangible products are the reports and papers that describe these ideas. This photograph (Figure 1) will give you some idea of the magnitude of a research laboratory's writing.

The stack at the left is one year of progress reports from just one of the five divisions at our Detroit laboratories. The middle stack is one month's output of similar reports from our two most prolific divisions. And for comparison of sheer amounts of reading matter, the Manhattan telephone directory is shown on the right.

Now some of this writing is just for the record. But most of it is written with action in mind. The author wants a research project continued or expanded; a new product marketed; a new plant built or modified; and so on. A lot of key people must read these reports not only to keep up to date, but in order to make decisions. Not surprisingly, we found that poor organization and complex writing were greatly magnifying the time required to read these reports. Therefore, we decided to try to teach our scientists and engineers to write more clearly and concisely.

[3] Quoted by permission of the Stromberg-Carlson Corporation.
[4] T. C. Carron, *Training the Professional-Technical Employee* (Ferndale, Michigan: Research and Development Department, Ethyl Corporation, no date), pp. 1–3. Quoted by permission.

Figure 1   Magnitude of written material.

Of course writing is especially important in a research group such as the Ethyl Corporation Research Laboratories, which issued the material just quoted. In fact, it is common for people in research to remark that the end product of their work is a report. For example, L. R. Buzan, of General Motors Research Laboratories in Warren, Michigan, said simply that ". . . we consider new information our primary product." What about other areas of science and engineering, then?

Well, as we said earlier, it depends on all the circumstances, but here is some relevant evidence. First, let's consider the situation of a student who is preparing for entrance into a medical school. Concerning the place of the study of writing in such a program, Dr. Bryan Williams, a member of the faculty and Associate Dean for Student Affairs of the University of Texas Southwestern Medical School in Dallas, made the following comment:

> My standard advice to all premed clubs, individual students, premed counselors, anybody-who-will-listen is to learn to write coherently before coming to medical school. Instead of being advised to take more molecular biology and biochemistry, all students are advised to become more proficient with the pen.

Evidence about a second area is found in a study made by Westinghouse Electric Corporation. Seventy Westinghouse managers, including all managerial levels within the corporation, were interviewed on the subject of what they most needed and desired in reports. The following is a brief summary of what these managers, representing a cross section of the activities in this large corporation, had to say concerning the best general approach to the whole problem of report writing. This summary is reprinted from the pamphlet *What to Report* (1962), long in use at Westinghouse.

Approach to Writing

The point of attack in report writing is the analysis of the problem. An effective report, like any good engineered product, must be designed to fill a particular purpose in a specific situation. Effective design depends on a detailed analysis of the problem and an awareness of the factors involved. In analyzing your writing problem, you should ask the following questions:

What is the purpose of the report?
Who will read it?
How will it be used?
What is wanted?
When is it wanted?
What decisions will be based on the report?
What does the reader need to be told in order to understand the material?

Once this analysis has been completed and the influencing factors defined, you should evaluate your material and select that to be included in the report, using the analysis as a basis for your selection. The material should then be organized functionally within the requirements of the reporting situation just as an engineering design must be functional within its technical and environmental requirements. If the analysis of the problem was thorough, you are reader-oriented, and writing the report becomes a much easier task.

For a third item of evidence, we go back to the point of view of a research group—which, as the preceding item makes clear, is actually not significantly different from the point of view of a broad cross section of managers in industry. What follows appears in the Introduction to *Written Reporting Methods,* a manual prepared by and for the staff of the Alliance Research Center of Babcock & Wilcox.

The reports published by the Research and Development Division are the principal evidence of the quality of our performance and ability. Since we are judged by the contents of these reports, our writing ability must match our research ability in the laboratory. The production of effective reports, therefore, is a major part of our professional performance. Most of us have already spent many years preparing for this essential duty. No single manual can serve as a substitute for this necessary background. A manual is definitely required, however, to gain consistency in the form of our reports.

The next and last item, referred to earlier, must have touched a sensitive spot, to judge from the amount of interest it provoked. We hope you will enjoy it. The footnotes are reprinted exactly as they were in the original. The article is reprinted here from the June 1958 issue of *American Scientist.*

Mathmanship
by Nicholas Vanserg

In an article published a few years ago, the writer[1] intimated with befitting subtlety that since most concepts of science are relatively simple (once you understand them) any ambitious scientist must, in self protection, prevent his colleagues from discovering that *his* ideas are simple too. So, if he can write his published contributions obscurely and uninterestingly enough no one will attempt to read them but all will instead genuflect in awe before such erudition.

WHAT IS MATHMANSHIP?

Above and beyond the now-familiar recourse of writing in some language that looks like English but isn't, such as Geologese, Biologese, or, perhaps most successful of all, Educationalese[2], is the further refinement of writing everything possible in mathematical symbols. This has but one disadvantage, namely, that some designing skunk equally proficient in this low form of cunning may be able to follow the reasoning and discover its hidden simplicity. Fortunately, however, any such nefarious design can be thwarted by a modification of the well-known art of gamesmanship[3].

The object of this technique which may, by analogy, be termed "Mathmanship" is to place unsuspected obstacles in the way of the pursuer until he is obliged by a series of delays and frustrations to give up the chase and concede his mental inferiority to the author.

THE TYPOGRAPHICAL TRICK

One of the more rudimentary practices of mathmanship is to slip in the wrong letter, say a y for a r. Even placing an exponent on the wrong side of the bracket will also do wonders. This subterfuge, while admittedly an infraction of the ground rules, rarely incurs a penalty as it can always be blamed on the printer. In fact the author need not stoop to it himself as any copyist will gladly enter into the spirit of the occasion and cooperate voluntarily. You need only be trusting and not read the proof.

STRATEGY OF THE SECRET SYMBOL

But, if by some mischance, the equations don't get badly garbled, the mathematics is apt to be all too easy to follow, *provided* the reader knows what the letters stand for. Here then is your firm line of defense: **at all cost *prevent him from finding out!***

Thus you may state in fine print in a footnote on page 35 that V^x is the total volume of a phase and then on page 873 introduce V^x out of a clear sky. This, you see, is not actually cheating because after all, or rather before all, you *did* tell what the symbol meant. By surreptitiously introducing one by one all the letters of the English, Greek and German alphabets right side up and upside down, you can make the reader, when he wants to look up any topic, read the book backward in order to find out what they mean. Some of the most impressive books read about as well backward as forward, anyway.

But should reading backward become so normal as to be considered straightforward you can always double back on the hounds. For example, introduce μ on page 66 and avoid defining μ until page 86.* This will make the whole book required reading.

THE PI-THROWING CONTEST OR HUMPTY-DUMPTY DODGE

Although your reader may eventually catch up with you, you can throw him off the scent temporarily by making him *think* he knows what the letters mean. For example every schoolboy knows what π stands for so you can hold him at bay by heaving some entirely different kind of π into the equation. The poor fellow will automatically multiply by 3.1416, then begin wondering how a π got into the act

---

* All these examples are from published literature. Readers desiring specific references may send a self-addressed stamped envelope. I collect uncanceled stamps.– N. Vanserg

anyhow, and finally discover that all the while π was osmotic pressure. If you are careful not to warn him, this one is good for a delay of about an hour and a half.

This principle, conveniently termed pi-throwing can, of course, be modified to apply to any other letter. Thus you can state perfectly truthfully on page 141 that F is free energy so if Gentle Reader has read another book that used F for *Helmholtz* free energy, he will waste a lot of his own free energy trying to reconcile your equations before he thinks to look for the footnote tucked away at the bottom of page 50, dutifully explaining that what you are talking about all the time is *Gibbs* free energy which he always thought was G. Meanwhile you can compound his confusion by using G for something else, such as "any extensive property." F, however, is a particularly happy letter as it can be used not only for any unspecified brand of free energy but also for fluorine, force, friction, Faradays, or a function of something or other, thus increasing the degree of randomness, dS. (S, as everyone knows stands for entropy, or maybe sulfur). The context, of course, will make the meaning clear, especially if you can contrive to use several kinds of F's or S's in the same equation.

For all such switching of letters on the reader you can cite unimpeachable authority by paraphrasing the writing of an eminent mathematician;[4] "When I use a letter it means just what I choose it to mean — neither more nor less . . . the question is, which is to be master — that's all."

THE UNCONSUMMATED ASTERISK

Speaking of footnotes (I was, don't you remember?) a subtle ruse is the "unconsummated asterisk" or "ill-starred letter." You can use P** to represent some pressure difference from P, thus tricking the innocent reader into looking at the bottom of the page for a footnote. There isn't any, of course, but by the time he has decided that P must be some registered trademark as in the magazine advertisements he has lost his place and has to start over again. Sometimes, just for variety, you can use instead of an asterisk a heavy round dot or bar over certain letters. In doing so, it is permissible to give the reader enough veiled hints to make him *think* he can figure out the system but do not at any one place explain the general idea of this mystic notation, which must remain a closely guarded secret known only to the initiated. Do not disclose it under pain of expulsion from the fraternity. Let the Baffled Barbarian beat his head against the wall of mystery. It may be bloodied, but if it is unbowed you lose the round.

The other side of the asterisk gambit is to use a superscript as a key to a *real* footnote. The knowledge-seeker reads that S is — 36.7[14] calories and thinks "Gee what a whale of a lot of calories" until he reads to the bottom of the page, finds footnote 14 and says "oh."

THE "HENCE" GAMBIT

But after all, the most successful device in mathmanship is to leave out one or two pages of calculations and for them substitute the word "hence" followed by a colon. This is guaranteed to hold the reader for a couple of days figuring out how you got hither from hence. Even more effective is to use "obviously" instead of "hence," since no reader is likely to show his ignorance by seeking help in elucidating

---

** April fool. See what I mean?

anything that is obvious. This succeeds not only in frustrating him but also in bringing him down with an inferiority complex, one of the prime desiderata of the art.

These, of course, are only the most common and elementary rules. The writer has in progress a two-volume work on mathmanship complete with examples and exercises. It will contain so many secret symbols, cryptic codes and hence-gambits that no one (but no one) will be able to read it.

REFERENCES

1. Vanserg, Nicholas. *How to Write Geologese,* Economic Geology, Vol. 47, pp. 220–223, 1952.
2. Carberry, Josiah. *Psychoceramics,* p. 1167, Brown University Press, 1945.
3. Potter, Stephen. *Theory and Practice of Gamesmanship or the Art of Winning Games without Actually Cheating.* London: R. Hart-Davis, 1947.
4. Carroll, Lewis. *Complete Works,* Modern Library edition, p. 214.

2

Five Basic Principles of Good Technical Writing

Chapter 1 outlined the general subject and plan of this book and indicated the importance of writing in the professional work of the engineer and scientist. The present chapter is devoted to a highly condensed preliminary statement of five basic principles that will later be presented in detail. There are many more than five principles involved in good technical writing, but the five stated below are so important that they may be taken as a foundation on which further development rests.

1. Always have in mind a specific reader, real or imaginary, when you are writing a report; and always assume that this reader is intelligent, but uninformed.
2. Before you start to write, always decide what the exact purpose of your report is; and make sure that every paragraph, every sentence, every word, makes a clear contribution to that purpose, and makes it at the right time.
3. Use language that is simple, concrete, and familiar.
4. At the beginning and end of every section of your report check your writing according to this principle: "First you tell your readers what you're going to tell them, then you tell them, and then you tell them what you've told them."
5. Make your report attractive to look at.

You will find that these principles are involved in one way or another with practically everything that is said throughout the rest of this book.

3

Style in Technical Writing

Introduction

The concept of style is familiar in many areas of our lives, such as clothes, automobiles, and architecture. At the present time we even talk of life-styles. Despite this familiarity, style is a difficult concept to define, and this difficulty is much evident in discussions of the style of written language.[1]

Although style is often defined as the author's choice and arrangement of words, it takes on a much larger meaning for the technical writer because much technical communication is achieved by nonverbal means. Before becoming concerned with vocabulary and matters of sentence and paragraph construction, the technical writer needs to determine whether words are really the most effective means of conveying the intended information. Consider, for example, what has happened in communication involving the operation and orderly movement of motor vehicles. At one time it was customary to use words for all the information to be conveyed. Motorists were given information about the status of their vehicles by means of gauges marked "gasoline" and "temperature"; highway conditions were indicated by signs reading "cattle crossing," "sharp curve," "merging traffic." In recent years the prevailing practice has been to present such information graphi-

[1] For an excellent survey of concepts of literary style, see James L. Kinneavy, *A Theory of Discourse* (Englewood Cliffs, N.J.: Prentice-Hall, 1971), pp. 357 ff.

cally whenever possible. Thus, the motorist sees instruments marked with tiny gas pumps or thermometers, and highway signs showing pictures of cows, trucks on inclines, and actual shapes of curves and intersections.

While there may be some question as to whether the change has been beneficial in every instance, it should be obvious that the choice of verbal or nonverbal media must be governed by the writer's purpose and the specific communication situation. In this chapter we will deal with style as it applies to verbal language, but much of what we say about words, sentences, and paragraphs is applicable also to photographs, drawings, diagrams, graphs, tables, maps, and formulas when used in technical writing. More specific discussion of nonverbal elements appears in a subsequent chapter.

We are not here concerned with the concept of style in all its aspects, but only with its place in technical writing, and this limitation permits focusing upon a few relatively simple problems. Instead of exploring the whole idea of style, we'll turn at once to these practical problems.

This chapter is divided into two parts. Part 1 is concerned with the two issues of adapting a report to a particular reader and of achieving a "scientific attitude." Part 2 takes up more detailed problems, such as vocabulary and sentences.

PART 1

Reader Adaptation

The Importance of Reader Adaptation   What we here term "Reader Adaptation" is often also termed "Audience Analysis." We continue to use the former in this text in the belief that it is the more accurate term. A common misconception continues to exist that good writing can exist in isolation, and that its merits can be judged independently of the communication situation in which it is to function. While it is true that spelling, punctuation, and grammar have become sufficiently standardized that a piece of writing may be judged defective to the degree that it departs from accepted standards in these areas with no apparent justification, it is not true that a piece of writing is necessarily good to the degree that it does *not* depart from these standards. Writing can follow all the accepted conventions of spelling, punctuation, and grammar and still fail to achieve its intended purpose.

For technical writing in particular, two additional criteria must be met: (1) the writing must be accurate; that is, the writing must not overstate, understate, or distort the reality that is being represented; (2) the writing must be tailored to meet the needs, the interests, and the capabilities of its expected readers. Since the first of these two criteria is heavily dependent on the second, we will focus the following discussion on the problems of recognizing the needs, interests, and capabilities of the expected readers and on possible solutions to these problems.

Perhaps the easiest way to think about the problem of reader adaptation is simply to look back over some of the reading you have done and ask yourself how you felt about it. No doubt you have had the experience of enthusiastically starting to read a technical article about some subject you thought interesting, only to find the article either dull or incomprehensible. It doesn't necessarily follow that the article was no good, only that it was no good for you. As an extreme example of how things can go wrong in this respect, we might imagine the feelings of a small boy who has noticed a halo around the sun, asked his father what it was, and received a reply which began, "My boy, we are here dealing with a phenomenon involving spicules of ice having a uniform refracting angle of 60 degrees and yielding, at minimum deviation, a deflection of 22 degrees of the incident light."

Let's contemplate still another situation. Let's imagine that in a time of war an American naval vessel, a destroyer, has entered a combat area. While it is on patrol there, a crucial element in its electronic equipment breaks down. The man in charge of maintaining the equipment, a noncommissioned officer, turns to the equipment manual for help in finding the trouble. This manual is the only source of information or assistance he has. Will he find the manual useful? For some thoughtful and informed discussion by naval personnel of precisely this problem, please turn to Appendix C, pp. 497–499, beginning with the heading "Theory of Operation Section."

Finally, let's consider an example from the world of theoretical physics. The Nov. 16, 1973, issue of *Science* contains a review by physicist Bryce De Witt of *The Large Scale Structure of Space-Time,* a book written by S. W. Hawking and G. F. R. Ellis. In his review, De Witt said:

> The book is a masterpiece, written by sure hands. But it is a flawed masterpiece. The student who conquers it will be richly rewarded. But he will have to work unnecessarily hard. This is because the authors are unable to decide who their audience is.

The crucial implication of all these examples is that every aspect of an effective piece of technical writing is in some degree designed for a particular reader or group of readers. Another way of looking at this idea is to compare technical writing to other forms of writing, such as poetry, or even to an art form such as music. People sometimes write poetry or play a musical instrument to express their feelings. But who has ever seen a technical report written to express the author's feelings? We do not mean to say that no personal feeling is involved in all the activity that leads up to and perhaps even culminates in a technical report. As a matter of fact, the feeling is often intense, as revealed in very different ways in such remarkable books as Michael Polanyi's *Personal Knowledge,* and James Watson's *The Double Helix.* Our point about the purpose of technical writing can be put very simply: (1) poetry may be written and musical instruments may be played just for the fun of doing it, but (2) technical writing of the kind we are concerned with in this book is done primarily for the purpose of helping somebody else do something. This point is neatly summarized by J. A. Hutcheson, then a vice-president at Westinghouse, now retired: "Too few engineers realize that everything they write is used at some time by someone to help him make a decision." Mr. Hutcheson's comment happens to be directed to engineers, but his point applies equally well to most of the writing done by scientists or technical writers of any kind. For example, should college students be given guidance in selecting a personal computer? See "Using Personal Computers at the College Level" on pp. 212–216.

You will probably agree that it is wise to design your writing for your particular reader or readers. Nevertheless, you may be wondering if this kind of problem is really of any concern for a beginner. After all, you will be writing for colleagues who are equally well trained, and for experienced supervisors. True, in a way; but a look at the facts introduces some surprises.

Most writing done by students at almost every level of their formal education is done for readers quite different from those normally encountered outside the educational system. Since the writing is done primarily for teachers, a peculiar relationship exists that is rarely encountered elsewhere. With the exception of some specialized composition courses, almost all school writing — from elementary school assignments to Ph.D dissertations — involves just one of two possible audiences: either a reader who has already acquired specialized knowledge in the subject matter being discussed or a reader who needs no specialized background because the subject matter lies in the area of general knowledge. Thus science papers are written for science teachers, literature papers for literature teachers, engineering reports for engineering professors, medical case studies for certified medical instructors, and personal experience papers for composition teachers. None of these readers has much direct need to learn anything from the writer in regard to the subject being discussed.

Realizing this lack of need, students can very easily fall into the habit of writing to impress rather than to inform because they sense that their purpose is not to provide information about the subject but to provide information about the writer's command of the subject — which is quite a different matter.

We do not wish to attack the educational system for allowing this to happen; there are practical reasons for the development. But the fact remains that many highly trained people begin their professional careers with little or no experience in dealing with the types of readers they must actually write for.

When the writer is seen as a possible source of useful information rather than as a student who is to be trained and evaluated, there are very few instances in which a reader's primary purpose is to evaluate the writer's understanding of either subject matter or the manner of presenting it. In "real world" writing, the teacher – student roles are reversed. The reader becomes the person to be taught, and the writer is responsible for the teaching. Outside the educational system, readers of technical writing can be expected to be much more interested in the subject being discussed than in the writer who is discussing it, and they very rarely will be school teachers or college professors. Who, then, will they be?

Some technical writing experts have classified audiences on the supposition that a writer can more easily meet the needs of a great many individual readers by understanding the basic characteristics of just a few groups. Thomas E. Pearsall, for example, did some excellent pioneering work in this field by recognizing that most readers of technical information can be classified as laymen, executives, experts, technicians, or operators when grouped according to their relationship to the subject matter being discussed and their reasons for reading about it.[2] Other authorities have developed report worksheets or audience profile charts to identify the probable background, capabilities, motives, and attitudes of the intended reader. Regardless

[2] Thomas E. Pearsall, *Audience Analysis for Technical Writing* (Beverly Hills, CA: Glencoe Press), 1979.

of the method used, there simply is no substitute for the ability to put yourself in the reader's position and imagine what that person would want to know, would need to know, and would be able to understand. Only when the writer is aware of these three limitations is he or she in proper position to know what to say and how to say it.

A second surprise is the degree of difficulty that the writer may have in realizing how much he or she has learned that the intended reader has not learned. Technical writers normally are people who have acquired extensive specialized knowledge, and it isn't easy to distinguish between specialized knowledge and generalized knowledge. A medical student of our recent acquaintance could not believe that anyone should be allowed to graduate from college without being able to distinguish between medial and lateral movement of the blood or between inferior and superior vena cava, and a graduate student in mathematics felt his composition instructor could easily become conversant with fifth degree polynomials, and the properties of closure, commutativity, and associativity by a few quick references to a selected bibliography of texts used in his advanced courses in mathematics.

The supervisor or manager whose opinion of your writing is most important to you may not have been trained in your field at all. A little commonsense reflection makes this point meaningful. Suppose you are a mechanical engineer and you are a member of a team designing and building a prototype of an electric-powered automobile. What field of training will your section chief represent? Mechanical engineering? Electrical engineering? Chemical engineering? Here is what Westinghouse has to say to its own people on this subject of how to write for your supervisor, in the study *(What to Report)* described in Chapter 1. The italics appear in the original.

> *Usually* the management reader has an educational and experience background different from that of the writer. *Never* does the management reader have the same knowledge of and familiarity with the specific problem being reported that the writer has. *Therefore, the writer of a report for management should write at a technical level suitable for a reader whose educational and experience background is in a field different from his own.* For example, if the report writer is an electrical engineer, he should write his reports for a person educated and trained in a field such as chemical engineering, or mechanical engineering, or metallurgical engineering.

This statement goes a little further than we'd be willing to go in this book. That is, circumstances vary from one group or company to another, and we wouldn't want to adopt as a dogmatic rule the conclusions stated in the quotation. Nevertheless, this quotation does reveal the great practical importance of reader adaptation, and we most strongly and urgently endorse the principle that you should seldom write with the assumption that your reader is informed about your particular subject. The document from which this quotation was taken goes on to make the interesting assertion that, whenever possible, highly complex material should be put into an appendix. This idea opens up the whole subject of how organization can be used to help resolve problems of style.

Elements Requiring Adaptation   What elements of a report should you work on in adapting it to a given reader? Broadly speaking, the answer to this question is simply, as we said earlier—*everything.* It is difficult to think of any element in technical writing that is not influenced in some way by the intended reader. This fact will take on added significance as your experience with technical writing continues.

In this chapter on style we want to call attention to the three particular elements of style that should be given *first* attention in adapting your writing to your reader. These three elements are vocabulary, sentence length and structure, and organization. Our purpose at this time will be merely to indicate the importance of these elements and to prepare for later discussions that will be both greater in scope and presented from a different point of view.

As far as vocabulary is concerned, the problem is usually simple. Just don't use words your reader won't know. If you feel you must use a word that may not be familiar, define it. Definitions are taken up in Chapter 5, and problems of word choice in Part 2 of this chapter. For the present, we leave the matter of vocabulary with this thought: In many years of experience we have never heard of a complaint that the vocabulary in a report was too simple.

Sentences, the second element, should be relatively short, and should be simple in construction. What does "relatively" mean? That is, relative to what? Well, here is an example. Suppose you are writing a set of directions on how to assemble a piece of equipment and your reader is someone who dropped out of high school in the tenth grade. You'd better use only simple declarative sentences and restrict their average length to not more than 10 or 12 words. On the other hand, if you are writing about a chemical process and your reader is a Ph.D. in chemical engineering, you can probably get by with sentences that, on the average, are almost twice as long as well as more complex in structure. See p. 45 for further comment.

It may seem odd to speak of adapting the organization of the principal parts of a report to a reader. And it may seem even odder to claim, as we did earlier, that organization can be helpful in solving stylistic problems. Probably most people think of the organization of a piece of writing as flowing naturally from the content. For example, the "natural" organization of a report of a scientific experiment would be to explain the purpose of the experiment, to describe the experiment, and to end by presenting the conclusions reached. In technical writing, on the contrary, conclusions are often presented at or near the beginning. One of the advantages of this arbitrary way of organizing the report is that it helps solve a stylistic problem.

The stylistic problem we have in mind is that of writing for a group of readers among whom some have a much greater familiarity with the subject than others. In this situation, if the writer uses technical terminology, some readers won't understand what is being said. On the other hand, if the writer keeps explaining the terminology, those readers who are familiar with the subject may feel their time is being wasted. What can the writer do?

One of the best solutions to this problem is to begin the report with a brief summary. In this summary, the principal findings and conclusions are presented in language intelligible to the readers who are least familiar with the subject. These people will read only the summary. Consequently, the writer can use more technical terminology in the main body of the report, which will be read only by people having the necessary technical background. Let's consider a concrete example of the stylistic problem just described, and then return briefly for another look at the kind of solution suggested.

Let's assume that the report to be written will be of interest to readers who are well educated but not trained in the field being discussed. An oceanographer, let's say, is writing a report and has just completed this sentence:

> Two small commercial propellant-embedment anchors, nominally rated at 5- and 10-kip capacity, were investigated.

This sentence actually appears on page 34 of a 49-page booklet entitled *NCEL Ocean Engineering Program,* issued by the U.S. Naval Civil Engineering Laboratory in Port Hueneme, California. The purpose of the booklet is apparently to acquaint the educated public with the broad outlines of this laboratory's work. And for such readers it is a fascinating booklet, containing, among other things, a two-page summary of the laboratory's part in the search for some nuclear bombs accidentally dropped into the ocean off the coast of Spain, as well as brief descriptions of three proposed designs for manned underwater stations in which scientists could live at a depth of 6000 feet (1.8 km) in the ocean. The sentence just quoted is concerned with the problem of anchoring down such underwater stations.

Now, in this context, should all intended readers of this booklet be expected to understand the sentence quoted above? Probably not. Some would likely have trouble with that sentence because they lacked familiarity with the terms "propellant-embedment anchors" and "kip."

The author of the NCEL booklet evidently felt the same way because he provided this explanatory sentence:

> The propellant-embedment anchor is a self-contained device similar to a large-caliber gun consisting of a barrel, a recoil mechanism, and the projectile which is the anchor.

Fine. All is clear now. But what does "kip" mean? The author never did say. Why didn't he? Perhaps because he was beginning to wonder how his page was going to look if he added still another explanatory sentence.

In brief, here is an illustration of a kind of stylistic difficulty that is sure to result in annoyance for one reader or another. Some readers won't need the explanation of "propellant-embedment anchor" and will feel the passage is wordy. They might not need an explanation of "kip" either, but see the essay entitled "Mathmanship" at the end of Chapter 1. (As a matter of fact, a kip — kilopound — is a force of a thousand pounds.) In contrast, other readers won't understand the initial sentence when they first read it, and won't be interested in how many "kips" there are anyhow. Probably what these readers would like to have is something resembling the following:

> Upon reaching the ocean floor at a depth of as much as 6000 ft, the station will be secured in place by anchors fired like projectiles from the barrel of a specially designed gun.

A brief summary of the main points of the entire booklet, written in simple language like that of the sentence above, would be a real kindness to the uninformed readers. And thus a stylistic problem would be solved through use of an organizational device.

As you probably know, a routine part of a great many industrial and scientific reports is an initial summary or abstract. Often, although not always, this initial section is intended to help readers who are unfamiliar with the subject. As a matter of fact, many different terms besides "abstract" and "summary" are used for the kind of initial section just described, terms such as "results," "conclusions," "findings," "epitome," "digest." These terms don't necessarily all mean the same thing, but sometimes they do. As a matter of fact, some companies develop a number of elaborate forms for the organization of their reports, each form to be used for certain specified purposes or occasions, and these elaborate forms may begin with a formidable series of epitomes, results, findings, conclusions, and so on. These forms are often called "kinds of reports," and given names such as "Partial Report." There is no harm in these practices. They are easy to become accustomed to, and are sometimes useful. Further information about kinds of reports and their formats can be found in Sections IV and V.

We have commented on the reader who desires a summary at the beginning because the detailed text would be difficult. It is also worth remembering the advice Westinghouse gave about putting difficult material into an appendix, another helpful device. The Company was thinking particularly about readers at the managerial level.

What about the group of readers who are thoroughly familiar with the subject in general but who nevertheless want to read only the abstract? These are people who feel they need to know only the conclusions reached in the report and who want to learn what these conclusions are as quickly as possible. These readers want to be informed about the work, but don't need details; some of them might be at the managerial level too. Perhaps this situation raises a question. Should a beginning writer save some precious time by working hard on only the introductory sections of a report and getting out the rest in a hurry? Well, it might be risky.

The kind of problem we are getting into here is so important that we want to pause for a careful comment on it. We have pointed out that some, or even many, of the readers of a report may look at only the introductory portions of it. It would be tremendously misleading, however, to imagine that the full text of a report is seldom read, or is seldom read by people in executive positions. As J. A. Hutcheson observed in the statement quoted earlier, reports are written to help people make decisions; and intelligent people don't make important decisions until they have examined all the facts available. One of the co-authors of this book remembers vividly his own first important (to him) effort at writing a technical report. When the report had been submitted, his boss summoned him to a conference about it. "Your conclusions," his boss began, "do not adequately reflect the content of your report." And he went on to analyze in mortifying detail the fact that the conclusions really did not indicate the full significance of what had been presented in the report.[3]

In later chapters the writing of the initial sections of technical reports will be discussed and examples shown. At present, our purpose is simply to point out ways in which a writer can adapt the style of a report to a given reader by working on vocabulary, sentences, and organization.

[3] For a discussion of reader analysis and style in relation to recommendation reports, see pp. 28–29 and 250–252.

The Communication Situation   One other aspect of the general problem of adapting the report to the readers who will use it remains to be discussed. This is the need for analysis of what we'll call the communication situation.

So far, what we have said about readers has been confined virtually to consideration of how much they would know about a given subject. Obviously, in analyzing readers' needs and interests, a writer has a lot more to think about than just the state of their knowledge. The readers we are contemplating are, after all, real live people working in a complex situation in which they have responsibilities, ambitions, worries. When looked at in this larger context, the report itself takes on new dimensions. It is the sum of all these additional elements concerning both the reader and the report that we have reference to in the phrase "communication situation."

The scope of what we will have to say on this particular subject will extend beyond what is usually thought of as style, which is our principal subject. We believe you will feel, as we do, that the practical problems of style in technical writing can be understood only in the context of the total situation.

Every communication situation is unique, and adapting to it requires two distinct steps: first, perceiving it accurately, and second, making sensible decisions as to what to do about it. In a general way, it is easy to itemize the elements in terms of which any communication situation in technical writing can be perceived. On the other hand, almost the only means of discussing sensible decisions is to appeal to experience. We will therefore list and comment briefly on the elements of the communication situation, and then present a few examples.

As seen from the writer's point of view, the elements in the communication situation can be broken down into two categories, as follows:

1. Users of the report
   (a) Who they are
   (b) How much they know about the subject of the report
   (c) What their responsibility is concerning any action taken on the basis of the report
   (d) What their probable attitude is toward the conclusions and recommendations of the report
2. Uses of the report
   (a) As a guide for action
   (b) As a repository of information
   (c) As a long-term versus a short-term aid
   (d) As an aid under specified conditions

Points 1(a) and 1(b) have already been discussed in preceding pages and are repeated here only for the sake of the completeness of these categories. Points 1(c) and 1(d) are probably clear without explanation, but they do need to be illustrated, and will be in a moment. The elements listed in category 2 are perhaps less clear.

First, with respect to the elements in category 2, the uses of the report, let's consider the difference between 2(a) and 2(b). This difference may be seen by contrasting two imaginary reports, one recommending that production be started on a new additive for lead-acid storage batteries and the other explaining that efforts to find an effective additive have failed. The latter kind of report is put away in the files as a repository of information, useful as a safeguard against wasteful duplication of the work at some future time. (An abstract of just such a report describing unsuccessful research on a storage battery additive may be found on p. 75.) The recommendation report, unlike the one just mentioned, is intended as a guide for action, or at least as a way of initiating action. Of course a recommendation report may also serve as a repository of information, but that is not its primary function.

Some reports have a very short useful life; others, an indefinitely long one. A field report on the repair of a customer's equipment may be of interest only until the repairs have been completed, whereas a report of successful basic research may have permanent value. A report of the latter sort should be written in such a way that it will still be comprehensible many years later. Files of old field reports are often referred to by designers and engineers for clues to design weaknesses, and we don't mean to suggest that clarity is unimportant in such reports. On the other hand, the comprehensiveness of a report on what was done in a basic research project would be only an unnecessary burden in most "short-term" reports. This difference between the two kinds of reports is the subject of 2(c).

The fourth category, 2(d), the report as an aid under specific conditions, is concerned with the adaptation of the report to the immediate circumstances in which it will be used. For example, "trouble-shooting" manuals designed as an aid in emergency repairs on a piece of equipment are likely to be quite different from manuals designed for routine maintenance. Further illustration may be found on pp. 493 – 494, where differences between naval and civilian use of electronics service manuals are discussed.

As we said, these ideas about the elements of the communication situation are rather simple and obvious, once they have been pointed out. It is only when writers don't think about them at all that foolish things can happen. On the other hand, a good deal of imagination is required to take full advantage of the possibilities afforded by an accurate perception of these elements.

One very common, and often amusing, example of what happens when people simply don't think about these elements is that they apply some "standardized" way of organizing a report to a situation it doesn't fit. One of the authors of this text once examined several dozen proposals written by the technical staff of a large company in which this difficulty appeared. A proposal, as you may know, is a report in which a company attempts to sell its services to some other company or agency for the carrying out of a specific project. (See Chapter 14.) Since proposals are means of obtaining contracts, they are extremely important, and they should be tailor-made for every project sought. In our example, however, as the examination of the proposals written by the technical staff continued, it became evident that in every single proposal the major headings were identical. These headings were "Introduction," "Items and Services to Be Supplied," "Technical Approach," "Management Plan," "Reliability and Quality Assurance," and "Conclusion." For many of the proposals, these headings were in fact appropriate. For some, they most decidedly were not. One proposal, for instance, was concerned with a research study project in which nothing was to be delivered to the contracting company except a report containing the results of the study. In this proposal, nevertheless, the heading "Items and Services to Be Supplied" appeared as always — followed by the single word "None"!

How can imagination be used to adapt a conventional report to an unusual communication situation? The best answer, we believe, is not an abstract explanation but a concrete illustration.

We know an engineer who came to the conclusion that the company he worked for should invest a considerable sum (more than $100,000) in some new environmental test equipment. His study of work loads and of present and probable future commitments on contracts, and his estimate of company growth all led him to believe that money would be saved in the long run by replacement of the present equipment. The problem he then confronted was how to get approval for such an expenditure from his superior, the general manager of the company. This problem was especially forbidding because the general manager was a conservative man, a man who was rather negative about spending large sums of money. He would certainly take a hard look at the facts. On the other hand, the engineer believed the general manager respected his ability and judgment.

With all this in mind, the engineer wrote a 50-page report recommending purchase of the new equipment. In it, he provided detailed support of every aspect of his recommendation. To take the place of the usual introductory summary, however, he wrote a 2½-page section designed especially for his particular communication situation. These 2½ pages were made up principally of a series of questions and answers, each question typed entirely in capital letters. Underneath each question appeared the answer to it. He was thus attempting an adaptation that involved style in the conventional sense as well as organization.

The engineer's recommendation was approved. As a matter of fact, the general manager told us he thought the engineer had saved him a lot of time and energy by concentrating the most important data at the beginning of the report.

We certainly do not suggest that every recommendation be presented in the way this one was. Indeed, the point of our story is that almost any report can benefit from a sensible analysis of the particular circumstances in which it will be received and used. The questions the engineer presented in his first 2½ pages included the following: Why is new environmental test equipment needed? How much will it cost? How soon will it pay for itself? Where can space for it be found? What will be done with present equipment? How long will the proposed equipment serve company needs? What is its probable longevity? As you see, these are really just commonsense questions. They probably first appeared in the form of entries here and there in rough lists the engineer jotted down of things that must be included in his report. Such lists are the natural first step in the process of organizing a body of information. (See p. 54 for further discussion of the use of questions in outlining a report.) The crucial next step is to arrange the items in such lists into logical groups, and the third is to organize these groups into a formal outline. The success of the report we have just described can be attributed in large part to the skill with which the engineer presented at the beginning of his report a grouping of information that was well adapted to his particular, unique, communication situation.

The subject of organizing a body of information is discussed further in Chapter 4, "Outlines and Abstracts."

As a last example of adapting a report to a particular situation, we have in mind a case somewhat resembling the one just described, but one in which a very different report was written. Again, the following circumstances are actual.

An engineer was assigned to the job of surveying the layout of an assembly line in a manufacturing plant, and after making a survey, came to the conclusion that the entire assembly plant should be redesigned and reequipped for more efficient operation. The cost of such a change would be very great, not to mention the loss of income caused by the considerable amount of "down time." He was convinced, however, that the changeover would ultimately result in a significant saving of time for product assembly, would reduce the number of defective items, and—in short—would result in a marked decrease in unit cost, thus saving the company a great deal of money.

Unfortunately, the management was hostile to any increase in capital outlay, particularly in view of its current high unit cost. He therefore decided that a report which defined the problem and immediately suggested a solution might very well so antagonize the management that they would pay little attention to those parts of the report that proved his case. The solution was to begin with a careful survey of those production problems that were already recognized by the management and were known to be causing concern. The next step was to present solutions to those problems. In other words, the engineer's organization was essentially Socratic; he organized his body of information in such a way that his readers were virtually compelled to accept each new conclusion as they went along. This report, too, was successful.

As we have described this incident, the importance of the style of the report, in the conventional sense, appears much subordinate to that of the organization in achieving an effective adaptation to the situation. The fact is, however, that these two factors must work together. Were you to examine the report, we believe you would agree that its tone of detached and thoughtful analysis was well designed to harmonize with the kind of organization chosen.

The two examples we have just presented are both related primarily to points 1(c) and 1(d) in the list of the elements of the communication situation. That is, they are primarily related to the areas of the reader's responsibility and probable attitudes. Perhaps you have already noticed that, altogether, we have spent a good deal of time discussing the problems associated with category 1, the users of the report, and relatively little discussing the problems of category 2, the uses of the report. We have allotted the time in this way because the problems associated with category 2 are much simpler and more easily defined. We do not say they are less important than the problems of category 1—only that they are less complex. Sometimes it is wise to remind ourselves, especially when the excitement and pleasure of a new achievement or challenge in science or engineering absorbs our interest, that the most complex thing known to exist in the universe is a human being. "Users" are human beings.

In summary, we have discussed adapting the style of a report to, first, the state of the reader's knowledge of the subject, and second, the total situation in which the reader examines and uses the report. We have presented the discussion primarily in terms of vocabulary, sentences, and organization.

A last general comment at this point concerns the style of this textbook. We have tried to select a style adapted to our readers and to our communication situation. Since much of what we have to say consists of advice about writing problems rather than information about technical matters, the style of this book is more casual than would be found in technical writing. For example, our occasional use of contractions (like "we'll") would not be acceptable in a technical report. So please remember that this casual approach is not one we'd use when writing a technical report—but we hope you'll find it easy to read and suitable for your needs.

We will now leave the subject of reader adaptation and turn to a discussion of the scientific attitude in technical writing. The following quotation will serve as a transition to this new subject. It is taken from a booklet issued by the Whirlpool Corporation[4] for its own employees. It is transitional because it combines the subject we have just been considering, reader adaptation, with concern about point of view—which is one aspect of our next general subject, the scientific attitude.

> In report writing use your own style, not an imitation of anyone else's. Write the report from an objective point of view so as to focus the reader's attention on the subject and not on you; an objective viewpoint does not necessarily mean an impersonal viewpoint. Avoid personalities because this will have little value with the passage of time. Prefer the concrete expression to the abstract, the direct statement to the indirect. Give definite dates and values whenever possible. *KEEP YOUR READER IN MIND WHEN WRITING THE REPORT AND USE LANGUAGE HE CAN UNDERSTAND.*

The capital letters in this last sentence are found in the original.

The Scientific Attitude

The Traditional View   While you must be careful to adapt your writing to your reader, you must also adapt it to yourself. Technical writing is a communication *to* somebody *from* somebody. What sort of person do you want your reader to feel that you are?

Many years ago the answer to this question was both simple and, as we look back at it, rather strange. In writing about science, said the answer of 50 years ago, the author should apear to be no sort of person at all, simply a pure intelligence. As Albert Einstein wrote, "When a man is talking about scientific subjects, the little word 'I' should play no part in his expositions."[5] This point of view has subsequently been considerably modified. The reasons for this change are complex, and there are differences of opinion about them. To avoid becoming involved in the details of a big subject, we will merely say a few words about what has happened.

[4] *Style Manual for Project Reporting* (Benton Harbor, Mich.: Research and Engineering Center, Whirlpool Corporation, 1968), p. 19.
[5] Albert Einstein, *Essays in Science* (New York: The Philosophical Library, 1934), p. 113.

Historically, the change in the character of technical writing has been roughly paralleled by certain changes in science, or at least by the emergence of certain kinds of new questions in science. These questions are about the relationship between an experimenter in science and the events the experimenter observes. In classical physics it was usually assumed that the experimenter observed certain physical events and that the experiment — and even the report of the experiment — was concerned with nothing except the "objective" physical events. In modern physics, as no doubt you know, this assumption is often questioned. Now it is often asserted that the experimenter is subtly and importantly involved in the content of observations of physical events. A characteristic statement of this point of view is made by F. S. C. Northrup in his introduction to Werner Heisenberg's *Physics and Philosophy:*[6]

> Quantum mechanics, especially its Heisenberg principle of indeterminacy, has been notable in the change it has brought in the physicist's . . . theory of the relation of the experimenter to the object of his scientific knowledge.

Accompanying this new point of view there has been a tendency to doubt the old idea that science should be considered a monolithic, impersonal method. Instead, recognition has been accorded to the importance of the mind and temperament of the experimenter. No criticism of science is implied by this change; the intent is only to suggest a more thoughtful look at what actually happens. An example is seen in P. W. Bridgman's *The Way Things Are.*[7] At one point in his book, Dr. Bridgman, a winner of the Nobel prize in physics, devoted a couple of pages to a rather surprising subject — an explanation of why he decided to use the first-person singular pronoun in his writing.

Dr. Bridgman's unexpected interest in pronouns brings up again our initial question of how you present yourself to your reader. The use of personal pronouns has a lot to do with the impression a writer conveys. Let's look at a part of what Dr. Bridgman says:

> Insistence on the use of the first person . . . will inevitably focus attention on the individual. This, it seems to me, is all to the good. The philosophical and scientific exposition of our age has been too much obsessed with the ideal of a coldly impersonal generality. This has been especially true of some mathematicians, who in their final publications carefully erase all trace of the scaffolding by which they mounted to their final result, in the delusion that like God Almighty they have built for all the ages.

Now, should you — like Dr. Bridgman — use the first-person pronoun in your technical writing? Would such a stylistic practice be consistent with a properly scientific attitude? Let's examine the second question and return later to the first.

What *Is* the Scientific Attitude?   What is consistent with a properly scientific attitude depends upon how the scientific attitude is defined. As may be inferred from the evidence just looked at, the scientific attitude used to be associated with impersonality, exclusion of emotion, objectivity. Such a view leads to a style of writing in

---

[6] Werner Heisenberg, *Physics and Philosophy* (New York: Harper & Row, 1958), p. 4.
[7] P. W. Bridgman, *The Way Things Are* (Cambridge, Mass.: Harvard University Press, 1959), p. 5.

which these qualities are easily maintained, and it is not surprising that writing done in the third person, passive voice, came to be thought of as the scientific style. (We will illustrate this style soon.) On the other hand, those people who believe that the scientific attitude does not have to exclude feeling, that it is not completely impersonal, will at least to some degree accept Dr. Bridgman's opinion that there is no harm in using personal pronouns.[8]

Actually, we think it would be a mistake to attach great significance to the question of whether a personal pronoun should or should not be used. What matters, we think, are honesty about the facts, care in obtaining and evaluating the facts, dignity and restraint in manner. Probably you would agree that these are the kinds of qualities that make up a good scientific attitude. Suppose, for instance, that someone were to write a sentence like this: "There can be no doubt that this product is infinitely superior to all the others on the market; as a matter of fact, the others are worse than useless—they are shoddily made and placed on the market, it would appear, by an unscrupulous group of shysters." This sentence is written in the "scientific style" of third person, passive voice, but surely it does not express an acceptable scientific attitude. It might be effective journalism, but a scientist would say it lacks an honest concern about the facts, it lacks dignity, it lacks restraint. A scientist might write, "According to the criteria under consideration, test data show this product to be markedly superior to the others tested."

Here are a couple of additional examples. In the first, the conclusion is expressed merely as a personal opinion (the "weak" version); in the revised version the conclusion is somewhat objictified by reference to the facts.

> *Weak.*   In conclusion, it would be a mistake to relocate the factory.
> *Better.*   In conclusion, the evidence shows it would be a mistake to relocate the factory.

The initial version in the next example is not exactly true; for a homeowner wanting maximum insulation in a space that is, say, only 3½ in. (8.89 cm) in depth, styrofoam would be a better, although a more expensive, choice than fiber glass. The revised version corrects the unscientific and misleading implication of the initial version.

> *Weak.*   For the homeowner, fiber glass is a better choice than styrofoam for insulation.
> *Better.*   For the homeowner, fiber glass is effective insulation and is cheaper than
>   styrofoam; if space is restricted, however, it should be remembered that
>   almost twice as great a thickness of fiber glass as of styrofoam is needed for
>   a given resistance to the transmission of heat.

Scientific Attitude and the Choice of Person   Having considered these examples of the meaning of the scientific attitude, we return to the question of whether such an attitude is violated by the use of a first-person pronoun. The answer is that the scientific attitude is not achieved by either the use or the avoidance of a particular pronoun. Rather, it is achieved through the qualities mentioned earlier:

[8] For a further discussion of this point of view, see the following article by physicist Gerald Holton: "Science and New Styles of Thought," *The Graduate Journal,* 7 (Spring 1967), 399–422.

honesty, care in handling facts, dignity, and restraint in manner. Looked at in this way, there could be no objection to saying, "I believe I have succeeded in showing that it would be a mistake to relocate the factory." This point of view may be seen, for example, in the authoritative *Council of Biology Editors Style Manual*,[9] which asserts that

> The first person *(I, we)* is natural for relating what you did. The second (*you* expressed or *you* not expressed but understood) is convenient for giving directions. The third (*he, she, it, they,* or a substantive) has definite advantages for telling what happened.

Nevertheless, our advice is not to use first-person pronouns at all unless you are sure such a practice is acceptable in your organization. Some organizations seem to have been influenced by the historical trends we have mentioned; others have not. The following comment from the style guide used in the Stromberg-Carlson Corporation[10] is fairly representative of the opinion you are most likely to encounter:

> Personal pronouns are not to be used in Stromberg-Carlson Practices. However, personal pronouns can be used in other types of literature (such as Attendant's Operating Instructions) to avoid awkward or stilted sentence structure.

"Practice" is the name of a kind of report found at Stromberg-Carlson. We should add that it is doubtful that the author of the rule quoted above really intended to include the pronoun "I" among those approved; but for an example of the use of "I" and "my" in a report, see pp. 191–212.

To be as concrete and practical as we can in our advice on this subject of the personal element in style, we'll put it into the form of comments on the six illustrative sentences that follow. These six sentences represent the range of choices you have in deciding how personal to let your style become, so far as use of pronouns is concerned.

First Person Singular, Active Voice
1. I got surprising results from the three tests I made.

First Person Plural, Active Voice
2. We got surprising results from the three tests we made.

Third Person (Singular or Plural), Active Voice
3. The laboratory staff got surprising results from the three tests they [or "it"] made.
4. The three tests gave surprising results.
5. The writer got surprising results from the three tests made.

Third Person (Singular or Plural), Passive Voice
6. Surprising results were given by the three tests.

Until recently, the style represented by choice 1 was traditionally considered

---

[9] Committee on Form and Style of the Council of Biology Editors, *Council of Biology Editors Style Manual*, 3d ed. (Washington, D.C.: American Institute of Biological Sciences, 1972), p. 5.
[10] "Stromberg-Carlson Technical Literature Style Guide," *Stromberg-Carlson Practices* (Issue 3, October 1974; Section 00-001-00), p. 3.

unacceptable in formal reports, and in most other technical writing as well. The style of sentences 2–5 was acceptable (with the natural exception that "we" in choice 2 was acceptable only if it referred to a group, not just a single author). And the style of choice 6 (third person, passive voice) was widely regarded as *the* scientific style, as we said earlier.

If your organization has no objection to your doing so, you can let your style become somewhat "personalized" through the use of first-person pronouns; but remember these cautions: (1) Too many uses of "I" or "my" may cause your reader to decide you're conceited. Too many personal pronouns can become irritating. (2) There's a difference between reporting on original research you've carried out entirely by yourself and reporting on rather routine work done as a member of a group. In the latter circumstance the advice quoted earlier from the Whirlpool Corporation seems particularly sensible: "Write the report from an objective point of view, so as to focus the reader's attention on the subject and not on you."

Quite apart from the considerations just noted, however, the four styles represented by the six sentences listed above are not all equally effective. The style of choice 5 ("The writer . . .") would seem awkward and artificial in most circumstances. Similarly, the style of choice 2 ("We . . .") surely would seem pompous except where this pronoun actually referred to two or more people. Among the other four sentences (1, 3, 4, 6), the principal distinction, as far as effectiveness is concerned, seems to lie in the choice between the active and passive voices. And here is a problem that requires a little thought.

Scientific Attitude and the Choice of Voice   It is often argued that the active voice is the more vivid and effective, and comparison of sentences 4 and 6 undoubtedly lends support to this contention. Here they are again:

> The three tests gave surprising results.
> Surprising results were given by the three tests.

The first sentence has the obvious advantage of saying in six words what the second says in eight. And on the purely subjective grounds of how we happen to like the two sentences, we ourselves much prefer the first. But let's look at another example.

> Discussion at the meeting of the Board of Directors clearly revealed the weakness of plan X. Plan Y was adopted.

The second of these two sentences is passive, yet it seems to do very well. You could write the active, "They adopted plan Y" — but it's no shorter, and the referent of "they" is disturbingly vague. "The Board adopted plan Y" is still one word longer than the original, and perhaps you will feel that the repetition of "the Board" is slightly unpleasant. We ourselves will stick to the passive, "Plan Y was adopted."

These examples seem to point to two conclusions concerning this rather subjective problem of choosing between the active and passive voices. First, thoughtless use of the passive voice may unquestionably produce writing that is both wordier and less vivid than need be. And second, crisp and effective sentences can certainly be written in the passive. We're sorry we don't have any rules for you to go by, but it really isn't a very difficult problem if you just use your head.[11] The best time to attend to this stylistic problem is probably in the revision rather than the first draft of your writing.[12]

Style and Sexist Language

Style and Sexist Language   A problem requiring special attention has developed in recent years concerning the use of the masculine pronouns, "he, his, and him." As you no doubt know, there has been a strong protest, related to the women's liberation movement, against the use of a masculine pronoun in sentences in which the referent might actually be either masculine or feminine. An example of such a sentence is, "No citizen should neglect his duty." Although this problem may appear in any sort of writing, it is especially likely to appear in technical writing during the description of a process, where it is common to find sentences such as the following: "The operator then connects the cable. He does so by using a special tool." Like the citizen in the previous example, the operator might be either a man or a woman. At present, many of the important publishing companies have prepared special style sheets for their authors with regard to this problem of the masculine pronoun, but, as far as we are aware, nobody claims to have found a good solution. One universal recommendation by publishers is that "he or she" be substituted for "he" when necessary—but that this awkward phrasing not be used too often. Similarly, of course, "his or hers" would be substituted for "his," and "him or her" for "him." But as the publishers point out, this is indeed an awkward kind of solution.

Often the problem can be avoided very easily merely by using neutral nouns and the plural form of pronouns rather than the singular. For example, instead of writing, "Every man must wear his protective clothing while he is working in designated areas," you could write, "All employees must wear their protective clothing while working in the designated areas."

In summary, our suggestions concerning the problem of personal pronouns in technical writing are the following: First, make sparing use of "I" or avoid it altogether. Second, use "We" only where the referent is obviously appropriate. Third, with rare exceptions avoid using "The writer . . ." One point at which, in principle, using "The writer . . ." is effective is in the description of a process where the phrase becomes transformed into "The operator . . ." or some such term (see Chapter 7). Choose thoughtfully between the active and passive voice, on the basis of the immediate context. Finally, do what you can to prevent offense by using language that does not imply sexual bias.

[11] In the fourth edition of the *Council of Biology Editors Style Manual* (1978), the third edition of which is quoted on p. 33, authors are encouraged to "Use the active voice except where you have a good reason to use the passive. . . . Avoid the 'passive of modesty,' a device of writers who shun the first person singular. 'I discovered' is shorter and less likely to be ambiguous than 'it was discovered.' When you write 'experiments were conducted,' the reader cannot tell whether you or some other scientist conducted them" (p. 21).

[12] For further discussion of voice (and mood) see pp. 130–132 in Chapter 7.

PART 2

We come now to more detailed problems of style as found in sentences, vocabulary, and paragraph structure.

Making Sentences Say What You Mean

Besides giving attention to the needs of the reader and maintaining an objective manner, the technical writer must try hard to achieve accuracy and clarity. A great deal of bad writing results from the writer's failure to think carefully enough about what sentences actually say. Perhaps this fault is a habit of mind as much as anything; usually, when people are shown that sentences they have written don't make good sense, they recognize the difficulty at once and are puzzled over not having noticed it before. However this may be, one of the essentials of learning to write well is certainly the development of a habit of critically analyzing one's own sentences.

The kind of bad writing we are concerned with here is illustrated by the following passage:

> A problem usually arising in the minds of laymen considering solar heating for their home is the glare which might result from the use of large areas of glass. Actually, however, *just the reverse* has been found to be true. Large windows, while admitting more usable light, *produce less* than several small openings.

The italicized phrases don't convey the meaning the author intended. Just the reverse of what? Produce less what?

Here is a second illustration of the same sort of bad writing:

> The greatest problem which is found due to using large panes of glass is caused by the fact that glass is an excellent conductor of heat.

There are a lot of unnecessary words in this sentence, but that isn't the worst blunder in it. The worst blunder is the statement that glass is an excellent conductor of heat. The young man who wrote these words, a student of architectural engineering, knew very well that glass is a poor conductor, but he had a picture in his head of a lot of heat coming in through the window pane, and he knew the picture was correct. So he wrote down some words that had to do with the transfer of heat and was satisfied. What he was satisfied with was the picture in his head, which was a good picture. He paid little attention to the words. When he was later shown the words, he saw at once that they were wrong.

To avoid mistakes of this kind, put aside a piece of writing for as long as you can after finishing the first draft. Leave it until you can see the words instead of the pictures in your head. For some people, reading aloud is a help in spotting faulty passages. Ultimately, of course, everything depends on using words that mean precisely what you want to say.

Precision in the Use of Words

Precision in the use of words requires the technical writer to have an exact knowledge of the meaning of words, to avoid words that — in a given context — are vague, to leave out unnecessary words, to use simple words wherever possible, to avoid overworked or trite words, and to avoid technical jargon.

Knowing What Words Mean   Unfortunately, many words are used incorrectly in technical writing. We list below a sampling of those commonly confused or misused. Reference to a good dictionary or to books like Fowler's *A Dictionary of Modern English Usage*, Evans' *Dictionary of Contemporary American Usage*, and William and Mary Morris' *Harper Dictionary of Contemporary Usage* will help you with them — and many others like them.

ability/capacity

adjacent/contiguous — *touching*

advise/tell, inform

affect/effect

alternative/choice — *many*

among/between — *2*

anticipate/expect — *unavoidable*

apparent/obvious/evident — *evidence*

appreciate/understand

assume/presume *presumptions*

assure/insure/ensure *to make certain*

balance/remainder — *what's left*

bimonthly/semimonthly — *every 2 wks*

conclude/decide — *end doubt*

continual/continuous — *without ceasing*

deteriorate/degenerate — *living things*

effective/effectual — *capable of producing an effect*

encounter/experience — *thru*

essentially/basically

few/less

events become known

filtrate/filter — *out*

indicated/required   *observing*

infer/imply — *under surface meaning (speaker)*   *reader— gets another meaning*

liable/likely   *answerable*

maximum/optimum — *the best*   *biggest*

oral/verbal — *by word (oral or written)*   *aloud*

percent/percentage — *a part of a whole.*

perfect/unique — *one of a kind.*   *w/o flaw*

practical/practicable — *can be done.*   *sensible*

preventative/preventive   *don't use*

principal/principle — *article of faith.*   *money   main reason*

proportion/part *usually use*   *acting*

reaction/opinion — *thinking*   *new*

replace/reinstall — *old thing redone*   *w/ respect*

respectfully/respectively *in order mentioned*   *shorting*

target/objective   *bias*

theory/idea, view, opinion

transpire/occur

universally/generally — *maybe some exceptions*   *always*

waste/wastage — *loss by erosion, decay*

Avoiding Vague Words   Most often, however, precision of meaning is lost, not through outright error in the use of terms but by the use of words that, although not incorrect, do not convey the exact meaning demanded. For example, words like "connected," "fastened," or "attached" are used instead of terms that more accurately denote the nature of a connection—terms like "welded," "soldered," "bolted," and "spliced." Of course there is nothing wrong with any of the words mentioned; it is in the way the words are used that trouble may develop.

Leaving Out Unnecessary Words   Words that serve no useful purpose should be rigorously weeded out of your reports during the process of revision. A comprehensive discussion of the ways in which words can be, and are, unnecessarily used lies far beyond the scope of this chapter. We will discuss only a few ways. The positive principle we want to establish is probably demonstrated as well by a few examples as it would be by a great many. This principle is simply that you should take a hard look at every word in a sentence to make sure that it is there for a good reason.

To start with, various pitfalls are associated with the use of abstract words like "nature," "character," "condition," or "situation." (Please remember that it is not the word itself but the incorrect use of the word that we are criticizing.) Consider the following sentence: "The device is not one of a satisfactory description." What does the word "description" contribute to the sentence? Nothing at all. The sentence might better have read: "The device is unsatisfactory." Here are some other examples:

> The principal reason for this condition is that the areas which were indicated for street purposes were not intelligently proportioned. [*Better:* The principal reason for poor traffic flow is that the streets were not intelligently laid out.]

> An easy example for explanation purposes would be a shunt-type motor. [*Better:* A shunt-wound motor is a good example.]

In both original sentences the word "purposes" is used unnecessarily. In the following sentence the word "nature" is ineptly used:

> The soldering proved to be of an unsatisfactory nature. [*Better:* The soldering proved to be unsatisfactory.]

Finally, here is a sentence in which "position" is at fault:

> With this work now completed, the plant is in a position to proceed with work on the new product.

What revision would you make of this sentence?

A second common source of trouble is the use of modifying words that look fine at first but actually mean little or nothing. Examine the list below and then notice in your reading how often they turn out to be meaningless.

appreciable	fair
approximate	negligible
comparative	reasonable
considerable	relative
definite	sufficient
evident	suitable
excessive	undue

These are all good words when they are used with a concrete reference. But consider the following examples:

This newly developed machine proved to be comparatively efficient. [This sentence is not meaningful unless we know the efficiency of the machines with which comparison is made.]

Water-flooding effected a substantial increase in production. [This means little without specific amounts.]

The voltage regulator must definitely be checked at periodic intervals. [*Better:* The voltage regulator must be checked at periodic intervals. *Or:* It is important that the voltage regulator be checked at periodic intervals.]

But what is a periodic interval? The time should be stated. A last illustration follows:

Research personnel have made appreciable progress in solving this problem. [*Translation:* We haven't found out anything yet, but we have several ideas we're working on.]

A third source of unnecessary words is the use of pointlessly elaborate prepositions and connectives. The sentences below illustrate this problem:

Greater success has been enjoyed this year than last *in the case of* [by] the engineering department.

This problem is *in the nature of* [like] one encountered years ago.

Our reports must be made briefer *with a view to* ["to" is enough] ensure more successful research-production cooperation.

This recorder has been installed *for the purpose of providing* [to provide] a constant check of volume changes.

Many phrases and clauses used in introducing the main idea of a sentence are unnecessary, and they are often pompous-sounding and stilted as well. Study these examples:

It is perhaps well worth noting that the results of this study show that plant efficiency is low. [If the main idea the author wants to communicate is that "plant efficiency is low," the elaborate introductory clause is a waste of words. The clause can be justified only if the writer wants to emphasize the idea that "it is worth noting" that plant efficiency is low. "Perhaps" surely serves no useful purpose in either case.]

It will be observed that test specimen A is superior to test specimen B. [Possibly all the author wanted to say is that "test specimen A is superior to test speciman B." If so, the introductory clause is unnecessary. If the author really wanted to say that the superiority of A to B will be *observed,* then the sentence was all right.]

There is no inherent fault in the introductory clauses used above, or in others like them (such as "it will be noted," and "consideration should be given to"), but fault does lie in saying more than is meant and in using a great many words to say what could be said more emphatically and clearly with a few.

A comprehensive list of wordy, redundant phrases found in technical writing would make a book by itself. We conclude these remarks on wordiness with a miscellaneous list of frequent offenders.

absolutely essential (essential)
actual experience (experience)
aluminum metal (aluminum)
at the present time (at present, now)
completely eliminated (eliminated)
collaborate together (collaborate;
    "together" is unnecessary in
    many phrases, such as "connect
    together," "cooperate
    together," and "couple
    together")
during the time that (while)
few in number (few)
in many cases (often)
in most cases (usually)
in this case (here)
in all cases (always)
involve the necessity of (necessitates,
    requires)
in connection with (about)

in the event of (if)
in the neighborhood of (about)
make application to (apply)
make contact with (see, meet)
maintain cost control (control costs)
make a purchase (buy)
on the part of (by)
past history (history)
prepare a job analysis (analyze a job)
provide a continuous indication of
    (continuously indicate)
range all the way from (range from)
red in color (red)
stunted in growth (stunted)
subsequent to (after)
through the use of (by, with)
true facts (facts)
until such time as (until)
with the object of (to)

Words are used unnecessarily in many more ways than those we have pointed out, but the problem of avoiding unnecessary words is always to be solved in basically the same way: by thinking about what each of the words in a sentence is contributing to the meaning.

Using Simple, Familiar, Concrete Words Probably nobody would deny the wisdom of avoiding unnecessary words, but beginning technical writers are often reluctant to admit that simple and familiar words should be chosen in preference to "big" words. In fact, they may resent such a practice because it denies them the free use of the technical vocabulary they have been at such pains to acquire. Furthermore, they may feel that substituting simple words for technical terms will inevitably result in a loss of precise meaning, or even a loss of dignity and "professionalism."

Many years ago, Thomas O. Richards and Ralph A. Richardson,[13] both of General Motors Research Laboratories, pointed out a curiously interesting fact:

> *We have never had a report submitted by an engineer in our organization in which the explanations and terms were too simple.* [Italics ours.] We avoid highly technical words and phrases and try to make the work understandable, because we know that even the best engineer is not an expert in all lines. . . . Most reports err in being too technical and too formal.

These men were not talking about writing for people without any technical back-

---

[13] Thomas O. Richards and Ralph A. Richardson, *Technical Writing* (Detroit: General Motors Corporation, 1941), p. 4.

ground, someone like a stockholder or a director, but about writing for other technical people.

A large company of builders and contractors declared that one of the essential qualities of a good report is that it be clear, concise, and convenient, and added that "the use of technical words should be limited as far as possible to those with which the prospective readers are familiar." The Tennessee Valley Authority manual on reports has as one of the criteria in its report appraisal chart the question, "Is the language adapted to the vocabulary of the reader?" In 1945, E. W. Allen,[14] of the United States Agricultural Research Administration, made a comment that possibly reflects the tremendous sense of pressure of the years of World War II, but it remains good advice today.

> . . . it is necessary to understand and keep in mind the point of view of those it is desired to reach . . . it is not enough to use language that *may* be understood—it is necessary to use language that can not be misunderstood. . . . The style of the technical paper should be simple, straightforward, and dignified.

The list below provides a few examples of the problem these people were talking about. Most of the terms in the left-hand column are perfectly good words, and they are the best words in certain contexts. But if you mean "parts," why say "components"? Or if you mean what may be written as either "name" or "appellation," why not take the simpler word? Unless you have a good reason don't substitute

initiate	for	begin
disutility	for	uselessness
compensation	for	pay
conflagration	for	fire
veracious	for	true
activate	for	start
ramification	for	branch
verbose	for	wordy

H. W. Fowler wrote sensibly and wittily of this problem in *A Dictionary of Modern English Usage,* in such articles as "Love of the Long Word" and "Working and Stylish Words."

On the other hand, don't ever sacrifice precision for simplicity. Some ideas can't be expressed in simple language, and there's no use trying.

Making Familiar Comparisons   All effective technical writing in one way or another requires new knowledge to be connected to old knowledge. This means that the writer must establish connections between information the reader already possesses and new information the writer can supply. The most effective method for doing this has been to use familiar comparisons. Parables in the Bible and other ancient writings provide ample proof that this method has been universally recog-

[14] E. W. Allen, *The Publication of Research* (Washington: U.S. Agricultural Research Administration, 1945), p. 4.

nized by great teachers as a primary tool of communication. In an age when the diversity and complexity of technology are sufficient to bewilder even brilliant minds, the need for familiar comparisons becomes particularly acute. Obviously the writer who lacks a clear awareness of who the intended reader is to be is in no position to know what comparisons will be familiar and what will not, since the comparisons must be drawn from the world that the specific reader is familiar with. To state that an object is the size and shape of a .50 caliber machine-gun cartridge, for example, would be an effective comparison if the reader can be assumed to be familiar with machine-guns. For someone lacking that familiarity, the object might be better compared to a scaled-down space rocket.

The larger the expected audience is, the more diverse it is likely to be assumed to be and the greater the need for the comparisons to be based on common knowledge. Robert Jastrow, for example, in writing an astronomy text geared to a high school level audience, uses terms that any high school student would know to explain the immense size of the Solar System:

> Let the sun be the size of an orange; on that scale of size the earth is a grain of sand circling in orbit around the sun at a distance of 30 feet; the giant planet Jupiter, 11 times larger than the earth, is a cherry pit revolving at a distance of 200 feet or one city block; Saturn is another cherry pit two blocks from the sun; and Pluto, the outermost planet, is still another sand grain at a distance of ten city blocks from the sun.

> On the same scale the average distance between the stars is 2000 miles. . . . An orange, a few grains of sand some feet away, and then some cherry pits circling slowly around the orange at a distance of a city block. Two thousand miles away is another orange, perhaps with a few specks of planetary matter circling around it. That is the void of space.[15]

Of all the familiar comparisons used in technology, perhaps the most frequent is the analogy used to explain electricity and electrical circuitry to the layman. In the standard analogy, electricity is compared to water in a pressurized system: the wiring is compared to water pipes, voltage is compared to the pressure of the water in the system, amperage is spoken of as "current" to compare it to the rate the water moves in the system, resistors and capacitors are compared to valves that can regulate the pressure and the rate of flow. The basic assumption here is that a person who needs or wants a basic understanding of electricity is very likely to possess a basic understanding of plumbing. Obviously the analogy would be relatively worthless if the purpose is to enable uneducated natives of a primitive culture to take proper safety precautions while working around electrical apparatus. In this instance, one might do better to compare the behavior of electricity to the behavior of good and evil spirits!

Avoiding Overworked Words and Phrases   Some words and phrases are used so often that they seem to be second nature to technical writers. Although such trite words and phrases are not necessarily wrong, their frequent use makes them

---

[15] Robert Jastrow, *Red Giants and White Dwarfs: Man's Descent from the Stars.* New Edition (New York: W. W. Norton & Co., 1979), p. 28.

tiresome to discriminating readers. Moreover, such terms are likely to be pretentious and wordy. Since the beginning technical writer may have difficulty in recognizing trite words and phrases, take our word for it that the words and phrases we list below are overused. Keep alert in avoiding them—and dozens of others like them.

activate (begin)
approach (answer, solution)
appropriate (fitting, suitable)
assist (help)
cognizant authority (proper authority)
communicate (write, tell)
consider (think)
demonstrate (show)
develop (take place)
discontinue (stop)
effort (work)
endeavor (try)
facilitate (ease, simplify)
function (work, act)

implement (carry out)
indicate (point out, show)
investigate (study)
maximum (most, largest, greatest)
on the order of (about, nearly)
optimum (best)
personnel (workers, staff)
philosophy (plan, idea)
prior to (before)
subsequent to (later, after)
terminate (end, stop)
transmit (send)
utilize (use)
vital (important)

Avoiding Technical Jargon   In writing technical documents for readers who lack a thorough familiarity with the subject matter, you should avoid shoptalk or technical slang. Such terms may be clear to workers in your scientific or technical field, they may be colorful, and they are usually natural and unpretentious; but they will not serve your purpose if they are not known to your readers. The list below suggests the kind of term we mean:

breadboard (preliminary model of a circuit)
call out (refer to, specify)
ceiling (limit)
know-how (knowledge, experience)
megs (megacycles)
mike (micrometer, microphone, microscope)
optimize (put in the best possible working order)
pessimize (deliberately put in poor working order)
pot (potentiometer)
state of the art (present knowledge)
trigger (start, begin)
-wise (added to many terms like budget, production, design)

One other word that deserves special mention is "data." The problem with this familiar word is not what it means but how it is pronounced and whether it is singular or plural. There are three acceptable pronunciations: dayta, datta, and dahta. The first is the preferred pronunciation (that is, the commonest). And the word may be used as either singular or plural. This word is of Latin origin, and in Latin the word "data" is plural and "datum" is singular; but usage has made "data is" acceptable—with the exception that not everyone agrees. Some people still prefer to think of data as plural. There's no great point in fussing about it, especially if the person who demands "data are" is your boss. You should, of course, be consistent in your treatment of this troublesome word, and make your treatment appropriate to the context.

Sentence Structure and Length

Good technical writing calls for a natural word order, simple sentence structure, and fairly short sentences.

The normal, natural order of elements in English sentences is (1) subject, (2) verb, and (3) object or complement. Each of these elements may be modified or qualified by adjectives or adverbs. The normal position of adjectives is in front of the terms they modify. Adverbs usually appear before the verb, but often after. This order of parts should generally be followed in your sentences for the sake of clarity and ease of reading. Furthermore, subject and verb should usually be close together. Naturally, departure from these patterns is occasionally desirable to avoid monotony.

The following sentences illustrate some typical word orders:

Natural Order
1. The machine was designed for high-speed work.

Natural Order with Modifying Words and Phrases
2. This 90-ton, high-speed machine was efficiently designed to provide the motive power for a number of auxiliary devices.

Inverted Order
3. Remarkable was the performance of this machine.

Periodic Order
4. When these tests have been completed and the data has been analyzed, there will be a staff meeting.

The order of sentences 1 and 2 is usually preferred to that of the other two. In sentences 3 and 4 the principal subject is not clear until near the end of the sentence. Periodic and inverted sentences may certainly be used occasionally, but most of your sentences should be in the natural order.

So far we have been concerned with the effect of word order on the readability of sentences. Closely related is the type of sentence structure employed. In general, simple sentences should outnumber the other kinds: complex, compound, and complex-compound. You will recall from your study of composition that a simple sentence contains only one clause and that a clause is a group of words containing a subject and a predicate. Examples 1, 2, and 3 above are simple sentences. A complex sentence contains an independent clause plus one or more dependent clauses. A compound sentence contains two or more independent clauses. A complex-compound sentence contains two independent and at least one dependent clause.

Complex

1. When all other preparations are made, the final step may be taken. [The introductory clause here functions as an adverb and is dependent upon the main clause for its full meaning.]

Compound

2. The first stage of this process can be completed under the careful supervision of the shop personnel, but the second stage must be directed by trained engineers. [The compound sentence consists of two statements linked by a conjunction.]

Complex-Compound

3. If this process is to succeed, the first stage can be completed under the careful supervision of the shop personnel, but the second stage must be directed by trained engineers. [Here a qualifying dependent clause precedes the first main clause. Additional qualifying phrases and clauses could, of course, be added, further complicating the sentence.]

Reading is slowed by too large a proportion of complex and complex-compound sentences. What is too large a proportion? We wish we could answer that question with a precise figure, but we can't. The writer must have a sense of proportion—and we do intend that word to mean two things: a percentage and a balance or harmony.

You should also be careful about the length of your sentences. The amount of difficulty a person experiences in reading a given text is positively correlated with sentence length and number of syllables per word. Research indicates that the average sentence length should probably not exceed 20 words. Of course this does not mean that every sentence should be limited to no more than 20 words. Nor is it necessary to avoid all words of more than three syllables. Technical subject matter often requires the use of a complex technical vocabulary and the expression of complex ideas. But if you should discover that your sentences are long and your words have many syllables, the chances are that you can simplify. And always keep in mind the range of your reader's familiarity with your subject. For an interesting illustration of the practical application of these principles, see Appendix C, p. 498. Here you will find that the suggested maximum average sentence length is 25 to 30 words. This length seems to us a little too much, but no certain knowledge exists on this point.

Following is an example of how a difficult job of reading can be made easier by simplifying the sentences. For this example, we are indebted to the Ethyl Corporation Research Laboratories at Detroit, Michigan. It was used in a course in writing provided for their staff. First, here is the original passage:

> Although it is recognized that the question of soap content versus lubricating efficiency is a controversial subject in the grease industry, it is believed that the long record of eminently satisfactory lubrication performance, frequently under adverse conditions where no other grease was adequate, is sufficient evidence that high-soap content is not a detriment insofar as barium greases are concerned. Further, it is felt that a comparison between different types of greases solely upon the basis of soap content is rather a pointless argument unless proper cognizance is taken of the differences in the molecular weight of the bases, of the ultimate

effectiveness of the greases and lubricating bearing surfaces under service conditions, and of the various factors of composition that radically modify the oil-thickening action of the different soaps.

Here is the way this material was rewritten in the Ethyl Corporation's course in technical writing:[16]

The grease industry has long debated the effect of soap content on lubricating efficiency. Still, a long record of highly approved performance should show a high soap content is no drawback in barium greases. Often they have succeeded where other greases failed.

Comparing greases by soap content is rather pointless anyway unless such questions as these are answered:

1. How do molecular weights of their bases differ?
2. How well does the grease lubricate bearing surfaces in service?
3. What else in the grease might alter the oil-thickening action of the soaps?

The revised version has 88 words, the original 130; sentences are much shorter; and there are fewer polysyllabic words. We think the revised version is a lot easier to understand.

It is quite as possible to go to extremes in the use of short, simple sentences as in the use of complex sentences. If you go too far in the use of simple sentences, you may find yourself writing something like this:

He did not do well with the company at first. Later he managed to succeed very well. Finally he became president of the company.

This is bad writing because there is no use of subordination in it. All the ideas are given the same weight. Linking the three sentences together with simple conjunctions — "but later," "and finally" — would eliminate the unpleasant choppy effect, but what is really needed is subordination of one idea, something like that in the following complex sentence:

Although he did not do well at first, he was later very successful, finally becoming president of the company.

The word "although" subordinates the first clause. Such a word is called a subordinating conjunction. Some other words that will serve this function are: after, because, before, since, in order that, unless, when, where, while, why.

In general, then, the best policy is to make most of your sentences simple in structure and natural in order, but to vary the pattern enough to avoid unpleasant monotony and to provide proper emphasis.

Paragraph Structure and Length

Typically, a paragraph begins with a sentence (the topic sentence) that presents the idea to be developed. The other sentences of the paragraph develop, support, and

[16] T. J. Carron, *Training the Professional-Technical Employee* (Detroit: Research and Development Department, Ethyl Corporation), p. 4.

clarify this central idea. But, as a matter of fact, you have probably observed that this topic sentence may appear anywhere within the paragraph. It may appear in the middle, or it may appear last, as a summary or generalization based on material already presented. Sometimes it doesn't appear at all, in so many words, but is implied. The requirements of technical style being what they are, we urge you to follow the tried practice of placing the topic statement first in the paragraph, or, at the very latest, just after whatever transitional sentences appear. The technical writer doesn't want a reader to be in suspense as to what the subject is.

Compare the following two versions of a paragraph from a Shell Oil Company manual.[17] Version A is the original; version B is our revision, for the purpose of illustration.

Version A
These instructions are not designed to cope with exposure environment where highly corrosive vapors are encountered, although the paints recommended do have substantially good corrosion-resistant properties for normal plant tank farm conditions. Where such environments are encountered, special coatings may be required, such as vinyls, chlorinated rubber, Epon resin vehicle materials, or standard and other special paint systems applied to sprayed zinc undercoatings. In these cases proprietary brands may be used until open formulations are available. Experience in the field and the use of exposure test panels, pH indicators, and other methods will determine whether it will pay to apply the more expensive corrosion-resistant coatings. Special corrosion problems should be referred to the Atmospheric Corrosion Committee for investigation. On the other hand, the instructions, specifications and formulations contained in this manual are designed to cope adequately with exposure environments existing in the general run of tank farms where hydrocarbons and the less corrosive chemicals are stored.

Version B
The instructions, specifications and formulations contained in this manual are designed to cope adequately with exposure environments existing in the general run of tank farms where hydrocarbons and the less corrosive chemicals are stored. They are not designed to cope with exposure environment where. . . . [This version continues by completing the first sentence in Version A and concludes at the end of the next-to-last sentence in Version A.]

The main idea (the topic sentence) in version B is stated at the beginning so that readers will know without delay just what the object of the discussion is. It is true that readers need to know what will not be covered, but it is more important for them to know what will be covered by the discussion. In version A readers do not find this out until the very end of the paragraph. Verson B is the better of the two.

Two considerations govern paragraph length: unity of thought and eye relief for the reader. Since the paragraph is defined as the compositional unit for the development of a single thought, it may seem to you that length should be governed entirely by requirements of the development of the thought. And in theory, that's right. A simple, obvious idea, for example, might not take much development — perhaps no more than two or three sentences. A complex and highly important idea might, according to this line of reasoning, require a large number of sentences, perhaps covering several pages.

[17] From *Protective Coating Manual*, p. 2. Reprinted by permission of the Shell Oil Company.

Long paragraphs, however, do not permit easy reading. If there is no break in an entire page, or in more than a page, the reader's attention flags; it becomes difficult to keep the central idea in mind. Since long, unbroken sections of print repel most readers, the writer should devise paragraphs so that such sections will not occur.

An unfortunate, but common cause of many long paragraphs is that writers use paragraphs to convey material that is better suited for some other vehicle of communication. Few writers would make the mistake of arranging a personal shopping list in paragraph form, but they frequently overlook the advantages of the list and the many possible choices of graphic presentation when they need to present a long series of items, measurements, or specifications.

Breaking up discussion so that the reader's eye is given some relief does not demand violation of basic principles of paragraph development. It does not mean that paragraphing should be merely arbitrary. Nevertheless, the writer has a good deal of freedom in deciding what will constitute a unit of thought. An idea containing several parts or aspects may be broken up, with the sentence that originally stood as a topic sentence for a long paragraph serving as an introductory statement to a series of paragraphs. Let's consider a hypothetical case. Suppose a writer had written:

> For a brief explanation of the meaning of the term "skip distance" in radio communications, we must first turn our attention to the phenomena of the ground wave, the ionosphere, and the sky wave.

Suppose further that this sentence stood as the topic sentence and that it was followed by a description of the three phenomena, all in the same paragraph. The paragraph would run quite long, too long for comfortable reading. The solution would be simple. Instead of one long paragraph, there could be three shorter ones, one on each phenomenon. The original topic sentence could serve as an introductory, transitional paragraph, perhaps with the addition of another sentence something like this: "Each of these phenomena will now be described in detail." In other words, the writer can manage organization so that the material can be divided into conveniently small units.

When you desire an especially forceful effect, try using one or more very short paragraphs.

To sum up, remember that all sentences in a paragraph must be about the same topic, but also remember that paragraphs should not be too long. Try to have one or more breaks on every page of your report.

Summary

1. Technical writing style is distinguished by adaptation to the reader, by attention to the communication situation, by observance of the scientific attitude, and by certain conventions in writing symbols, numbers, and abbreviations.
2. Most organizations expect reports to be written in the passive voice, but other possibilities are useful.

3. It is highly desirable to develop a habit of looking critically at sentences to make sure that they express precisely the ideas they were intended to express.
4. Words and phrases must be used with precision.
5. Clarity and ease of reading are improved by moderately short sentences and paragraphs.
6. The organization of both sentences and paragraphs should usually be natural, with main ideas appearing near the beginning.

Sentences for Revision I

The sentences below are not so succinct or clear as they might be. Rewrite them for greater conciseness, without omitting essential content.

1. They ~~evidenced a surprisingly uniform communality of attitude to the effect~~ *agreed* that t~~he most vital area of training was~~ the develop~~ment of~~*ing* military skills and courtesy~~,~~ *were the most important training.*
2. ~~It is seen that~~ there ~~are~~ five output voltages from the analog computer. ~~These voltages~~ are proportional to the yawing velocity of the fighter aircraft.
3. ~~Prior to the conductance of these tests,~~ *Before* condensed moisture should be removed *dry* from the equipment by ~~either~~ inverting or tilting, ~~which~~ever ~~is more compati-ble with its configuration.~~ *depending on its ___* *Condensation should free.*
4. Numer~~als are used to~~ *bes* identify the ~~various~~ adjustment screws ~~provided~~ on the panel ~~located~~ inside the door ~~of the equipment.~~ *of*
5. Poor living ac~~commodations give promise of incrementing~~ *quarters* *increase* the negative side of the morale ba~~lance so far as~~ *we* *the* new personnel ~~are concerned.~~
6. ~~It is~~ expected to co~~mplete the full~~ integration ~~of~~ these new units into the system ~~as a whole~~ by early ~~in the~~ next month.
7. It would ~~seem desirable to terminate the prior~~ *be better to stop* *using the old* process and i~~nitiate the new one~~ *use a* ~~if optimum results are to be secured.~~ *for best results.*
8. The proposed program is ~~intended for the utilization of~~ foresters ~~who are in the employ of the United States government~~ *U.S.* *to ensure best* ~~in seeing to it that~~ fire prevention ~~is~~ *fire prevention* *^carried out* ~~with optimum~~ results.
9. This diagram indicates that there are twenty-one instrument servomechanisms in the control room which do the necessary computing for the system.
10. Due to the many and varied applications a system of this type may have in the immediate future, it is felt that techniques should be utilized which will give the system the maximum amount of versatility and reliability.
11. ~~On the basis of past history,~~ *As in the* it ~~is expected by~~ management t~~hat great progress~~ *expects* ~~will be made by~~ personnel ~~in providing a solution~~ *solve* to these problems ~~in the near~~ *soon.* ~~future.~~
12. ~~The~~ first th~~ing that~~ *we* must ~~be done~~ is edit the report.
13. Personnel of the purchasing department must prepare a cost estimate for the purpose of making it possible to make a purchase.
14. ~~In most case~~*usually*s the installation of a monitoring device that provides continuous indication of deviations from the normal will permit the reduction of shutdowns.

15. Whether or not these ~~anticipated~~ operations to correct errors in procedure enable the staff to cooperate ~~together~~ more efficiently, it is intended that they be inaugurated without undue delay by reason of communication difficulties.

16. Enclosed ~~herewith~~ is a list of important essentials that should be subject to coverage in the next conference ~~dealing with the matter of~~ absenteeism.

17. ~~It is to be~~ hoped that work to be scheduled will not involve the ~~necessity of any undue~~ overtime work in ~~the neighborhood of~~ the holiday period.

18. ~~In this quite unique design,~~ labels have ~~been provided for the purpose of~~ identifying each ~~of the various~~ controls.

19. ~~There is a city-owned pier running out from this land which is used by~~ a marine repair firm. uses a public pier.

20. ~~From a cleaning point of view,~~ these valves are ~~relatively good.~~ cleaning

---

Sentences for Revision II

The following sentences are too short and "choppy." In each instance combine the short sentences into one well-constructed longer sentence. In doing so, you may need to reword some of the short sentences, and to rearrange their sequence.

1. We finished buffing the parts. We gave them a primer coat.

2. These reports show that the tests have been disappointing. The authors don't say so.

3. The work will be finished. The results will be reported. The decision as to whether to continue will be made.

4. Now we have the necessary equipment. The project can be started. The expected postponement will be unnecessary.

5. Centralized control would lead to expansion of the management division. The management division is too large. An alternative solution is available. This solution will now be explained.

6. The flow of water in this stream has been reduced to a trickle. Property owners are complaining. The water is used for irrigation.

7. Some want to spend additional money on pollution control. Some want to spend additional money on health care. There isn't much money.

8. Let us consider the possible consequences. Our problem is whether to invest more money. Our information is inadequate.

9. The problem is to determine the depth of the well. It is discussed briefly in the report. The problem is more complex than the author indicates. It must be solved at once.

10. A good technical writing style does not make use of many extremely short sentences. A good technical writing style does not make use of many extremely long sentences.

Effective Word Choice Exercise

In the sentences below, you are offered some word choices. Sometimes the choice is between an acceptable ("correct") word and an unacceptable ("incorrect") word; at other times, the choice is between a word with precise and suitable meaning, and one that is less precise and suitable. Rewrite these sentences and be prepared to justify your choices. You may need to consult a good dictionary or one of the books on usage, such as the Evans' *Dictionary of Contemporary American Usage,* or the Morris' *Harper Dictionary of Contemporary Usage.* These books contain discussions of words that are closely related in meaning and which are frequently misused. In making your choices, you should assume that the sentence is to appear in a formal, written document (speech allows greater freedom — or lenience). *& colloquial expressions*

1. Ms. Brown will (accept/except) the invitation.
2. The man paid all interest and part of the (principal/principle).
3. Government taxing and spending seriously (affect/effect) the economy.
4. The next step is to (filtrate/filter) the fluid.
5. The machine has (degenerated/deteriorated) through overuse.
6. A new design is not (indicated/required).
7. They did not (consider/think) many changes should be made.
8. This successful engineer proved to be most (ingenious/ingenuous).
9. The acid finally (eroded/corroded) the pipes.
10. The project engineer (informed/advised) the staff that no overtime work would be required.
11. (Oxidization/oxidation) should be prevented.
12. The chief engineer paid them a fine (complement/compliment).
13. Close supervision seriously (affects/effects) our success.
14. The time (passed/past) quickly.
15. The manager's remarks (infer/imply) that the engineer's report wasn't believable.
16. The geologists spent the (remainder/balance) of the time working.
17. The committee selected a (sight/site/cite) for the meeting.
18. A large (percentage/percent) of this report is useless.
19. Our supervisors are (continually/continuously) trying to help us.
20. We would like to (devise/device) a new (devise/device) for this purpose.
21. The secretary sent the memorandum to (its/it's) destination.
22. The president is worried about the (economic/economical) situation.
23. The task was divided (among/between) five engineers.
24. A few days' work destroyed all their (allusions/illusions).
25. We did not think a holiday was (likely/liable/apt) to be given.
26. They did not believe they needed any (council/counsel).
27. All the subordinates were (aggravated/irritated) by them.
28. The engineer's plan was (practical/practicable), but it was not (practical/practicable).
29. The department has (all ready/already) met its responsibilities.
30. The manager did not know how to (adopt/adapt) the report to departmental purposes.
31. They would not agree to the plan: they were (disinterested/uninterested).
32. There were (fewer/less) people at this meeting than at the last.
33. The designer's (implicit/explicit) instructions were written in detail.

colloquial – don't use.

34. The chief engineer was (enthused/enthusiastic) about the plan.
35. Now employed in Houston, the cashier was (formally/formerly) in Washington.
36. The supervisor was chosen to (administrate/administer) the program. *don't use*
37. He gave us (oral/verbal) instructions.
38. The mixture was said to be (inflammable/flammable). *mean same* – *preferred*
39. The specifications called for (bimonthly/semimonthly) reports. *either*
40. We must proceed, (irregardless/regardless) of difficulties. *not a word* *referring to people.*
41. The new system was designed to (ensure/assure) success.
42. We thought the decision (equitable/equable).
43. Although she did not say so, we (inferred/implied) that she was pleased.
44. The committee asked him for his (opinion/reaction).
45. We asked for a written (estimation/estimate).
46. Many such (incidence/incidents) have occurred.
47. The plan outlined in the report had (obvious/evident/apparent) merit.
48. Although this design was not greatly different from others proposed, it had some (unique/unusual) features.
49. We will (utilize/use) this material in our report.
50. (More than/Better than) a million dollars were spent on the project.
51. The worker (fixed/repaired) the damage.
52. He (acquainted/told) us (with) the facts.
53. She described her (approach/solution) to the problem.
54. Plans for this project must be (finalized/completed) soon.
55. The supervisor (concluded/decided) to make some changes.
56. We do not (envision/expect) any difficulty in completing the job.
57. The project engineer did not (consider/think) that any changes should be made.
58. The results were (nowhere near/not nearly as) good. *colloquial*
59. The scientists declared that they (appreciated/understood) the problem.
60. This research project does not need to be carried (further/farther).
61. Our (target/objective) for the quarter was an increase in production.
62. In solving the problem, he (assumed/presumed) nothing.
63. The investigators (encountered/experienced) many difficulties.
64. The device proved to be most (effectual/effective).
65. Should it be (desirous/desirable) to obtain a steady tone rather than the interrupted signal, more work will have to be done.

refers to person what a person feels.

4

Outlines and Abstracts

Introduction

Outlines and abstracts are much alike in one respect — both are highly condensed statements of, or descriptions of, the content of a piece of writing. For this reason they are taken up together in this chapter. In some respects it would be more logical to discuss only outlines at this point and to defer consideration of abstracts until after examination of various types of reports. Such a sequence of study can easily be managed simply by skipping the section on abstracts in this chapter, for the present, and returning to it later. On the other hand, you may find that study of abstracts in direct relationship to the study of outlining will be helpful in clarifying the basic concepts of organization in technical writing, and will provide a good background for the later examination of various techniques of writing and various types of reports.

Abstracts are written solely for the convenience of the reader. Outlines, on the other hand, serve both reader and writer. They serve the writer as an analytical device for studying other people's writing, and also as a means of organizing his or her own writing. In technical writing, outlines serve the reader in the form of the table of contents of a report, and also as the system of subheads within the report. It is worth remembering that an outline of almost everything you write on the job will be presented to your readers in the form of the table of contents and the subheads.

We will discuss outlines first, then abstracts, and finally introductory sum-

maries. An introductory summary is a combined introduction and abstract, as will
be explained later.

Outlines

Kinds of Outlines   There are three kinds of outlines that are useful to a technical
writer: sentence, topic, and question. In a sentence outline each entry is a complete
sentence. Conversely, in a topic outline each entry is a phrase or a single word; no
entry is in the form of a complete sentence. These two outlines are of use to both the
reader and the writer. If you will now turn to pp. 70 and 72, you will find examples
of these two kinds of outlines. The third kind, the question outline, is of almost no
use to the reader, but it can be very useful to the writer, especially in the early stages
of the writing.

   The sentence outline has one important advantage over the topic outline, but
it also has at least one important disadvantage. The advantage is that in making a
sentence outline, the writer is forced to think out each entry to a much greater
degree than for the topical form. In a topical form a writer might say merely,
"Materials"; in the sentence form it would be necessary to say something like, "The
materials required are seasoned white pine, glue, and whatever finish is desired."
The greater thoroughness of the sentence outline lessens the possibility of ambiguity and vagueness in the thought. It also means, on the other hand, that the sentence
outline is more difficult and time consuming to write than the topical. The sentence
outline is an excellent analytical device for studying the organization of a given
piece of writing. The topic outline, however, is probably more practical as a guide for
writing, although opinions differ about this. It is not a good idea to combine the two
forms. There is nothing greatly wrong with such a combination, but it does indicate
an inconsistency in the logical process that might appear later as an inconsistency in
the text itself.

   The question outline has its greatest value during the exploratory stages of the
writing process. Initially, it is just a list of all the questions the writer foresees a need
to answer somewhere in the course of the writing project. As the list grows longer, it
should become obvious that some questions have more in common with each other
than with others in the list. In a thoroughly developed question outline, some
questions will bear on the process of writing, others will pertain directly to the
subject matter to be discussed. By grouping and subgrouping related questions, a
plan for the writing process and a plan for the structure of the paper should both
begin to emerge. Let us point out here that this type of outline is not a specific stage
in the writing but is a process that continues up to the stage of the final draft of the
paper. Once significant questions have been raised, the writer can concentrate on
obtaining and presenting significant answers. As the answers become available they
should replace the questions. After all the questions have been answered, the question outline should have been transformed into a sentence or a topic outline that can
now be used to meet the needs of the reader.

   All outlines are based fundamentally on a combination of two simple principles. The first is the arithmetic principle of division; the second is the principle of classification. In addition to these two principles, we'll discuss the signal-to-write aspect of outlines, make some suggestions about how to prepare an outline, and conclude with a few examples.

The Arithmetic Principle   Since outlining is, from one point of view, a method of dividing, it naturally conforms in a certain degree to the principles of arithmetic. Let x equal the entire subject to be divided, or outlined. Then $x = \text{I} + \text{II} + \text{III} + \cdots n$. In turn, $\text{I} = \text{A} + \text{B} + \text{C} + \cdots n$, and $\text{A} = 1 + 2 + 3 + \cdots n$, and so forth. Please understand that this is more than an analogy. It is a principle that not only can be, but also should be, applied to every outline you write, to test its logical soundness. For instance, we might consider the following simple example from the outline on p. 70. This outline is taken from a report on the subject of sanitation in isolated construction camps. We are considering only part I.B here.

 I. . . . .
  A. . . . .
  B. Stopping the spread of these diseases by breaking the cycle of transmission
   1. Removing or destroying the breeding places of insects and rodents
   2. Killing the adult insects and rodents

This might be rewritten in the following form:

> Stopping the spread of these diseases by breaking the cycle of transmission = Removing or destroying the breeding places of insects and rodents + Killing the adult insects and rodents. Therefore,

$$B = 1 + 2$$

After thinking about this equation, however, you may object and say that it is not necessarily valid. That is, if all adult insects and rodents were destroyed, it would be pointless to worry about their breeding places. The equation would therefore be reduced to $B = 2$.

This objection is certainly justified here, but nevertheless it is justified only if we think of our subject matter as actually being arithmetic rather than outlining. But we do not want to think of outlining as being identical with arithmetic.

The two propositions we want to make are the following. First, the principle that $\text{I} = \text{A} + \text{B} + \cdots n$, and so on, should be used to test the logical soundness of any outline. Second, as the first proposition indicates, outlining does conform to a certain degree with the fundamental principles of arithmetic. But now we must emphasize that this conformity between outlining and arithmetic is indeed only to a certain degree; it is by no means complete. What we need to think about, then, is how the first proposition can be true if there is a difference between outlining and arithmetic. What is the difference?

We are not entirely joking when we say that this question will be easy to answer for anyone who has tried to kill all the adult insects and rodents around a construction camp. In the practical world of outlining technical reports, we remember what formidable antagonists insects and rodents have proven to be.

To tell the honest, unvarnished truth, perhaps we should rewrite our equation like this:

$$\begin{bmatrix} \text{The partly successful} \\ \text{efforts of the sanita-} \\ \text{tion engineer to pre-} \\ \text{vent an outbreak of} \\ \text{disease} \end{bmatrix} = \begin{bmatrix} \text{Trying very} \\ \text{hard to kill} \\ \text{the adult insects} \\ \text{and rodents} \end{bmatrix} + \begin{bmatrix} \text{Trying very} \\ \text{hard to} \\ \text{destroy} \\ \text{breeding} \\ \text{places} \end{bmatrix}$$

In summary, the principles of division and addition can be used to test the logical soundness of an outline, but they must be used within the practical world of the subject matter of the outline rather than just within the abstract world of arithmetic.

Should this summary leave you with the feeling that there must be some loose philosophical and logical ends lying around, we heartily agree. The whole problem shades off into fascinating but exceedingly puzzling questions. You might enjoy pondering it from the point of view of a little book by Ernest Nagel and James R. Newman entitled *Gödel's Proof.* In practical terms, however, all you have to do is to make sure that the subdivisions of any part of your outline add up in a commonsense way to the point you are trying to make in that specific part.

The discovery that the equation might take the form of B = 2, as noted above, calls attention to a question that is often asked. What is wrong with having only a single entry under some heading in an outline? For a particularly simple example of this problem to think about, let's turn to a different subject. Let's say we are writing an outline on the subject of power sanders and have got as far as the following entries (dotted lines indicate material omitted for brevity):

I. Introduction
   A. ....
   B. ....
   C. ....
   D. ....
II. Types of sanders
   A. Vibrating
      1. Straightline
      2. Orbital
      3. Combination
   B. Belt
      1. Straightline

Having got this far, we come to an abrupt halt because we realize there is no further entry to put under "Belt" — no number 2. Since all belt sanders are straightline sanders, there is simply no other subdivision that will here go with "Straightline" in the way "Orbital" did in part A. Consequently, it is pointless to have the heading "Straightline" in part B. So all we can do is cancel this heading and continue with the outline, as follows:

I. Types of sanders
   A. Vibrating
      1. Straightline

```
      2. Orbital
      3. Combination
  B. Belt
  C. Disk
  D. Drum
```

This is just another way of expressing the idea that the fundamental principle of outlining is division. And nothing can be divided into fewer than two parts. If there is an A, there must also be at least a B; if there is a 1, there must be at least a 2; and so on.

But, you may be thinking, suppose that for my particular reader I need to do only two things, as far as belt sanders are concerned. First, I need to point out that the belt sander is one of the types to be considered; second, I must explain that there is a smaller range of choices of grits in sanding belts than there is in sanding sheets. Why not simply write the following?

```
I.  Types of sanders
    A. Vibrating
       1. Straightline
       2. Orbital
       3. Combination
    B. Belt
       1. Limitation of choices of grits
    C. Disk
    D. Drum
```

You are wondering, in other words, if there aren't some situations in which it is sensible to have a single subdivision. Yes, if the writing required is brief, and if the outline is just a list of reminders about facts with which the writer is thoroughly familiar, the preceding outline might be all right. As we further ponder the question of whether it would really make good sense to set up an outline like the one above, however, we come to a helpful insight about how outlines work. We think you will probably decide that the sort of outline shown above would be all right only when the subject was so simple, the amount of writing required so little, and your familiarity with the subject so great that you scarcely needed an outline at all.

Let's imagine we have begun writing a discussion of power sanders, using the outline above as a guide, and have finished with part A and are ready to start on part B.

We sit there staring at the word "Belt." We are supposed to write something. What should we say? Apparently we are supposed to say something we have stored away in our mind about Belt, except that what we say must not be about the choice of grit. As the separate entry on the choice of grit shows, our discussion of grit comes second, after our discussion of Belt.

In short, our outline instructs us to say at least two things about belt sanders, and it reminds us explicitly what one of these things is, but it does not remind us of what the other, or others, may be. When writing about an extremely simple subject, this sort of vagueness may not create much trouble, but you can imagine what happens when the subject becomes so complex that the writer has trouble keeping it all clear. That is a very different situation.

What entries should the outline have had, then? As an absolute minimum, we'd say, something like the following would prove helpful (we're considering only part B).

> B. Belt
>   1. Special characteristics of belt sanders
>   2. Limitation of choices of grit

As a final comment on the arithmetic principle in outlining, we should consider how this principle affects the physical form the outline takes on paper. Various forms have been worked out, the objective in each case being to devise a physical arrangement that will help reveal the fundamental logical structure of division and subdivision. Tables of contents are often set up in a similar way. The following form is most commonly used for outlines. The dotted lines represent the text.

I. ....
....
  A. ....
  ......
    1. ....
    ....
      a. ....
      ....
        (1) ....
        ....
          (a) ....
          ....
          (b) ....
        (2) ....
      b. ....
    2. ....
  B. ....
II. ....

Observe the following points: (1) periods are used after symbols (that is, numbers or letters) except when the symbol is in parentheses; (2) in an entry of more than one line, the second line is started directly beneath the beginning of the first sentence or word entry; (3) the symbol of a subdivision (A, 2, and so on) is placed directly beneath the first letter in the entry of the preceding highest order; (4) periods are placed at the end of complete sentence entries, but not after topic entries; and (5) lines are usually double-spaced.

One other aspect of form calls for attention: the need for parallel grammatical structure in the sentence or topic entries. This may sound like an unimportant matter, but our own opinion is that carelessness in this respect is like the tiny fissure on an exposed slope of earth that, if not attended to, may become a badly eroded gully. Parallelism is explained on pp. 467–468.

The Principle of Classification   Once a subject has been divided into parts, the parts have to be brought together into logical groups. It might seem at first that the principles of arithmetic already considered would be sufficient for this purpose, but, as we'll see in a moment, this assumption proves to be false. More exactly, the principle we will be speaking about is only one element within the general subject of classification, but it is the most fundamental element — it is the principle of selecting and controlling the actual basis of the classification. The whole subject of classification will be taken up later, in Chapter 8.

Classification is the gathering together of things into logical groups. To form a logical group, a basis is needed on which to determine what belongs in the group and what doesn't. Suppose, for example, we wanted to organize all football players into logical groups. One way of doing so would be on the basis of position played: all quarterbacks in one group, all tackles in another, and so on. Another basis on which the players could be grouped would be the number of years of experience each has had in playing on an organized team. And, of course, many other bases could be chosen.

To see how the principle of a basis of classification enters into organizing a report, you might think about what is involved in something like a long report for your course in technical writing. Let's imagine, for example, that you had decided to write a report entitled "Considerations Involved in Choosing Which Musical Instrument to Learn to Play." Probably many ideas would soon come to mind. One part of the report would no doubt be concerned with a review of the types of instruments there are to choose among. Another part would be devoted to the relationship between certain special abilities and the choice of an instrument. Large hands are desirable for a piano player or a stringed instrument player, for example, but are not necessary for a trumpet player. A violinist needs a good sense of pitch, but an organist does not. Still another part of the report might be about the special advantages offered by the various instruments. For a person who wanted to be a soloist, the oboe would be a poor choice; for one who wanted to play jazz, the trumpet would be a good choice. Perhaps, then, our first rough attempt at an outline might look something like this:

I. Introduction
II. Types of instruments
III. Special abilities and the choice of an instrument
IV. Special advantages and disadvantages offered by individual instruments
V. Summary

This doesn't look too bad. The arithmetic works out all right; the topics to be discussed add up to about what a person would want to know in deciding which instrument to play.

Next, let's develop part II, the first substantive part of the outline.

II. Types of instruments
A. Woodwinds
B. Strings
C. Brasses
D. Percussion

Part II passes the arithmetic test. It is true that $II = A + B + C + D$. Nevertheless, closer inspection reveals that we have in fact got into a little trouble — and the trouble is one of the commonest kinds found in organizing reports. What we have done wrong, of course, is to use two different bases of classification to create one group of things. The group of things we are creating is "types of musical instruments." One basis of classification we've used for this purpose is the material out of which the instruments are made (woodwinds, brasses, and — in a sense — strings). The other basis of classification is how the sound is produced (percussion is the most obvious instance). Let us therefore eliminate material as a basis and limit ourselves to the other basis — how the sound is produced. Now we have an acceptable grouping:

    II. Types of instruments
      A. Wind
      B. Strings
      C. Percussion

The principle involved in this change is that *only one basis can be used for a given group.*

Naturally, we would later subdivide wind instruments into woodwinds and brasses. The basis of this subdivision is "material," and since it is the only basis used for this subgroup, it is perfectly acceptable here.

The organization of any piece of writing, except for the shortest, requires arranging the parts into logical groups; consequently, the selection of bases of classification inevitably becomes involved. Awareness of this simple fact can prevent a great deal of wasted time and confusion.

A special problem arising out of the role of bases of classification in outlining is that a writer sometimes becomes trapped by his or her logic. For an illustration of this curious fact, let's go back to an outline we were playing with earlier.

    II. Types of sanders
      A. Vibrating
        1. Straightline
        2. Orbital
        3. Combination
      B. Belt

The basis of classification used for the sanders within part A is the kind of motion made by the tool's sanding surface. On some vibrating sanders the motion is straight back-and-forth; on others it is orbital; and on still others ("combination" sanders) the user can select either motion by moving a switch. But all this availability of choice does not hold true of a belt sander, in which the sanding surface is literally a belt that runs over rollers. So there are no such subdivisions under "Belt."

A writer who is really trapped by logic may write page after page under a topic like "Belt" with never a heading to help the reader along. Because the basis of classification used for vibrating sanders doesn't work for belt sanders, it is apparently assumed that no subdivision is possible. Incredible as it may seem, we once saw a long proposal, for a contract representing a very large amount of money, in which at one point there was a stretch of over a hundred pages without a single subhead to help the reader. The people who prepared this proposal had been trapped by exactly the same logic you saw in our simple example. Of course there were actually a great many subtopics within those hundred pages, and what the writers should have done was to figure out what these subtopics were and set up subheads for them.

This bit of reminiscence brings up one other curious fact about the organization of reports — at least we believe it's a fact. We have become convinced through personal experience that very often when people complain that there are a lot of bad sentences in a report, the real trouble lies in the organization. Perhaps their situation resembles that of a man who complains of a pain in his leg but who is told upon examination that his backbone has got a little disorganized. We don't mean to say that poorly written sentences are not a common problem. They certainly are. Nevertheless, our personal experience is that the best place to start, in hunting for the trouble in a report people are complaining about, is not with the sentences but with the organization.

Previously, we remarked that the outline of a technical report becomes the table of contents and the system of subheads in the finished report. Let's take a brief, summary look now at outlines from the point of view of some examples of tables of contents. The advantages of this point of view are the perspective it gives on the value of informative headings, on the benefits of having neither too few nor too many headings, and on the confusion created by a disregard of the sort of logic we have just been considering.

As you look at the examples that follow, remember that technical reports, even very long ones, almost never contain an index. For assistance in finding a particular passage in the text, the reader must rely on the table of contents.

One difference between the function of an outline and that of a table of contents is that the headings in an outline are primarily reminders to the writer about things he or she already knows. If the headings in a table of contents are not self-evident sources of information, on the other hand, they aren't much good. A heading like "Vibration Tests," for instance, is not nearly so useful in a table of contents as something like "Procedures for Vibration Tests."

Sometimes the question of whether a heading is sufficiently informative becomes involved with the problem of how to set up the subheads. A common example is found in that kind of report called a "proposal" (see Chapter 14). In a proposal, the most important single section is usually one concerned with how to solve some problem. All too often this section of a proposal will have only a heading like "Technical Discussion," or "Detailed Description," or "Technical Approach." Such a general heading may be defensible if the discussion is brief; it is not very informative at best, however, and if the discussion is long, the table of contents becomes unnecessarily complex. Thus version 1 below is improved if presented in the form of version 2.

Version 1

   III. Technical discussion
     A. Buoy electronics package
       [Nine Arabic-numeral subheads appeared here in the original.]

   B. Shipboard receiving station
      [Three subheads.]
   C. Experience in supplying related equipment
      [Three subheads.]
   D. Reliability
   E. Quality assurance
   F. Testing

Version 2

   III. Buoy electronics package
        [The nine Arabic-numeral subheads now become A through I.]
   IV. Shipboard receiving station
       [Subheads A, B, and C.]
    V. Experience in supplying related equipment
       [Subheads A, B, and C.]
   VI. Reliability
  VII. Quality assurance
 VIII. Testing

Here is another example, this one containing too many entries.

   A. Experience..14
      1. Sonar 14
      2. Seismic 14
   B. Personnel..14

Why not just write it this way?

   A. Experience in sonar and seismic work...............................14
   B. Personnel..14

So brief a discussion as section A represents is scarcely in need of having its subdivisions labeled by separate headings. (On the other hand, it is very important that the text itself be clearly divided, probably by separate paragraphs for each topic and a transitional phrase or two.)

The following table of contents reveals a defect in logic.

   II. Circuit redesign
       A. Preamplifier
       B. Modulator and demodulator
       C. Ripple and lead networks
       D. Servoamplifier
       E. Packaging

Clearly, the entry "E" should appear as "III." In terms of our previous discussion of the application of the principles of arithmetic to outlining, the heading "Circuit redesign" is less than the sum of "A" through "E." Here is another illustration of the same problem.

    V. Data-transfer subsystem
       A. Introduction
       B. System operation

      C. Airborne unit
        1. Data-processing unit
        2. Transmitter
        3. Power supply
      D. Ground unit
        1. Translator
        2. Telemeter receiver
        3. Subcarrier discriminator
        4. Countermeasures considerations

Here, obviously, the last item, 4, should appear as "E" rather than 4, since "Countermeasures considerations" are not a component of the ground unit.

    Our last example is difficult to categorize. Does it represent a failure of logic or simply a careless redundancy?

    III. Technical discussion
      A. Data-processing techniques
        1. Time-shifting linear addition
        2. Cross-correlation
      B. Time shifting
      C. Characteristics of correlator filter

Something looks very odd here, with one of the subheads under "A" reappearing as "B." Perhaps the general heading "Technical discussion" could have been deleted and the following setup used, with appropriate subheads under "A" and "B":

    III. Data-processing techniques
      A. Time shifting
      B. Cross-correlation

The above excerpts from tables of contents are all drawn from actual reports.

The Signal-to-Write Aspect of Outlines   The entries in an outline can be divided into two kinds: signals-to-write and titles. We again return to the outline about sanders for an illustration. One version we looked at in that outline was the following:

    II. Types of sanders
      A. Vibrating
        1. Straightline
        2. Orbital
        3. Combination
      B. Belt
      C. Disk
      D. Drum

Upon reaching the entry "Belt" in this outline, the writer would naturally begin discussing the subject of belt sanders. That is to say, the entry "Belt" is a signal to the writer concerning what should be written about next. In another version, however, this entry was subdivided as follows:

    B. Belt
        1. Special characteristics of belt sanders
        2. Limitation of choices of grits

As we look at these last three headings, we may observe still another fact about how an outline functions. We now have two explicit reminders of what to say about belt sanders. But what about the word "Belt" itself? What do we say concerning it? Absolutely nothing. The fact is that the word "Belt" has become merely a title.

This idea can be seen more clearly when we divide all entries in an outline into two kinds: titles, and signals-to-write. A title is any entry that has subdivisions under it. A signal-to-write is any entry that is not subdivided. In other words, if we are using the outline above as a guide for writing, we will not write anything about "Belt" but instead will begin by writing about "Special characteristics of belt sanders."

Let's follow this idea one step further, beginning by adding some entries to the outline.

    B. Belt
        1. Special characteristics of belt sanders
            a. Speed in removing material
            b. Danger of removing too much material in fine finishing work
        2. Limitation of choices of grits

According to the principle that any subdivided entry becomes a title, B.1. is now a title and no longer a signal-to-write. The first subject to be written about in part B is now B.1.a., "Speed in removing material," which is one of the special characteristics of belt sanders.

But now we may begin to feel apprehensive about a new problem. Isn't there danger in starting off a whole new part of our report (part B) by going at once to such a detail as B.1.a.? What if "a" itself had been subdivided, thus becoming a title, and our first signal-to-write had been B.1.a.(1)? Wouldn't following such principles in writing a report result in a text that would appear to be a mass of unorganized detail? It certainly would. And we believe this is in fact one of the commonest trouble spots in the writing of reports.

The fundamental problem we are confronting might be put this way. Since the signal-to-write aspect of outlines tends to lead the writer into beginning a new part of a report by discussing a mere detail, what can the writer do to make sure the details are integrated into a smooth-flowing report? The sudden appearance of a detail that seems unrelated to the preceding text is naturally puzzling.

One answer is that a special kind of signal-to-write can be put in the outline as an indication of the problem and as a suggestion of what should be done. These special signals-to-write that we are thinking about, as you may already have guessed, are called "introductions," "transitions," and "conclusions" or "summaries." It is unusual to see the word "transition" in an outline, but of course "introduction" and "conclusion" or "summary" are common. We haven't yet discussed these particular devices, but they are taken up in detail in Chapters 10, 11, and 12.

Another answer to the question of how to integrate the details into a smooth-flowing report is this: As writers acquire skill, they begin to use the "titles" in an outline themselves as signals-to-write — but in the second sense in which we used this term rather than the first. The titles (like "Belt") are understood not as signals to write directly about the subject matter of the report, but as signals to write an introduction or a transition. These ideas will be clarified in the chapters just mentioned.

How to Make an Outline   Let's distinguish carefully between two things: (1) using the principles of outlining as a help in organizing material about which a report must be written, and (2) writing the report itself, with the outline as a guide. Of these two things, the first is the subject we are now concerned with: how to make an outline.

It is hard to discuss this subject without implying one idea that is actually ridiculous. This ridiculous idea is that the minds of intelligent people work in an orderly process just as an outline is developed. Nothing could be further from the truth. Of course the purpose of an outline is to establish an orderly relationship among a group of facts or ideas. And intelligence could almost be defined as the ability to perceive relationships. Nevertheless, intelligent thought processes appear to be infinitely more complex and more varied in structure than is even remotely implied in the concept of an outline. Indeed, when you are instructed to prepare an outline for a paper you are to write, you may very well feel a distaste for doing so. You may feel that, logical though it appears to be, an outline doesn't have very much to do with how your mind really works. And you are quite right. It doesn't, except in very limited and explicit ways. These ways do not — apparently — include the most fundamental insights about how to present material.[1]

Typically, when people start making outlines, they have already reached a few conclusions. One is that they have so much material to write about that if they don't make an outline they'll get mixed up. The second is that they must give some attention to problems of the sort discussed in the section "Reader Adaptation" in Part I of Chapter 3. As a matter of fact, people often decide to make an outline only after they have actually started writing and have discovered they really are beginning to get mixed up. We see nothing particularly objectionable in going about the job in this fashion. If the writer doesn't have sense enough to stop writing before getting in too deep, however, a lot of time may be wasted.

At any rate, you will most often begin an outline with a general idea already in mind as to how you want to present your material. Before going very far with the outline you can clarify this general idea by writing a statement of purpose. That is, write a sentence beginning. "The purpose of this report is . . . ." If you can't finish the sentence, you need to think some more about what you are really trying to do.

The first thing to do with respect to the outline is to make lists of the topics that must be discussed. The great thing here is not order but inclusiveness. Don't bother about the sequence in which you list the topics you are to discuss; just don't omit anything.

---

[1] For a discussion of some relevant issues in the philosophy of science, see Michael Polanyi, *Personal Knowledge* (New York: Harper & Row, 1962), especially sections 7 and 8 in Chapter 8.

you have written down all the topics you can think of, the next step is to
t major groups you will have. These major groups become the Roman
numeral divisions in the outline. For example, if you decide you want a division on
equipment, "Equipment" becomes the heading for that division; and you then look
through your lists to find all the entries that pertain to this topic and jot them down
under the heading "Equipment."

As the outline nears completion, you can start testing the logic according to the
principles described and illustrated throughout the preceding discussion.

Summary of Discussion of Outlines

1. Outlines used as guides in technical writing are almost always in topical form,
   and it is in this form that they appear in tables of contents and as subheads in
   reports.
2. The two fundamental principles involved in outlines are division and the basis of
   classification.
3. A writer may become trapped by the logic of classification.
4. All entries in an outline function either as titles or as signals-to-write; the titles,
   however, may signal the need for transitions.
5. Outlines do not reflect the most basic creative intellectual processes, but are
   rather a means of organizing the results of these processes.

Abstracts

An abstract is a short description, or a condensation, of a piece of writing. It is a
timesaving device. Naturally, it is a device that is highly popular with executives.
The person whose opinion of your report matters most may read only the abstract
of it.

We will identify the two types of abstracts, note the advantages of each, and
then make some remarks on how to write an abstract.

Types of Abstracts   One type of abstract, the descriptive, tells what topics are
taken up in the report itself, but mentions little or nothing about what the report
says concerning these topics. This type of abstract is illustrated on p. 72. The advantages of a descriptive abstract are that it is easy to write and is usually short; a serious
disadvantage is that it contains little information.

The other type of abstract, illustrated on p. 73, is sometimes called "informational." The informational abstract gives the gist, or essence, of a piece of writing; it
includes the most significant material in the writing. It is the report in miniature.
Instead of learning merely that such and such topics are taken up in the report, we
are told something of what the report has to say about these topics. The advantage of
an informational abstract is that it provides much more information than does a
descriptive abstract. Of course it is harder to write, and it may be a little longer than
the descriptive type. Except where brevity is of special importance, however, there
can be no question as to the superiority of the informational type.

How long should an abstract be? A good rule of thumb is to make it as short as you can, and then cut it by half. Some people say it should be about 5 percent of the length of the report. In industrial reports, an abstract rarely exceeds one page.

It is illuminating to look at these differences from the point of view of the preparation of abstracts for a great published series such as *Chemical Abstracts.* The following comments about such published abstracts are contained in a paper presented by John C. Lane at the 138th National Meeting of the American Chemical Society:[2]

> . . . let's look briefly at two extremes [in types of abstracts]. The purpose of Chemical Abstracts had been defined as "preparing concise summaries . . . *from the indexing point of view."* Here the goal is to provide a timeless reference tool for chemists. Speed in reporting must necessarily be sacrificed for the sake of comprehensive coverage of the world's literature and for thorough indexing and cross-indexing. Thus, the time it takes an abstract to appear averages three to four months. But the value of these abstracts continues for years. In fact, from the literature searcher's point of view, their value increases with time — as their use is facilitated by annual and decennial indexes.

> At the other extreme, Current Chemical Papers, which has replaced British Abstracts, cuts appearance time to only a matter of weeks, but at a sacrifice in comprehensiveness. It is primarily an indexed listing of titles. This is practical only because our Chemical Abstracts fills the need for a comprehensive abstract journal written in English.

Usually, the abstract written to appear as part of a report circulated within a company or other organization resembles the first type that John Lane refers to. That is, it is essentially informational. Examples may be found at the end of this chapter.

In concluding these remarks on types of abstracts, we must point out that most abstracts are not exclusively either descriptive or informational, but a combination of both. This is perfectly all right. As a matter of fact, the first sentence in the descriptive abstract on p. 231 is more nearly informational than descriptive. Writing an abstract invariably presents a problem in compromising between saying everything you think you ought to and keeping it as short as you think you ought to. Descriptive statements here and there in an informational abstract often help solve this problem. Sometimes the term "epitome" is applied to a very short informational abstract in which only the most important facts or ideas are presented, and the term "abstract" is reserved for a longer, more detailed statement. Whatever the terminology you encounter, you have fundamentally two sets of conflicting variables to balance: brevity versus detail, and description versus information.

Suggestions about Writing Abstracts   The best single suggestion we can make about writing an abstract is to have a well-organized report to begin with. Having that, you simply write a brief summary of each one of the major divisions of the report. To help avoid getting tangled up in the details in the text, you may find it a good idea to write the abstract from the outline or table of contents rather than

[2] John C. Lane, "Digesting for a Multicompany Management Audience," Abstracts, 138th National Meeting of the American Chemical Society, Sept. 1960. Quoted by permission.

from the text itself. This advice is based on the assumption that you are preparing an abstract of a report you yourself have written and that you are therefore thoroughly familiar with the content. It is often difficult for an author to compress into the brief space available in an abstract all the thinking that has gone into a long report. Working just from the outline or table of contents may encourage a sense of balance and perspective.

If you are preparing an abstract of a report someone else has written, it is helpful to underscore the key sentences in the report before beginning the abstract. You can think of the task of writing the abstract as a problem in bringing together into one brief space the content of these key passages. Of course you might find the same method helpful when abstracting a report you had written yourself.

In any event, begin by writing a sentence that focuses the reader's attention on what you think is the principal idea communicated by the report. If you have difficulty deciding what that idea is, you may find some assistance in Chapter 10, which is concerned with the writing of introductions. Examples of good beginning sentences will be found in the illustrative materials at the end of the present chapter.

Once you have your beginning sentence, each additional sentence should constitute a development of the principal idea expressed in that beginning sentence.

As will be seen in the section "Illustrative Material," there is a tendency for abstracts of reports of research work to have a rather specialized form. Emphasis falls on the problem and the scope of the attack on the problem, on findings, on conclusions, and on recommendations. In contrast, reports on other kinds of subjects may lead naturally to abstracts containing virtually none of these subtopics.

In form, the abstract is usually set up as a single paragraph, double-spaced, on a page by itself. It should be written in good English; articles should not be omitted; and no abbreviations should be used that would not be acceptable in the body of the report. A special effort should be made to avoid terminology unfamiliar to an executive or any reader who is not intimately acquainted with the work. With the exception noted in the next section, the abstract should be regarded as a completely independent unit, intelligible without reference to any part of the report itself.

Introductory Summaries

Abstracts are sometimes called "summaries," so it is easy to guess that an introductory summary is a combination of introduction[3] and abstract. It isn't exactly a combination, however, in the way that H_2 and O make water; it is rather a joining together, as a handle and a blade make a knife — it's still easy to identify both parts.

There are really two kinds of introductory summaries. One is an ordinary abstract put at the top of the first page of the text of a report. The only thing introductory about it is the fact it is the first thing the reader sees. Since this is just a matter of what name you want to call an abstract by, we will say no more about it.

In the second type of introductory summary, special emphasis is given to the introductory portion. The idea back of this is to show clearly at the outset how the

[3] For a discussion of the elements of an introduction, see Chapter 10.

project being reported fits into the whole program of which it is a part. If the report itself is short, there may be no further introductory material. In longer reports there is likely to be a formal introduction following the introductory summary. There is always a temptation, however, to let the introductory summary do the whole job, even when a separate formal introduction is definitely needed.

The introductory summary reproduced below was written in the Pulp and Paper Research Institute of Canada. This organization customarily circulates such summaries separately from the reports to which they are introductory, much as many abstracts are published independently of the reports to which they refer. At the end of this introductory summary you will see a reference to the report concerned. The introductory summary could equally well have been printed as the initial section of this report.

The first three paragraphs of the introductory summary are the introductory part.

Delignification Using Pressurized Oxygen

Concern for environmental protection is making it necessary for kraft and sulphite mills to reduce their emissions both of noxious gases and of liquid effluents. Increasingly stringent legislation on the permissible levels of mill emissions is leading to very high costs for the installation of non-productive pollution control equipment. It would be a better long-term solution to this problem if new processes can be developed which do not create noxious effluents. One possible new process is to pulp wood in alkali under oxygen at moderately high pressure.

In recent work at the Institute to develop a commercially viable pulping process using oxygen as the delignification agent, our first aim has been to define the optimal range of the process variables in which pulping can be best achieved.

Consequently, the effect of several process variables (temperature, oxygen pressure, nature and charge of the alkali, liquor-to-wood ratio, gas-to-liquor volume ratio) on the kinetics of delignification of spruce wood has been studied, as well as their effect upon the physical characteristics of the resulting pulps.

The following results have been obtained:
— Delignification of wood by oxygen and alkali begins only above 140°C;
— The practical range of pulping temperature has been found to be between 140 and 160°C. In this range, increasing the temperature increases dramatically the rate of pulping but does not affect the ratio of the rates of dissolution of lignin and polysaccharides;
— The rate of pulping increases as the oxygen pressure increases up to a limiting value, where it reaches a plateau. The limiting value has been found to be of the order of 400–500 psi at 150°C;
— Optimum results were obtained for a 9–11% charge of alkali on wood, expressed as Na_2O, irrespective of the nature of the anion;
— Under optimal conditions, a high yield (50–55%) of a readily defiberisable pulp (0–7% Klason lignin) of medium brightness (50%) can be obtained from black spruce;
— Homogeneous pulping has been found to depend heavily on the efficient mass transfer of oxygen from the gas phase to the solution, and then on the efficient transfer of dissolved oxygen into the wood structure;

—The pulps obtained have physical properties quite similar to those of sulphite pulps. The individual fibres seem to be strong, undamaged, but inelastic.

This research has shown that oxygen pulping offers the opportunity of pulping softwoods to high yields of relatively bright pulp, with a low chemical consumption and drastically reduced pollution problems. Further research will aim at finding ways of optimizing the mass transfer phenomenon, at elucidating the mechanism of the reactions involved and at finding a suitable pre- or post-treatment to improve the physical characteristics of the pulps.

A detailed report on this study is being issued as PPR 75, "Delignification of Wood Using Pressurized Oxygen—I. A. Preliminary Study" by J. J. Renard, D. M. Mackie, H. G. Jones, H. I. Bolker and D. W. Clayton. If you would like to have a copy, please complete and return the attached reply card.

Illustrative Material

The materials in the following pages are divided into two parts.

The first part is made up of five exhibits, including an entire chapter, "Insect and Rodent Control," of a report entitled *Sanitation Requirements for an Isolated Construction Project,* which was written by Jerry Garrett while he was a student at the University of Texas at Austin. The first four exhibits in Part I are all based on this one chapter. The second part is taken from a booklet on report writing prepared for use in General Motors Research Laboratories at Warren, Michigan.

Part I   The five exhibits in this part are:

1. A topic outline
2. A portion of a sentence outline
3. A descriptive abstract
4. An informational abstract
5. "Insect and Rodent Control"—a chapter from *Sanitation Requirements for an Isolated Construction Project*

(Perhaps we should remark that all poisons mentioned in the materials included in his section are used under carefully controlled conditions and that some of them— for example, DDT—cannot legally be used at all except by special permission.)

Topic Outline

    I. Introduction

       A. Flies, mosquitoes, and rats as the vehicles of infection for ten widespread diseases

         1. Flies

           a. Mechanical transmission of disease

           b. Intestinal diseases they transmit

             (1) Typhoid

             (2) Paratyphoid

             (3) Dysentery

          (4) Cholera

          (5) Hookworm

     2. Mosquitoes

        a. Transmission of disease by biting

        b. Diseases they transmit

          (1) Malaria

          (2) Yellow fever

          (3) Dengue

     3. Rats

        a. Transmission of disease through harboring fleas

        b. Diseases they transmit

          (1) Plague

          (2) Typhus

  B. Stopping the spread of these diseases by breaking the cycle of transmission

     1. Removing or destroying the breeding places of insects and rodents

     2. Killing the adult insects and rodents

II. Breeding control

  A. Introduction

  B. Flies

     1. Breeding habits

     2. Control measures

        a. Sewage disposal

        b. Removal of manure

          (1) Time limit

          (2) Storage bins

          (3) Compression

        c. Destruction of all decaying organic matter

  C. Mosquitoes

     1. Differences from flies

        a. Greater difficulty in control of breeding places

        b. Small percentage that carry disease

     2. Disease-transmitting mosquitoes

        a. Female *Aedes aegypti*

          (1) Transmission of yellow fever and dengue

          (2) Breeding in clean water in artificial containers

        b. *Anopheles quadrimaculatus*

          (1) Transmission of malaria in southern United States

          (2) Habit of biting at night

          (3) Breeding in natural places

             (a) Preference for stationary water

             (b) Protection afforded by vegetation and floating matter

     3. Control measures

        a. Removing water

        b. Spreading oil on stationary water

  D. Rats

     1. Lack of direct ways to control breeding of rats or their fleas

     2. Prevention of breeding in specific areas

        a. Building rat-resistant houses

        b. Keeping rats from food

III. Adult control
    A. Flies
        1. Screens
        2. Traps
        3. Baits
           a. Fish scraps
           b. Overripe bananas
           c. Bran and syrup mixture
        4. DDT
    B. Mosquitoes
        1. Screens
        2. Larvae-eating minnows
        3. Poisons
           a. DDT
           b. Pyrethrum
    C. Rats
        1. Importance in property destruction as well as in disease
        2. Poisons
           a. Barium carbonate
           b. Red squill
           c. 1080
           d. Antu
        3. Trapping
        4. Fumigating

A Portion of a Sentence Outline

I. The fact that flies, mosquitoes, and rats transmit ten diseases makes it important that these insects and rodents be destroyed by preventing them from breeding or by killing adults.
    A. Flies, mosquitoes, and rats transmit ten widespread diseases.
        1. Flies transmit five intestinal diseases.
           a. Flies are mechanical carriers of diseases.
           b. They transmit typhoid, paratyphoid, dysentery, cholera, and hookworm.
        2. Mosquitoes transmit three diseases.
           a. Mosquitoes spread diseases by biting.
           b. They transmit malaria, yellow fever, and dengue.
        3. Rats transmit two diseases.
           a. Rats transmit disease through harboring fleas.
           b. They transmit plague and typhus.
    B. The spread of the diseases listed above can be stopped by breaking the cycle of transmission.
        1. The breeding places of insects and rodents can be removed or destroyed.
        2. The adult insects and rodents can be killed.

Descriptive Abstract

Ten widespread diseases that are hazards in isolated construction camps can be prevented by removing or destroying the breeding places of flies, mosquitoes and rats, and by killing their adult forms.

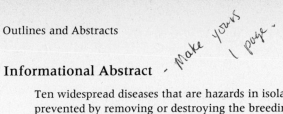

Informational Abstract

Ten widespread diseases that are hazards in isolated construction camps can be prevented by removing or destroying the breeding places of flies, mosquitoes and rats, and by killing their adult forms. The breeding of flies is controlled by proper disposal of decaying organic matter, and of mosquitoes by destroying or draining pools, or spraying them with oil. For rats, only the indirect methods of rat-resistant houses and protected food supplies are valuable. Control of adult forms of both insects and rodents requires use of poisons. Screens are used for insects. Minnows can be planted to eat mosquito larvae.

A Chapter from Jerry Garrett's Report

IV. Insect and Rodent Control[4]

INTRODUCTION

Flies, mosquitoes, and rats are the vehicles of infection for ten widespread diseases. Flies, which are mechanical carriers, are responsible for the transmission of the intestinal diseases; i.e., (1) typhoid, (2) paratyphoid, (3) dysentery, (4) cholera, and (5) hookworms. Mosquitoes spread diseases by biting; they are vectors in the cycle of transmission of (6) malaria, (7) yellow fever, and (8) dengue. Rats are the reservoirs of (9) plague and (10) typhus, but the rat's fleas are the vehicles of transmission.

There is but one way to stop the spread of these diseases, and that is to break the cycle of transmission. The best way to do this is to get rid of the insects and rodents, and the most effective method of getting rid of them is to remove their breeding places by good general sanitation. The only alternative is to kill the adults. Positive steps which may be taken in these operations are discussed below.

BREEDING CONTROL

As pointed out above, if there are no insects or rodents the diseases which depend on them for transmission must vanish. It is certainly cheaper and simpler to destroy their breeding places than to try to kill billions of adults only to find more billions waiting to be killed.

Flies
One characteristic of the fly makes it particularly susceptible to breeding control. The fly always lays its eggs in decaying organic matter, preferably excreta or manure. Three stages in the life of the fly — the egg, larva, pupa — are spent in the manure. A minimum of eight to ten days is spent here before the adult emerges. Therefore, the measures are relatively simple. First, there should be proper sewage disposal; i.e., the flies are never permitted to come into contact with human excreta. Secondly, all animal manure should be removed within four or five days, or in other words, before pupation takes place. The manure should either be placed in fly-proof storage bins or tightly compressed so that the adult fly cannot emerge after pupation. The final breeding control is to destroy all decaying organic matter such as garbage by either burying it two feet deep or burning it.

---

[4] "Insect and Rodent Control" is Section IV of *Sanitation Requirements for an Isolated Construction Project,* by Jerry Garrett.

Mosquitoes

It is not as simple to control the breeding places of the mosquito as it is to control those of the fly. But it can be done! First, it must be realized that there are many kinds of mosquitoes and that only a few are disease vectors. Still they must all be killed to be sure the correct ones are dead, and they are all important as pests anyway. The female *Aedes aegypti* is the vector for yellow fever and dengue; this mosquito breeds only in clean water in artificial containers. In the southern section of the United States (the chief malaria area in the United States), the malaria vector is the *Anopheles quadrimaculatus,* a night biter, which breeds in natural places, particularly where the water is stationary and where there is vegetation and floating matter to protect the eggs, larvae, and pupae.

Therefore, the best way to prevent the breeding of mosquitoes is to remove all water in which they breed by draining or filling pools, and removing or covering artificial containers. However, since the construction project is only temporary, the operators will be interested in the most economical measures rather than the most permanent. Artificial containers must still be covered, but it might be cheaper to spread a film of oil over all the natural, stationary water rather than to try to drain it or fill in the low spots.

Rats

There are no direct ways to control the breeding of rats or their fleas, but sufficient control can be exerted to make them take their breeding elsewhere. This is done by building rat-resistant houses and by preventing the rats from reaching food.

Adult Control

Flies

Houses should be screened to keep the flies from getting to food. Then, traps such as the standard conical bait trap should be distributed. The most attractive baits, as established by experiment, are fish scraps, overripe bananas, and a bran and syrup mixture. DDT may be used effectively to leave a residual poison for flies.

Mosquitoes

If a house is well screened, the mosquitoes cannot get into the house to bite their victims. Advantage can be taken of the mosquitoes' natural enemies by stocking waterways with minnows which eat the larvae. Poisons which may be used against mosquitoes are DDT and pyrethrum.

Rats

Besides carrying diseases, the rat of course destroys much property. Usually, however, the construction project operator need be concerned with rats only to the extent that they endanger his workers' health. Poisons which may be used against rats are barium carbonate, red squill, 1080, and antu. Other effective means of getting rid of rats are by trapping and fumigating.

Part II   The material that follows is taken from a rather typical booklet of instructions prepared for the technical staff of General Motors Corporation.[5] We do have one half-hearted complaint to make about the abstract (Figure 1) that appears as part

---

[5] Robert F. Schultz, *Preparing Technical Reports,* Research Publication GMR-427. (Warren, Mich.: Research Laboratories, General Motors Corporation.)

Abstract	
Chemists have long sought an electrolyte additive that would improve battery performance, especially at low temperatures. Ten organic additives were investigated to learn more about their effects on the rechargeability of an experimental lead-acid cell at 5°F. In each set of trials a different additive (0.001% by weight) was mixed with the sulfuric acid in the cell. The cell was then charged and discharged at eight current intensities. The additives included adipic acid, azelaic acid, p-phenolsulfonic acid, hydroquinone, 1-naphthol 5-sulfonic acid, phthalic acid, sucrose, tartaric acid, alpha–naphthol, and 8-hydroxyquinoline.	The "why" What you are trying to do Scope
Adipic acid resulted in the best high-rate charge acceptance; however, the level reached was only 96% of that achieved without any additives. Alpha-naphthol, the worst, reached a level of only 20%. In addition to the obvious conclusion that these additives do not improve rechargeability, the following trend appears valid: the higher the recharge voltage, the lower the charge acceptance.	Findings Conclusions

Figure 1 Sample abstract.

of this material. We don't think the first sentence is effective. See if you feel that something like the following would be better:

> Tests of ten electrolyte additives revealed that none of them improved battery performance. Ten organic additives . . .

Our complaint is only half-hearted because we are fully sympathetic with any researcher's reluctance to emphasize a negative result in a report. Could it be that this reluctance helped determine the form taken by the beginning of this abstract? However that may be, this is a rather typical example of an abstract of a research report. And, on the whole, it is well done. The marginal comments appear in the original, as does the box within which they are enclosed.

ABSTRACT

The abstract should present the report in a nutshell. Many readers, including management, may be too busy to read the entire report; therefore, the abstract might be the only chance to tell them what has been accomplished. Another reason for writing meaningful abstracts is that the Library is becoming increasingly dependent on them for cataloging information properly.

To communicate its message effectively, the abstract appears double spaced on a page by itself. It should not exceed 150 words.

Textbooks recommend that an abstract be informative, that it convey the essence of the report rather than merely amplify the title or describe the contents. In some cases, however, such as when a report consists of a procedure or a data compilation, a descriptive abstract is the only alternative. Following is a suggested guide for writing an abstract:

The problem (what you are trying to do, and, if not obvious, why)
The scope of your work
The significant findings or results
Any major conclusions
Any major recommendations

Suggestions for Writing

Opportunities for practicing the arts of abstracting and outlining are everywhere, and we suggest that you form a habit of noticing and taking advantage of them. For example, when you go to a public lecture, see if you can come away with a mental abstract of the lecture. Each time the speaker starts to discuss a new aspect of the subject — a new Roman numeral division — say over to yourself what seems to you the proper heading for it, and also for any divisions that preceded it. With only a little concentration you can leave the lecture with a mental outline of its contents. With this outline in mind, give a brief summary of the lecture to one of your friends, and you will have made an abstract, even though it isn't written down.

But practice in writing is essential too. Choose a magazine you enjoy reading, and outline and abstract an article you find interesting. It is particularly instructive to choose a journal whose articles are abstracted in one of the professional journals of abstracts listed on pp. 407 – 410, and later to compare your abstract with the one published in the journal.

Whenever it is feasible to do so, prepare an abstract of reports and exercises you yourself have written.

Finally, here are some outlines for you to try revising. Each of them violates one or more of the principles discussed earlier. There is no single right way to improve them, but all do need changes. Make the necessary changes, and indicate what principle required the change.

Houses

    I. Introduction
   II. Important types of residential architecture
      A. Modernistic
      B. American Colonial
      C. Greek Revival
      D. Georgian
      E. Adobe
      F. French Provincial
      G. Tudor
      H. Ranch

    I. Masonry
    J. Spanish
    K. Victorian
    L. Other
III. Cost
    A. Per square foot
IV. Summary

need a B here or no A

Head Coverings

    I. Introduction
   II. Purpose
    A. Protection
      1. From sun
      2. From rain
    B. Ornamentation
  III. Straw
  IV. Fabric
   V. Fur
  VI. Metal
 VII. Plastic
VIII. Summary

Writing Instruments

    I. Introduction
   II. Fountain pens
    A. Cartridge
    B. Bladder
    C. Plunger
  III. Ball-point pens
    A. Replaceable cartridge
    B. ~~The cheapest have a~~ nonreplaceable cartridge
  IV. Lead pencils
    A. Wood-encased
    B. Mechanical
      1. Choice of hardness of lead
    C. Colored lead
   V. Summary

Popular Games of the United States

    I. Introduction
   II. Strenuous athletic games
    A. Baseball
    B. Football
    C. Basketball
    D. Hockey
    E. Soccer
    F. Handball
    G. Tennis
    H. Volleyball
    I. Other

III. Nonstrenuous games of physical skill
    A. Golf
    B. Bowling
    C. Billiards
    D. Pool
    E. Foosball
    F. Video games
IV. Sedentary games
    A. Checkers
    B. Chess
    C. Games of chance
        1. Dominoes
        2. Roulette
        3. Bingo
        4. Card games
        5. Craps
    D. Other
 V. Summary

SECTION 11

Special Techniques of Technical Writing

Five techniques are of special importance in technical writing: definition, description of a mechanism, description of a process, classification, and interpretation. These techniques will be discussed separately in the five chapters that make up this section.

For emphasis, it is worth repeating that these techniques must not be considered as types of reports. Usually, several of them will appear in a single report. It would be exceptional to find an entire report, even a short one, containing only one of these techniques. For example, two or more techniques might be closely interwoven as a writer described the design, construction, and operation of a mechanism. The intermingling of these techniques, however, does not alter the basic principles of their use. And these basic principles can be studied most effectively by taking one technique at a time.

The treatment of these techniques will stress the practical rather than the theoretical, particularly in the chapters on definition and classification.

5

Definition

In this chapter on definition we have three specific objectives: (1) to clarify the problem of what should be defined in technical writing; (2) to suggest effective methods for defining what needs to be defined; and (3) to point out where definitions can be most effectively placed in reports.

What to Define

Before we can tackle the problem of *how* to define, we must think about *what* should be defined. It is not possible, of course, to set up an absolute list of terms and ideas that would require definition, not even for a specific body of readers, but it is possible and desirable to clarify the point of view from which the problem of definition should be attacked.

First of all, let's recall a rather obvious but extremely significant fact about the nature of language: as used in science, words are primarily labels or symbols for things and ideas. The semanticists—those who study the science of meaning in language—speak of the thing for which a word stands as its "referent." For instance, five letters of the alphabet, l-e-m-o-n, are used as a symbol for a fruit with which we are all familiar. In a sense, it is unimportant that these letters happen to be used, for the lemon would be what it is no matter what combination of letters was used to name it. This fact, however obvious, is an important one to keep in mind, for it often happens that a writer and reader are not in perfect agreement as to the referent for certain words. That is, the same word, or symbol, may call to the reader's mind a referent different from the one the writer had in mind, and thus communication may not be achieved. Or, more importantly for our purposes, a word used by the writer may not call to the reader's mind any referent at all. A reader who is familiar with banana oil may not have it called to mind by the technical term "amyl acetate" because the latter term is unfamiliar.

The relationships of words to the ideas and things for which they stand can become very complex, but without going into the problem of semantics any further we can discern a simple and helpful way of classifying words as they will appear to your reader. The words you use will fall into one of the following categories:

1. Familiar words for familiar things
2. Familiar words for unfamiliar things
3. Unfamiliar words for familiar things
4. Unfamiliar words for unfamiliar things

Each of these categories deserves some attention.

Familiar Words for Familiar Things   The only observation that need be made about the first category is that familiar words for familiar things are fine; they should be used whenever possible. To the extent that they can be used, definition is unnecessary. This might be dismissed as superfluous advice were it not for the fact that a great many writers appear to seek unfamiliar words in preference to everyday, simple terms. There is, as a matter of fact, a tendency for some people to be impressed by obscure language, by big words. Thus we find "amelioration" when "improvement" would do as well, "excoriate" for "denounce," "implement" for "carry out" or "fulfill." It scarcely needs to be pointed out that a "poor appetite" is not really changed by being called "anorexia." Nothing is ever gained by using, just for their impressiveness, words that may puzzle your reader.

Familiar Words for Unfamiliar Things   The words in this second category present a rather special problem to the technical writer. These are the everyday, simple words that have special meanings in science and technology.[1] Most of them may be classed as "shop-talk," or language characteristic of a given occupation. Because they are a part (often a very colorful part) of the language of a specialized field, it is easy to forget that they may not be a part of the vocabulary of the reader, at any rate not in the special sense in which they are used. Consider a term like "puddle." Everyone knows this word in the familiar sense, but not everyone knows that in the metallurgical sense it means a mass of molten metal. Or take "quench" in the same field. Quenching a metal by immersing it in water or oil bears some relation to quenching one's thirst, but it is a distant relationship.

---

[1] Our phrase "familiar words for unfamiliar things" does not cover all situations. Sometimes a well-known word is *unfamiliarly applied* to a well-known thing, and hence needs explanation. In anatomy, for example, the word "orbit" (a familiar word) means what most people call "eye socket" (a familiar thing).

Every field of engineering and science has a great many of these simple words that have been given specialized meanings. Examine the following list (you could probably add a number from your own experience):

apron: as on a lathe, the vertical place in front of the carriage of a lathe. This term is also used in aeronautics, navigation, furniture, textiles, carpentering, hydraulics, and plumbing, with different meanings in each field.
backlash: play between the teeth of two gears that are in mesh or engaged. Not quite the same thing the word would mean to a fisherman!
blooms: heaving semifinished forms of steel.
chase: iron frame in which a form is imposed and locked up for the press.
cheater: an extension on a pipe wrench.
Christmas Tree: the network of pipe at the mouth of an oil well. Also red and green lights in a submarine control room to show closed and open passages, and in fishing to describe multiple lures.
diaper: a form of surface decoration used in art and architecture consisting of geometric designs.
dirty: to make ink darker.
dwell: (of a cam) the angular period during which the cam follower is allowed to remain at its maximum lift; and, in printing, for the slight pause in the motion of a hand press or platen when the impression is being made.
freeze: seizing of metals that are brought into intimate contact.
galling: a characteristic of metals that causes them to seize when brought into intimate contact with each other.
lake: a compound of a dye with a mordant.

This somewhat haphazard list of terms — it could be extended at length — suggests the nature of the terms we have in mind. The reader may not confuse the everyday meaning of such terms with the technical sense they have in a particular report, but there is not much doubt that when encountering a term of this sort, the reader, unless a specialist in the field being discussed, will initially wrongly identify the word according to its everyday meaning. In any event, the writer must be alert to the need for defining such terms.

Unfamiliar Words for Familiar Things   A moment ago we condemned writers who prefer big and pretentious words for referents with which their readers are familiar. Such a practice should always be condemned if a simple, familiar term exists which means the same thing. But an unfamiliar word for a familiar thing may be used if there does not exist any simple, familiar term for it. Both convenience and accuracy justify it. Suppose you were writing on the subject of hydroponics. You can easily imagine addressing readers who know that plants may be grown without soil in a chemical solution, but who are unfamiliar with the technical term "hydroponics." Since there isn't a simple, familiar word for this process, you would scarcely want to give up the word "hydroponics" for an awkward, rather long phrase. Your solution is simple: you use the convenient term but you define it. Let's take another example. Suppose an electrical engineer were writing about special tactical electronic equipment making use of direct wave transmission. It is not likely that he or she would be satisfied to use the phrase "short wave" when dealing specifically with, say, the 300- to 3000-megacycle band. A more precise phrase is "ultrahigh frequency" (UHF). Similarly, a physician might prefer, in the interests of precise accuracy, the term "analgesic" to the simple word "painkiller."

You will have to judge whether your subject matter demands the use of such terms and whether they are familiar to your readers. If they are needed, or if they are justifiably convenient, and you decide that your readers do not know them, you should define them.

Unfamiliar Words for Unfamiliar Things This category, unfamiliar words for unfamiliar things, embraces most of those words that are commonly thought of as "technical" terms. They are the specialized terms of professional groups; big, and formidable looking (to the nonspecialist), they are more often than not of Greek or Latin origin. Terms like "dielectric," "hydrosol," "impedance," "pyrometer," and "siderite" are typical. We do not want to suggest that a static, precise list could be set up in this group, but since the reader's response determines the category into which a word falls, a great many of the terms that constitute the professional language of any special science or branch of engineering would, for the nontechnical reader, stand for unfamiliar things. These same words, however, when used by one expert in talking or writing to another expert would be familiar words for familiar things. It is important to remember, on the other hand, that the "nontechnical reader" does not necessarily mean the "lay reader," for even an expert in one field of science or engineering becomes a nontechnical reader when presented with technical writing in another field.

So far our interest has been in the problem of what needs to be defined. We can sum it up this way: you need to define (1) terms familiar to your reader in a different sense from that in which you are using them; (2) terms unknown to your readers, but which name things that actually are familiar to them, or at least things that can be explained simply and briefly in readily understandable, familiar terms; and (3) terms unfamiliar to the readers, and which name scientific and technical things and processes with which they are also unfamiliar. With these facts in mind about what to define, we can more intelligently consider the problem of how to define.

Methods of Definition

Before discussing the methods of definition, we want to remind you that insofar as it is possible to use simple, familiar terminology, the problem of definition may be avoided entirely. In other words, the best solution to the problem of definition is to avoid the need for it. When it is necessary, however, there are two methods or techniques that may be employed. The first may be described as informal; the second, as formal. The second takes two forms: the sentence definition and the extended or amplified definition. Each of these techniques has its own special usefulness.

Informal   Essentially, informal definition is the substitution of a familiar word or phrase for the unfamiliar terms used. It is therefore a technique to be employed only when you are reasonably certain that it is the term alone and not the referent which is unfamiliar to the reader. You must feel sure, in other words, that the reader actually knows what you are talking about, even if you give it another name. Thus, you might write " . . . normal (perpendicular) to the surface . . . " with the parenthetical substitution accomplishing the definition. Or "dielectric" might under certain circumstances simply be explained as "a nonconductor." Or "eosin" as "dye."

Instead of a single-word substitution, sometimes a phrase, clause, or even a sentence may be used in informal definition. Thus, dielectric might be informally explained as "a nonconducting material placed between the plates of a condenser," or eosin as "a beautiful red dye." Or you might use a clause, as "eosin, which is the potassium salt of tetrabromofluorescin used in making red printing ink." In very informal, colloquial style, you might prefer a statement like this: "The chemical used in making red ink and in coloring various kinds of cloth is technically known as 'eosin.' " Or, "When you use rubber insulating tape in some home-wiring job, you are making use of what the electrical engineer might call a 'dielectric.' "

Several general facts should be noted about such definitions. First, they are partial, not complete, definitions. The illustrations just given, for instance, do not really define dielectric or eosin in a complete sense. But such illustrations are enough in a discussion where thorough understanding of the terms is not necessary and the writer merely wants to identify the term with the reader's experience. Second, informal definitions are particularly adapted for use in the text of a discussion. Because of their informality and brevity, they can be fitted smoothly into a discussion without appearing to be serious interruptions. Third, we should note that when the informal definition reaches sentence length, it may not be greatly different from the formal sentence definition to be discussed in the next section. It lacks the emphasis, and usually the completeness, however, which may be required if a term defines an idea or a thing that is of critical importance in a discussion. In short, if you want to make certain that your reader understands a term, if you think the term is important enough to focus special attention on it, you will find the formal sentence definition, and perhaps the amplified definition or article of definition, more effective.

Formal Sentence Definition   We have seen that informal definition does not require the application of an unchanging, rigid formula; rather, it is an "in other words" technique — the sort of thing we all do frequently in conversation to make ourselves clear. With formal definitions the situation is different. Here a logically dictated, equation-like statement is always called for, a statement composed of three principal parts for which there are universally accepted names. These are the *species,* the *genus,* and the *differentia.* The species is the subject of the definition, or the term to be defined. The genus is the family or class to which the species belongs. And the differentia is that part of the statement in which the particular species' distinguishing traits, qualities, and so forth are pointed out so that it is set apart from the other species comprising the genus. Note this pattern:

Species =	*Genus* +	*Differentia*
Brazing is a welding process		wherein the filler metal is a nonferrous metal or alloy whose melting point is higher than 1000° F but lower than that of the metals or alloys to be joined.

Defined as a process, then, formal definition involves two steps: (1) identifying the species as a member of a family or class, and (2) differentiating the species from other members of the same class.

Don't let these Latin terms worry you. Actually, the process of working out a formal definition is both logical and natural. It is perfectly natural to try to classify an unfamiliar thing when it is first encountered. In doing so, we simply try to tie the thing in with our experience. Suppose you had never seen or heard of a micrometer caliper. If, when you first saw one, a friend should say — in response to your "What's this?" — that it is a measuring instrument, you would begin to feel a sense of recognition because of your familiarity with other measuring instruments. You still would not know what a micrometer caliper is, in a complete sense, but you would have taken a step in the right direction by having it loosely identified. To understand it fully, you would need to know how it differs from other measuring instruments like the vernier caliper, the rule, a gauge block, and so on. In all likelihood, therefore, your next question would be, "What kind of measuring instrument?" An accurate answer to this question would constitute the differentia. What you would be told would probably be something about the micrometer caliper's principle of operation, its use, and the degree of accuracy obtainable with it. Were it not for the fact that you had it in your hand, you would undoubtedly also be given a description of its shape, for physical appearance is a distinctive feature of the instrument. To be quite realistic about our hypothetical instance, we must admit that you would probably be told more than is essential to a good sentence definition. But if the essential distinguishing characteristics of the micrometer caliper were sifted out from all that was said, and put into a well-ordered sentence, your friend would have made a formal sentence definition — something like this, no doubt: "A micrometer caliper is a C-shaped thickness-measuring gauge in which the gap between the measuring faces is minutely adjustable by means of a screw whose end forms one face."

Natural as the process of identifying and noting the particular characteristics of something new may be, it must not be done carelessly. Let's take another look at some of the problems of handling the genus and differentia. The first step in the process of formal definition is that of identifying a thing as a member of a genus, or class. It is important to choose a genus that will limit the meaning of the species and give as much information as possible. In other words, the genus should be made to do its share of the work of defining. You wouldn't have been helped much, for instance, had your friend told you that a micrometer caliper is a "thing" or "device." If a ceramic engineer were to begin a definition of an engobe by saying it is a "substance," he or she wouldn't be making a very good start; after all, there are thousands of substances. A great deal more could be said by classifying the engobe at once as a "thin layer of fluid clay." With this informative beginning, it would be necessary only to go on to say that this thin layer of fluid clay is applied to the body of a piece of defective ceramic ware to cover its blemishes. Generally speaking, the more informative you can make the genus, the less you will have to say in the differentia. Another way of saying this is that the more specific you can be in the genus, the less you have to say in the differentia.

Care must be taken in carrying out the second step of the process of formulating a sentence definition. Here the important point is to see that the differentia actually differentiates — singles out the specific differences of the species. Each time you compose a statement in which you attempt to differentiate a species, examine it critically to see if what you have said is applicable *solely* to the species you are defining. If what you have said is also true of something else, you may be sure that the differentia is not sufficiently precise. One who says, for instance, that a micrometer caliper is "a measuring instrument used where precision is necessary" will recognize upon reflection that this statement is also true of a vernier caliper or, for that matter, of a steel rule (depending, of course, upon what is meant by "precision"). One way to test a statement is to turn it around and see whether the species is the only term described by the genus and differentia. Consider this example: "A C-shaped gauge in which the gap between the measuring faces is minutely adjustable by means of a screw whose end forms one face is a _____." "Micrometer caliper" fills the blank, and if the definition is correct, it is the only term that accurately fills the blank.

The foregoing discussion about methods can be reduced to the statement that an accurate limiting genus coupled with a precisely accurate differentia will always ensure a good definition. A few specific suggestions about particularly common difficulties should be added. Please regard the itemized points that follow not as guidelines to be memorized, but as possible sources of help in case of trouble. We have included a few suggestions for solving some particularly common difficulties.

Repetition of Key Terms. Do not repeat the term to be defined, or any variant form of it, in the genus or differentia. Statements like "A screw driver is an instrument for driving screws" or "A caliper square is a square with attached calipers" merely bring the reader back to the starting point. These examples may be so elementary as to suggest that this advice is unnecessary, but the truth is that such repetition is not at all uncommon.

There are, however, some occasions when it is perfectly permissible to repeat a part of the term to be defined. For instance, it would be perfectly permissible to begin a definition of an anastigmatic lens with "An anastigmatic lens is a lens . . . " if it could be assumed that it is the *anastigmatic* lens, and not all lenses, that is unfamiliar to the reader.

Qualifying Phrases. When a definition is being made for a specific purpose, a common practice in reports, limitations should be clearly stated. For example, an engineer might write, "Dielectric, as used in this report, signifies . . . " and go on to stipulate just what the term means in the present context. Unless such limitations are clearly stated (usually as a modifier of the species) the reader may feel — and rightly so — that the definition is inaccurate or incomplete.

Single-Instance (or Example) Definitions. In an amplified definition, as we will see in the following section, the use of examples, instances, and illustrations is fine; they help as much as any thing to clarify the meaning of a term. But the single instance, or example, is not a definition by itself. "Tempering is what is done to make a metal hard" may be a true statement, but it is not a definition. So it is with "A girder is what stiffens the superstructure of a bridge." In general, guard against following the species with phrases like "is when" and "is what."

Word Choice in Genus and Differentia. Try as much as possible not to defeat the purpose of a definition by using difficult, unfamiliar terminology in the genus and differentia. The nonbotanist, for instance, might be confused rather than helped by "A septum is a transverse wall in a fungal hypha, an algal filament, or a spore." And everyone remembers Samuel Johnson's classic: "A network is any thing reticulated or decussated, at equal distances, with interstices between the intersections."

Amplified Definition   Although brief informal definitions or sentence definitions are usually adequate explanations of the unfamiliar in technical writing, there are occasions when more than a word, phrase, clause, or sentence is needed in order to ensure a reader's understanding of a thing or idea. If you think that a sentence definition will still leave a number of significant questions unanswered in the reader's mind, then an amplified or extended definition is required.

Consider, for example, a medical term like "shock,' which often appears in news stories about accidents. A formal sentence definition goes like this: "Physiological shock may be defined as acute progressive circulatory failure in which defective tissue perfusion occurs." This is a helpful definition, but what causes shock? What are its symptoms? Answers to such questions call for an amplified definition. Here is what the author of the definition above said in the article from which the definition was taken[2]:

Shock, Physiological

Physiological shock may be defined as acute progressive circulatory failure in which defective tissue perfusion occurs. This definition is derived from the one constant feature of physiological shock, the failure of adequate blood flow through the capillaries, the smallest of the blood vessels. As a result of this failure, an inadequate supply of oxygen and nutrients is delivered to the tissues, and the removal of metabolic waste products from the tissues is incomplete.

All, or some, of the following features may be present: a raised pulse rate, each pulse being diminished in strength or "thready"; diminished arterial blood pressure; a cold, sweaty skin; rapid, deep respirations (breathing called air hunger in this context); mental confusion; dilated pupils; a dry mouth; and diminished flow of urine.

Many classifications of various types of shock have been developed, the most popular being one that relates the state to its cause; *e.g.,* hemorrhage (bleeding) or sepsis (invasion by pus-forming organisms). This classification has the advantage of possessing diagnostic and therapeutic implications, but unfortunately the cause of shock is often uncertain. Alternatively, the state may be called hypovolemic if it is associated with a reduction in blood volume or normovolemic if no blood-volume diminution has occurred. Shock can also be subdivided into "warm hypotension," if

[2] *Encyclopaedia Britannica,* 15th ed. (Chicago, Ill.: Helen Hemingway Benton, 1974), Vol. 16, p. 699.

the sufferer has lowered blood pressure and warm dry skin, and "cold hypotension," if the skin is cold.

Having completed this amplified definition, the author of the article on shock continued with a discussion of how to recognize a given type of shock, and how to manage shock. The continuation of the discussion is not reprinted here.

Below is a second example of an amplified definition. This one is more complex in structure. The amplified definition of the principal term, "closed loop," includes a classification ("regulator," "servomechanism"); and each of the terms in this classification is in turn defined. This author[3] also has a sense of humor; but perhaps that is indefinable.

informal defn —

All regulating systems are closed loop systems. Closed loop systems, consequently (regulating systems,) are common in our everyday lives. For example, while driving an automobile the driver judges the degree of pressure to exert on the steering wheel by considering such factors as road surface and curvature, speed of the automobile, foreign objects on the road, etc. The driver's eyes monitor these factors and, through the human system, control his limbs to allow safe travel. This is a closed loop system. However, if the driver were to close his eyes, the result would be an open loop system. An open loop system is one which does not allow for unpredictable necessary corrections and is not recommended while driving an automobile. This chapter will be concerned with closed loop systems which do not rely on the human for completing the loop.

There are three terms which are used to describe closed loop systems: regulator, servomechanism, and servo. A *regulator* is a closed loop system that holds at a steady level or quantity such elements as voltage, current or temperature, and often it needs no moving parts. A *servomechanism* is a closed loop system that moves or changes the position of the controlled object in accordance with a command signal. It includes some moving parts such as motors, solenoids, etc. The term *servo* has been generally adopted to mean either a regulator or a servomechanism and will be used as such in this chapter.

There is no single way to go about amplifying a definition. You must use your own judgment in determining how much, and what, to say. Examination of many definitions, however, does indicate that the following techniques often prove useful.

Further Definition. If you think that some of the words in a definition you have written may not be familiar to your reader, you should go on to explain them (some readers, for example, might like to have the word "hypotension" explained in the definition of shock).

Concrete Examples and Instances. Since sentence definitions are likely to be abstract statements, they do not contain concrete examples of the thing being defined. It helps, therefore, to give the reader some specific examples. As a matter of fact, this technique is probably the best of all.

Comparison and Contrast. Since we tend to relate — or try to relate — new things and experiences to those we already know, it helps to tell a reader that what you are talking about is like some familiar thing. Remember that the relationship must be

---

[3] *SCR Manual,* 4th ed. (Syracuse, N.Y.: Semiconductor Products Department, General Electric Co., 1967), p.263. Quoted by permission of the General Electric Co.

one of the unfamiliar to the familiar. If you were attempting to explain what a tennis racket is to a South Sea islander, it wouldn't help much to compare it to a snowshoe! On the other hand, it may be better to stress the differences between the things compared. See "Negative Statement" below.

Word Derivation. It rarely happens that information about the origin of a word sheds much light on its present meaning, but sometimes it does and the information is nearly always interesting. Take the term "diastrophism" for instance. It comes from the Greek word *diastrophe* meaning "distortion" and ultimately from *dia* (meaning "through") and *strephein* (meaning "to turn"). Thus, the word appropriately names the phenomenon of deformation, that is, "turning through" or "distortion" of the earth's crust, which created oceans and mountains. As you know, etymological information may be found in any reputable dictionary.

Negative Statement. Negative statement is mentioned in many books as a possible means of developing a definition. Sometimes it is called "obverse iteration," sometimes "negation," and sometimes "elimination." Whatever it is called, you should realize that you will never really get anywhere by telling what something is not. But in some cases you can simplify the problem of telling what something *is* by first clearing up any confusion the term may have in the reader's mind with closely related terms. You might, for instance, say that a suspensoid is not an emulsoid, but is a colloid dispersed in a suitable medium only with difficulty, yielding an unstable solution that cannot be re-formed after coagulation. An emulsoid is a colloid readily dispersed in a suitable medium that may be redispersed after coagulation.

Physical Description. We mentioned earlier that you could scarcely give a reader a very thorough understanding of a micrometer caliper without explaining what it looks like. So it is with virtually all physical objects.

Analysis. Telling what steps comprise a process, or what functional parts make up a device, or what constituents make up a substance obviously helps a reader. This technique is applicable to many subjects: a breakdown of a thing or idea permits the reader to think of it a little at a time, and this is easier to do than trying to grasp the whole all at once.

Basic Principle. Explaining a basic principle is particularly applicable to processes and mechanisms. Distillation processes, for instance, make use of the principle that one liquid will vaporize at a different temperature than another.

Cause and Effect. Magnetism may be defined in terms of its effects. In defining a disease, one might very well include information about its cause.

Location. Although of minor importance, it is sometimes helpful to tell where a thing may be found. Petalite, for instance, is a mineral found in Sweden and on the island of Elba; and in the United States at Bolton, Massachusetts; and Peru, Maine; vast deposits are located in Southern Rhodesia and South Africa.

The foregoing techniques do not exhaust the possibilities for amplifying a definition. Anything you can say that will help the reader comprehend a concept is legitimate. We have seen mention of authorities' names (in a definition article on the incandescent lamp, it would be natural to find Edison's name mentioned), history of a subject, classification, and even quotations from literature on a subject used to good advantage. Nor should every one of these techniques be employed in any given case, necessarily; often only a few of them would be pertinent. You will have to depend upon your own judgment to decide how much you need to say and what techniques are best suited in a specific situation.

Two organization patterns are possible for amplified definitions. The first pattern begins with the formal sentence definition and proceeds with supporting discussion. A glance at the definition of shock, given earlier, reveals that it is organized in this way. After the initial sentence definition, an explanation is given of the "one constant feature" of shock, symptoms are described, and some bases of classification of types of shock are presented. In a general way, this pattern of organization may be regarded as deductive in that it begins with a statement regarded as true and proceeds to the particulars and details. Altogether, it is a method to be preferred over the second, or inductive, pattern of organization, which places the sentence definition last as the conclusion to the evidence presented. The deductive method is preferred because there is no point in keeping the reader waiting for needed information. Where the inductive method is used, the issue is in doubt, in a sense, until the last sentence is reached.

Placing Definitions in Reports

Very often it is difficult to decide where to put definitions in reports. There are three possibilities: (1) in the text, (2) in footnotes, and (3) in a glossary at the end of the report, or in a special section in the introduction.

If the terms requiring definition are not numerous and require brief rather than amplified definition, it is most convenient to place explanatory words or phrases in the text itself as appositives (set off with commas or parentheses). If you are not sure whether your readers know a term, or if you feel that some readers will know it and some will not, it is probably best to put the definition in the form of a footnote with a numeral or some suitable designating mark or symbol after the word itself in the text. If placing definitions in the text would result in too many interruptions, especially for the reader who may know them, it is a good idea to make a separate list to be put into an appendix. If there are a number of terms of highly critical importance to an understanding of your report, they may be defined in a separate subdivision of the introduction of the report. An introduction to a report on, say, a bridge construction may contain a statement like this: "concrete, in this report, will mean . . . " with the rest of the statement specifying the composition of the mix.

The point of all this is that definitions should be strategically placed to suit your purposes and the convenience of your readers. Once you decide on the importance of the terms you use and the probable knowledge of your readers, you will find it easy to decide where to put the definitions.

Summary

1. Definition is needed when familiar words are used in an unfamiliar sense or for unfamiliar things, when unfamiliar words are used for familiar things, and when

unfamiliar words are used for unfamiliar things. The question of familiarity or unfamiliarity applies in all cases to the reader, not the writer.

2. Definitions may be either informal (essentially the substitution of a familiar word or phrase for the unknown term) or formal.

3. Formal definitions always require the use of a "sentence definition," which is composed of three principal parts: species, genus, and differentia.

4. Sometimes it is necessary to expand a formal definition into an article. An article of definition may be developed by either the deductive or the inductive method, the deductive being generally preferable.

5. Definitions may appear in the text of a report, in footnotes, in a glossary at the end of the report, or in a special section in the introduction. Their proper location depends upon their importance to the text and on the knowledge of the readers.

Illustrative Material

The illustrative material presented here consists of two items: the first is a generalized explanation of eyesight which appears in a pamphlet published by Bausch and Lomb, Soflens Division, for distribution to the public; the second is an amplified "definition" of land treatment published in *EXXON USA* and written by Denise Zwicker.

Insight into Eyesight

Each of your eyes is like a camera. Or more correctly, the camera is designed to operate like the human eye. Both the eye and the camera have an opening for light to enter, a lens or lens system, and a screen for registering an image of the visible world.

A simple camera with a small lens will take good pictures when the light is fairly bright and the objects photographed are at least several feet from the lens. But this isn't good enough for human vision. You need the ability to see and focus on dimly and brightly lit objects both at a distance and very close to your eyes. To accomplish this, the human eye has a complex lens system and a unique screen where the image is formed in the eye.

Rays of light coming from external objects enter each eye through a multi-layered, transparent dome at the front of the eye. This is the *cornea,* the first lens in the system, which begins the bending or refracting of light rays toward a point of focus on the retina, the interior back surface of the eye.

After passing through the cornea, the refracted rays of light travel through the *pupil* of the eye, the round black opening inside the colored ring of the iris. The *iris* acts like a circular window shade,

This explanation of eyesight (which may also serve as an illustration of a description of a mechanism) does not begin with a formal definition of an eye or an explanation of its purpose and general appearance since these are presumed to be obvious. Instead it begins with an analogy. The second sentence lists three major parts; in the second paragraph these three are reduced to two, since the "opening" is part of the lens system. This change is unnecessary and momentarily confusing.

The detailed parts of the lens system are not listed in the text, but are made clear in the sketch. Here the value of an illustration for brevity and clarity is evident.

Throughout the article, the parts are defined and their purpose explained, and analogy is used.

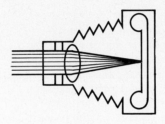

CAMERA

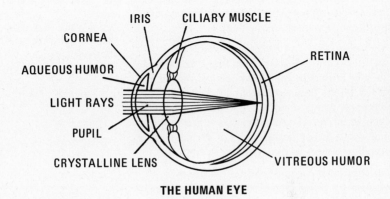

THE HUMAN EYE

enlarging or diminishing the size of the pupil so
that the proper amount of light enters the inner
chamber of the eye.

The rays of light which pass through the pupil
then enter the *crystalline lens* of the eye, the most
complex part of the system. Unlike camera lenses
which are hard and rigid, this lens is flexible, and
its shape can be altered by a circular muscle
attached to its circumference.

Light rays from distant objects are virtually
parallel. The crystalline lens focuses them on the
retina while the lens is flattened by the encircling
ciliary muscle. Rays coming from objects close to
the eye are divergent, and a stronger lens is
needed to refract or bend them to a focus. In this
situation, the ciliary muscle reduces its pull on the
lens to allow it to assume its relaxed (thicker)
shape. Thus the lens becomes more powerful and
bends the light rays to a greater degree. This
process is called *accommodation*.

After the crystalline lens has done its job, the refracted light rays pass through a gel-like substance called the *vitreous humor,* and arrive in focus at their final destination, the retina of the eye. The *retina* is a transparent membrane containing millions of tiny light sensitive receptors. Like the film in a camera, the retina reacts to the image projected upon it. Rather than record the image, the many sensations it receives are passed along to the brain through a single, complex optic nerve. The brain analyzes this data and so you "see."

Land Treatment: Safe Disposal of Petroleum Wastes[4]

In a world where "high technology" has become a cliche, it's hard to believe that some of a refinery's most important assets are its bugs.

That's right, *bugs.* In this case, microscopic creatures — mostly bacteria and fungi. You'll find the same critters in your flower bed or vegetable garden, where they work around the clock to decompose compost and fallen leaves, helping your plants to thrive. At a refinery, they function the same way — decomposing many of the wastes the refinery produces in its day-to-day operations.

This controlled decomposition of wastes by soil organisms is called land treatment, landfarming, biodisposal farming, and many other names. All describe the simple process of cultivating wastes into specially prepared soil plots. The wastes are consumed and transformed into various byproducts by the bacteria, fungi, and other organisms that exist naturally in the soil.

Today, about 200 industrial land treatment facilities operate in the United States, with the petroleum industry accounting for more than half. Others are operated by city governments, chemical producers, pulp and paper companies, food processors, and tanning companies. Land treatment facilities also operate in Canada, the United Kingdom, France, Switzerland, Denmark, Sweden, Finland, and the Netherlands.

Although land treatment is not a new waste-disposal technique, it has only recently come into wide use in the petroleum industry. Exxon, as one of the first oil companies to practice land treatment, has played a leading role in that increase.

Published in a widely distributed company magazine, this article begins with a lead which captures the reader's attention. The definition proper does not begin until the third paragraph.

---

[4] *Exxon USA,* First Quarter 1984, pp 18–21. To save space we have omitted illustrations and the last few paragraphs, including a final section on continuing research.

Exxon's experience with land treatment dates back to 1954, with the opening of a treatment facility at the Baytown (Texas) Refinery. Since that time, Exxon has successfully operated land treatment facilities throughout the United States, Canada, and Europe.

Of course, land treatment is only one of several waste-management options available to the petroleum industry. Other commonly used methods include landfilling, incineration, temporary surface impoundment, deep-well injection, and chemical treatment.

"A refinery generates about 100 different solid wastes, so it's logical that we'd use many different disposal methods," says Dr. W. M. Kachel, senior staff engineer for Exxon Research and Engineering Company. "However, land treatment and landfilling are used for about half of our wastes."

For many years, the majority of Exxon's refinery wastes were landfilled. Recently, the use of land treatment has increased to the point that it accounts for one-third of the total at Exxon Company, U.S.A. refineries.

LAND TREATMENT VS. LANDFILLING

Often, people confuse land treatment and landfilling. It's important to understand the difference.

In land treatment, oily refinery wastes are spread in a thin layer onto a designated plot of land and cultivated into the soil with farm equipment. Soil organisms break down and immobilize the wastes in the soil. The application/cultivation process can be repeated many times on the same plot since the soil organisms use the wastes as food—and maintain a continuing appetite for more.

Landfills, on the other hand, store wastes in constructed or natural excavations. When a landfill is full, it is covered with an impermeable material, such as compacted clay. The purpose of a landfill is not to decompose wastes, but to store them permanently.

Both land treatment and landfilling have their places—as do each of the other waste-management processes used by the petroleum industry. To select the best method, waste-management experts consider the type and volume of waste as well as the relative cost and efficiency of each method.

Note that a significant portion of this discussion of land treatment is devoted to differentiating it from landfilling, another way of handling the disposal of petroleum wastes.

For example, only combustible wastes with a relatively high thermal content are suited for incineration. Land treatment is effective only for wastes that will biodegrade and that are not highly volatile or toxic. Chemical fixation is difficult with wastes that contain organics. And landfilling is best restricted to inorganic wastes, since landfill capacity is limited and other methods often are more appropriate for organic wastes.

"Because most refinery wastes have a high water content, it's difficult to incinerate them without the use of auxiliary fuel, which increases the already high cost. And, since many refinery wastes are organic, chemical fixation isn't practical either," explains Kachel. "So, most of the time, we use land treatment or landfills."

Although the cost of each method varies widely from one region to another, land treatment generally is the least expensive — often half the cost of landfilling and one-quarter or less of the cost of incineration. Thus, as the safety and effectiveness of land treatment has been proved through the field experience of Exxon and other companies, its use by the petroleum industry has increased.

"Exxon land-treats two types of refinery wastes: oily and biological," says Kachel. "The oily wastes come from many operations, but mostly wastewater treatment. The biological wastes consisting mostly of the organic remains of microorganisms also come from our wastewater treatment facilities."

Some of these wastes are classified as hazardous by the U.S. Environmental Protection Agency (EPA) because they contain lead and chromium — metals that the EPA believes could prove toxic to people or the environment. A variety of processes are approved by the EPA for the managment of these wastes. However, as the EPA said in a 1983 report: "Land treatment is one alternative for handling hazardous waste that simultaneously constitutes treatment and final disposal of the waste . . . Many wastes currently being disposed by other methods without treatment could be treated and rendered less hazardous by land treatment, often at lower cost."

The difference is that land treatment not only stores the wastes, but also renders them less

hazardous by breaking down the organic portions
of the waste, ultimately producing carbon dioxide,
water, and soil humus. Both organic and inorganic
portions of the waste also become chemically and
physically attached to soil particles, preventing
potentially hazardous components from reaching
groundwater.

MANAGEMENT IS CRUCIAL

Yet, simple as it is, land treatment requires careful
management if it is to be effective.

In many ways, the management process resembles
farming. Just as farmers must cultivate, fertilize,
and water their plant seedlings, land-treatment
operators must supply oxygen, nutrients, and
water to the soil organisms that degrade the
wastes. In many cases, farm machinery is used to
apply and cultivate the wastes. Some land treatment operations even use field-rotation techniques
borrowed from agriculture.

The organisms that degrade oily wastes need
oxygen to survive. Oxygen also prevents the wastes
from becoming odorous. Therefore, an important
part of the land treatment process is periodic
tilling of the topsoil to mix oxygen with the soil
and wastes.

Another important ingredient is nutrients.
Nitrogen and phosphorous, for example, may not
be available in sufficient quantities in refinery
wastes. Commercial fertilizers can supply these
and other nutrients.

The acidity/alkalinity (pH) of the soil also must be
monitored. If the soil is too acidic, the microorganisms will not decompose the waste as effectively.
The proper pH level is one of several mechanisms
that help prevent lead and chromium from moving
through the soil to groundwater.

Soil microorganisms, like plants, also function best
when they have the right amount of water.
Because oily wastes usually contain plenty of
water, irrigation is seldom necessary at land treatment sites. Frequent tilling, with rest periods
between waste applications, permits excess water
to evaporate quickly.

Vegetation may be grown on inactive portions of
the site to reduce soil erosion and improve the

appearance of the site. However, no food-chain crops are grown and no livestock are raised on petroleum-waste land treatment sites.

ENVIRONMENTAL MONITORING

Land treatment facilities also follow strict monitoring procedures to verify that the wastes are decomposing and that potentially harmful components are immobilized in the upper layer of soil.

Before a land-treatment facility is constructed, samples are taken of the groundwater and any surface waters near the site. Chemical analyses of these samples set the standards against which post-treatment analyses will be compared.

Once land treatment begins, site personnel regularly analyze soil cores, soil water, and groundwater to calculate the waste-degradation rate and to quickly detect any movement of waste components below the approximately five-foot layer of soil that is designated as the treatment zone. If such movement is detected, prompt remedial measures can be taken.

Run-on water is diverted from the site. Runoff water is collected and reapplied to the treatment plot or, if necessary, treated in a wastewater treatment system before it is released to public sewer systems or surface waters.

Most petroleum-company land treatment sites are situated on property owned by the refinery that generates the waste. In many cases, the treatment site is included within the perimeter of the refinery itself. This proximity eliminates the costs and potential hazards of waste transportation — and enables refinery management to more closely supervise land treatment activities.

CLOSURE AND POST-CLOSURE

Most land treatment facilities can operate successfully for many years. But, eventually — depending on waste composition, application rate, or the life span of the facility that generates the waste — the time will come to close the site.

Closure plans, which are regulated by the EPA, address four questions:

How the movement of waste components will be
    controlled to prevent seepage into ground-
    water

How run-on waters will be diverted and runoff
waters collected and, if necessary, treated before they are released to sewage systems or
surface waters
How wind erosion will be controlled to prevent
the release of contaminants into the air
How vegetation growth will be controlled to
prevent possible contamination of the food
chain

This post-closure care may be required for 30 years
or more, depending on the results of routine
monitoring.

Suggestions for Writing

1. Often it is necessary in technical and scientific presentations to give the reader an informal or working definition of a term, a definition that makes the meaning clear without the formality of a complete sentence definition. Write informal definitions of the following terms: chronometer, clear-cutting (in forestry), catalyst, antimatter, nucleus (of an atom), fault (geology), cyclone (meteorology), merger (in business), tourniquet, virus, mean (arithmetic), stress (mechanical).
2. Write sentence definitions of the following terms: calculator (electronic), outboard motor (for a boat), term insurance, cold front (weather), guitar.
3. Assume that after a trip to some remote foreign country, you have been corresponding, in English, with a friend you made there. In one letter your friend asks about the meaning of the terms listed below; write brief amplified definitions of these terms: credit-hour (academic), freeway (a road), hamburger stand, World Series (baseball), primary election.
4. Write an amplified definition of a key term or concept in your major field of study. A central concept from one of the courses you are taking might be a good choice. We suggest 200 to 300 words.
5. If you are working on a research paper, prepare a glossary of the key technical terms that will appear in your report.

6

Description of a Mechanism

This chapter brings us to the second of the special techniques of technical writing —the description of a mechanism. In the strictest sense a mechanism is simply a piece of machinery, but the term has come to include almost anything in which some type of activity can be detected. The principles of the description of a mechanism apply to all types of mechanisms and to static objects as well. In general, the technique presented in this chapter applies to the description of things; the technique to be used for the description of actions is presented in the following chapter.

The general procedure in describing a mechanism is quite simple; in practice the chief difficulty lies in writing sentences that really say what you want them to say. There is no more fertile field for "boners."[1]

The three fundamental divisions of the description are the introduction, the part-by-part description, and the conclusion. Before discussing these divisions in detail, we'd like to remind you of two things. The first is that a description of a mechanism almost never constitutes an entire report by itself. For practice in the technique, it is wise to write papers devoted exclusively to description, but it should be understood that such papers will not represent the complete reports found in actual use. The second reminder is that what needs to be said in the description always depends on what the reader needs to know. For example, your reader might

[1] The following extract from a student paper suggests the possibilities: "The Dragoon Colts were issued to the army and sold to civilians equipped with shoulder stocks that locked into the butts to make short rifles out of them."

want to construct a similar device. This would require a highly detailed treatment. Or your reader might be chiefly interested in knowing what the device will do or can be used for, and need only a generalized description.

The Introduction

Because the description of a mechanism seldom constitutes an article or report by itself, the introduction required is usually rather simple. Nevertheless, it is very important that the introduction be done carefully. The two elements that need most careful attention are (1) the initial presentation of the mechanism, and (2) the organization of the description.

The Initial Presentation   At the beginning of a discussion of an unfamiliar mechanism, a reader immediately needs three kinds of information: (1) what it is, (2) what its purpose is, and (3) what it looks like.

The problem of identifying a mechanism for the reader is simply a problem of giving a suitable definition. If the reader is already familiar with the name of the mechanism and knows something about the type of mechanism it is, all you need do is write the differentia. For example, if you were about to describe some special type of lawn mower for an American reader, you would not need to define "lawn mower," but you would need to differentiate between the type you were describing and other types of lawn mowers with which the reader was familiar. Or, if the name of the mechanism to be described is unfamiliar to the reader, perhaps a substitute term will do. Suppose we write in a report, ". . . each of these small boats is equipped with a grains." How should we tell the reader what a "grains" is? We can do so very easily by writing, "A grains is a kind of harpoon." As you no doubt recall, both ways of clarifying what the mechanism is — that is, defining it — were discussed in Chapter 5.

The reader must also know the purpose of the mechanism. Often, an indication of purpose will appear as a natural part of the statement of what the mechanism is. For instance, to say that a grains is a harpoon indicates something about its purpose. To take another example, let's suppose we are writing a description of the Golfer's Pal Score-Keeper. Here, the purpose is suggested by the name itself. It is frequently desirable, however, to state the purpose of the mechanism explicitly. In writing about the Golfer's Pal Score-Keeper we might be more certain that its purpose had been made clear by stating that it is a small mechanical device that a golfer can use instead of pencil and paper for recording each stroke and getting a total. The purpose of a mechanism is often clarified by a statement about who uses the device, or about when and where it is used.

Finally, when the mechanism is initially presented, the reader needs a clear visual image of it. The most effective way to give a reader a visual image of a mechanism is to present a photograph of it — assuming the thing itself is not available for examination. A drawing would be second best. Our interest, however, is in creating the visual image with words. Photographs and drawings are more effective than words for this purpose, and should be used if possible, but expense, or the need for haste, or the lack of facilities often rule out the use of such visual aids. A very interesting example of the practical importance of the visual image may be found in Appendix C under the heading, "Information Presentation."

In the initial presentation of a mechanism, the visual image created by words should be general, not detailed. There will be time enough for details later on. Fundamentally, there are two ways of creating this general image. One is to describe the general appearance of the device; the other is to compare it with something that is familiar to the reader. You must be careful, of course, not to compare an unfamiliar thing with another unfamiliar thing. Reference to the Score-Keeper again suggests how illuminating a good analogy can be: "This device is very much like a wristwatch in size and general appearance." To this comparison might be added some such direct description as the following: "It consists of a mechanism enclosed in a rectangular metal case — 1⁵⁄₁₆ in. long (3.3 cm), ⅞ in. wide (2.2 cm), and ¼ in. thick (0.6 cm) — to which is attached a leather wristband."

Size and shape are often the conspicuous characteristics of a mechanism, but don't overlook other general characteristics, such as color, weight, material, if they will help the reader gain a more complete and accurate general understanding of the subject.

We remarked earlier that stating what a mechanism is constitutes a problem in definition. Now, in concluding these comments on the initial presentation of a mechanism, we should acknowledge that references to purpose and appearance are among the methods of making a definition that were discussed in the preceding chapter. It becomes apparent that there is a close similarity between acquainting readers with a mechanism new to them and acquainting them with a term or concept new to them.

Organization of the Description   It is possible to divide almost every mechanism into parts. Such division is an essential part of a detailed description. In the introduction to a description, a statement of the principal parts into which the mechanism can be divided serves two purposes. The first is that it is an additional way, and an important one, of giving the reader a general understanding of what the mechanism is. From this point of view, what we are saying here actually belongs under the preceding heading ("The Initial Presentation"). The second purpose is to indicate the organization of the discussion that is to follow. Since it is logical to describe the principal parts one at a time, a list of the principal parts in the order in which you wish to discuss them is a clear indication of the organization of the remainder of the description. The list of principal parts should be limited to the largest useful divisions possible. For instance, in the description of the A & J patented peeler (Illustration 2 in Illustrative Material), the principal parts are listed as the blade, the tang, and the handle. Later, to the degree possible, these parts are broken down into subparts. In practice, engineering assembly prints are valuable keys to identification of principal parts of a mechanism.

The order in which the parts are taken up will normally be determined by either their physical arrangement or their function. From the point of view of physical arrangement, an ordinary umbrella might be divided as follows: (1) the handle, (2) the metal framework that spreads the fabric, and (3) the fabric. The handle comes first because it supports the other parts. From the point of view of function, the fabric comes first, then the metal framework, and last the handle.

Finally, you should make sure that the list of principal parts is in parallel form. It is hard to make a mistake in this because the list will almost inevitably be composed of names — the names of the parts; nevertheless it might be well to check your list. The parts are usually named in normal sentence form, like this: "The principal parts of the peeler are (1) the blade, (2) the tang, and (3) the handle." But if the parts are numerous, it may be preferable to present them in the form of a list, as follows:

1. Part 1
2. Part 2
3. Part 3
4. Part 4

The Part-by-Part Description

The introduction being out of the way, and the mechanism logically divided into parts, we are ready to take up the description of the first part. But the fact is that now, as far as method goes, we start all over again, almost as if we hadn't written a line. For, what is the "part" but a brand-new mechanism? The reader wants to know what it is. So we must introduce it.

We have divided the peeler — say — into the blade, the tang, and the handle and are about to describe the blade. The first problem is to tell the reader what the blade is. The general procedure will be — as before — to define the part, to state its purpose, to indicate its general appearance (preferably by a comparison with an object with which the reader is familiar), and finally, if necessary, to divide it into subparts.

And what do we do with the subparts? The same thing exactly. In other words, the mechanism as a whole is progressively broken down into smaller and smaller units until common sense says it is time to stop. Then each of these small units is described in detail.

By this time you may have a mental image of a chain of subparts and sub-sub-parts stretching across the room with a detailed description glimmering faintly at the end. That certainly isn't what is wanted. Nevertheless, we do wish to emphasize the value of breaking down the mechanism into parts before beginning a detailed description. But, if the break-down procedure goes very far before you're ready to describe, it probably means that the principal part with which you started was too broad in scope. You need more principal parts. Although we urge the value of this system as a general policy, it is simply not true that all description must be handled in this way. Sometimes, for example, instead of giving a preliminary statement of *all* the subparts that will be described in a given section of the description, it is desirable not to mention a certain minor subpart at all except when you actually describe it.

"Described in detail" means careful attention to the following aspects of the mechanism:

Shape	Finish
Size	Color
Number of parts	Weight
Relationship to other parts	Hardness
Methods of attachment	Special properties
Material	

Each of these matters needn't be labored over mechanically, in the order stated, in every description. Which ones need attention, and what kind of attention, depends — as always — upon the reader and the subject. For instance, let's take the term "material" in the list above. The discussion so far has implied that the material of which a mechanism is constructed is not discussed until the mechanism has been divided into its smallest components. But if you were describing an open-end wrench made of drop-forged steel, it would seem unnatural to wait until you were taking up one of the smaller parts to let the reader in on the fact that the whole wrench was drop-forged steel.

The same line of reasoning can be applied throughout the description. There is no formula that will fit every situation. The important thing is to decide what information the reader needs, and to provide it in as nearly crystal-clear a form as you can.

The Conclusion of the Description

The last principal function of the description of a mechanism is to let the reader know how it works, or how it is used, if this hasn't been done in the general introduction. Emphasis should naturally fall upon the action of the parts in relation to one another. This part of the writing constitutes in effect a description of a process, usually highly condensed (see Chapter 7). If the operation of the mechanism is of minor importance or has been clearly established earlier, this section of the description can be devoted to a summary, or, in the case of short descriptions, be omitted entirely.

Summary of the Principles of Organization

The outline below indicates in a general way the organization of the description of a mechanism. As has been explained, the order of some of the topics listed and the inclusion or exclusion of certain topics depend upon the situation. This outline is to be taken as suggestive, not prescriptive.

DESCRIPTION OF A MECHANISM

    I. Introduction
       A. What the mechanism is
       B. Purpose

   C. General appearance (including a comparison with a familiar object)

   D. Division into principal parts

 II. Part-by-part description

  A. Part number 1

   1. What the part is

   2. Purpose

   3. Appearance (including comparison)

   4. Division into subparts

    a. Subpart number 1

     (1) What the subpart is

     (2) Purpose

     (3) Appearance (including comparison)

     (4) Detailed description

      (a) Shape

      (b) Size

      (c) Relationship to other parts

      (d) Methods of attachment

      (e) Material

      (f) Finish

    b, c, etc.—same as "a."

  B, C, etc.—same as "A."

 III. Brief description of the mechanism in operation

Some Other Problems

Style By far the most difficult problem in describing a mechanism is simply to tell the truth. The writer is seldom in any doubt as to what the truth is; he or she wouldn't be writing about an unfamiliar mechanism. But it is one thing to understand a mechanism and another to communicate that understanding to somebody else. Painstaking observation and careful attention to language are needed to ensure accuracy.

It is probably a mistake, however, to try to be perfectly accurate in the first draft of a description. Write it as well as you can the first time through, but without laboring the details; then put it away for as long as you can. When you read it over again, keep asking yourself if what the words say is what you actually meant. At especially critical points, try the experiment of putting what you have said into the form of a sketch, being guided only by the words you have written. Sometimes the results are amazing in showing how the words have distorted your intended meaning.

Whenever you see the letters "ing" or "ed" on a word, watch out for a booby trap (specifically, a dangling modifier). And make sure that every pronoun has an easily identified antecedent.

Finally, don't forget to watch the tense. Usually the entire description will be in the present tense. Occasionally it will be past or future. But almost invariably the tense should be the same throughout the description.

Illustrations   People who like to draw and do not like to write are often loud in argument as to the waste of writing anything at all when a drawing would do. We ourselves are rather sympathetic toward this attitude, but the trouble lies in deciding when the drawing will do.

First of all is the question of plain facts. Sometimes it can be difficult or impossible to show in a drawing how heavy an object or device is, what material it is composed of, how a device functions or how much tension is applied to a certain fitting (where a torque wrench might be used). Words are usually much better than drawings for such matters.

There is again a psychological problem. Some people seem to have a greater aptitude for comprehending things in verbal form than in graphic form, and vice versa, just as some people more readily comprehend the language of mathematics than they do the language of words, and vice versa.

Certainly, the wisest course is to use every means of communication at your command if you really want to make yourself understood. The corollary is to use discretion; you don't want to swamp your reader with either text or drawings.

One of the skills that technical writers need is that of effectively relating a written discussion to a drawing. In general, two possibilities are open. One is to print the name of each part of the device on the drawing; the other is to use only a symbol. In other words, if you were discussing the blade of a peeler, you might write, "The blade (see Figure 1) is. . . ." Or, if you had used only a symbol on the drawing, instead of the name, you might write, "The blade (Figure 1-A) is. . . ." If there is only one figure in the report, it need not be numbered. You could then write, "The blade (A) is. . . ."

Information about the form of drawings and other illustrations can be found in Chapter 20.

A problem that comes up in every description is how many dimensions to indicate, both in the text and on the illustration. A decision must be based upon the purpose of the description. If you anticipate that the description may be used as a guide in construction, then all dimensions should be shown on the drawing and a great many stated in the text.

Problems of Precision and Scope   Finally, we want to return to two problems that were mentioned briefly in the first three paragraphs of this chapter. One of these is what we called the problem of telling the truth about a mechanism. The other has to do with the amount of detail in a description.

What does telling the truth mean here? First, it means *seeing*. Second, it means *communicating*. We'd like to make a personal suggestion about this problem.

Don't be surprised if you find yourself having something of an emotional experience over your first attempt to describe a mechanism. If, for example, after working hard at your description, you show it to somebody (a friend, an instructor, a supervisor on your job), what happens next may assume the proportions of a disaster. The person to whom you show it may announce that he or she cannot understand it at all. If the person to whom you show your description is a friend who is too good-natured to risk hurting your feelings, you can challenge fate by asking your friend to make a drawing of the mechanism, relying on nothing but your written words for guidance. This is almost guaranteed to result in disaster.

We are only partly joking in using the word "disaster." A very real disaster can, in fact, occur. It does occur if, having run into trouble with your description, you decide that you are simply unable to write a good one. Writing a description of a mechanism is, like most other writing, merely a skill. It is not a divine gift. It requires practice. It requires practice in seeing, and in communicating.

Probably you will feel something of a shock if you discover you haven't really been seeing what you were looking at; and yet this discovery is an almost universal experience. Let's consider an example. In the "Illustrative Material" section at the end of this chapter, you will find a long, detailed description of a simple mechanism, a patented peeler. It is accompanied by illustrations so that the details of the writing can be checked. In describing the blade the author wrote, "the sharpening has been done by grinding the convex side of the slit. As a result, the convex side is flattened along the slit." Are these true statements? Not exactly. The first speaks of grinding the convex side of the *slit* rather than the convex side of the *blade*. The second at least implies that if a straightedge were laid across the area of the blade that was ground away, this area would prove to be uniformly flat. But this implication is misleading. Careful observation of the illustrations reveals that the flattened area on one side of the slit is set at an angle to the flattened area on the other side. Does this inaccuracy in the description matter? Yes, it probably matters a lot. First, these angles presumably are important in the way the blade functions, in the "bite" it takes. Second, the width of the slit — also important in the way the blade functions — is partly dependent on the angle at which the grinding is done.

Did the author fail to see this detail, or only fail to communicate it? There is no way of knowing. But it does seem likely that if the author had seen it, the description would have been written somewhat differently.

One suggestion that you may find helpful in training yourself to see what you are looking at is to verbalize. Talk to yourself (silently, preferably!). Instead of just looking at the blade of the peeler (for example), ask yourself questions, in words, like — "What is the shape of the cross section of this blade?" "Why does the slit down the center of the blade have this particular width?" Keep talking, and it will help you see.

We've been thinking about accuracy of observation. Let's shift attention to the closely related subject of successfully communicating what is seen. Now the reader becomes our chief concern. Here are two ideas you may find especially useful.

The first is to be very careful about the physical point of view or orientation from which you are asking the reader to look at the mechanism. Obviously, if you start describing one end of a mechanism and then suddenly begin to describe the other end, your reader is likely to become confused. An example of this need for control of orientation can be seen in the description of the handle of the peeler just mentioned. Further examples are found in the following interesting comments on problems of orientation by John Sterling Harris:[2]

---

[2] Quoted by permission of the author.

Some mechanisms have a natural orientation, some have a variety of natural orientations, and some have no natural orientation at all. Automobiles and airplanes have a natural front and back, top and bottom, right and left, but a rifle has one top in the firing position, and another top in the military *order arms* position. Such items as rivets, pliers, and sleeve bearings have no recognizable natural orientation. In description, it is nearly always best to use the natural orientation of the device, if the device has a natural orientation. For mechanisms with two or three natural orientations it is necessary to state clearly which is being used: "With the rifle in the firing position, you will see the sights on the top of the barrel. The blade sight is at the front or muzzle end. The V notch is at the rear or breech end." For mechanisms having no natural orientation, it is necessary to establish a temporary or arbitrary orientation: "To understand the parts of the can opener, hold the can opener in front of you with the key to your right and the turning axis of the key running from left to right. Then turn the longer end of the body of the can opener so that it points upward."

For such orientation, the X, Y, and Z axes used in mathematics are often useful; however, the writer needs to be sure that the reader understands what an X axis is. It may be preferable to refer to the three axes as vertical, horizontal and transverse. It is still necessary, however, to relate the device to the axes: "In this discussion, the length of the hammer handle will be called the longitudinal axis, the length of the head will be called the vertical axis, and the width of the head will be called the transverse axis." Without such explanation, the reader might assume that the length of the handle was the vertical axis and the length of the head the transverse. The result would be confusion. Incidentally, a review of basic terms of solid geometry can aid the writer who is going to describe a mechanism.

Specialized fields often have their own established set of orientation terms. For example, in anatomy there are such terms as dorsal and ventral, anterior and posterior, proximal and distal. To the specialist, such terms indicate direction as clearly as the sailor's topside and below, forward and aft, and amidships and abeam. However, using such terms outside their own field may prove unwieldy or incongruous. A sailor on watch can say, "Periscope, two points off the starboard bow." But an Air Force gunner can more naturally say, "Bandit, one o'clock high." The gunner conveys the same information as the sailor while adding data on relative altitude.

A related problem arises from confusing two-dimensional and three-dimensional figures. Thus it is not quite accurate to speak of a hockey puck as round (viewed radially it has a rectangular silhouette) but as cylindrical. A similar kind of problem arises in calling a chisel a pointed rather than an edged tool. Such ambiguities are especially likely to occur when you are working from orthographic drawings rather than from the mechanism itself.

The second suggestion we have to make about communicating accurately, a very brief one, has already been noted by implication. It is actually a means of testing the accuracy of your description after you have completed it. Simply make a drawing based entirely on what you have written, being careful not to draw anything you haven't put into words. In this way you can not only help yourself to detect omissions and ambiguities but to see the whole description more nearly as your reader will see it.

Finally, we must consider the problem of the scope of a description. All descriptions of mechanisms fall somewhere between two extremes. At one extreme is the detailed description of a particular individual mechanism, including consideration of its condition at the time of the description. At the other extreme is a generalized description intended to convey only a reasonable understanding of what the mechanism is. Imagine, for example, that someone is preparing a catalog of antique firearms. One item for the catalog is a rare blunderbuss worth thousands of dollars. It would be described with care, including its condition. Another item is a not-so-rare musket manufactured in the 1850s. The dealer has 20 of them. Here the description would probably not be of an individual weapon, but of the model in general. Between these two extremes, examples could be found of almost any degree of compromise between detail and generalization.

The problem confronted here is only a special form of the familiar problem of adapting your writing to your reader. All you need do is apply common sense. Ask yourself how much detail the reader wants, and what kind, and then be consistent throughout the description.

Illustrative Material

The following pages contain four examples of the description of a mechanism. The first,[3] which is concerned with a weight-equalizing hitch for travel trailers, is presented for the specific purpose of illustrating the value of a carefully thought-out analogy in the description of a rather complex mechanism.

The second and third illustrations are both descriptions of the same simple mechanism, a patented peeler. One is highly detailed, the other generalized.

The fourth illustration is a highly condensed description of a complex mechanism—the Pentax Super Program camera.

Illustration 1—Analogy

Engineering Principles of Weight-Equalizing and Sway Resistant Hitches
T. J. Reese

WHAT IS A WEIGHT-EQUALIZING HITCH?

A weight-equalizing hitch is a mechanical device usually consisting of three units:

1. A TOW BAR that is either welded or bolted to the rear of the frame of the car and extending forward within an inch [2.54 cm] or two of the rear axle housing. This tow bar is made of material that is strong enough to take the force imposed on it that would tend to spring or bend it. (Mr. J. R. O'Brien has outlined the typical

The passage quoted, which is only a small portion of the paper read at the SAE meeting, opens with a division of the weight-equalizing hitch into its principal parts. One fact, which is perhaps not entirely clear within this passage, is that the removable ball mount is attached to the tow bar, which in turn is either bolted

---

[3] Printed by permission of T. J. Reese. The material quoted here is excerpted from a paper read at a regional meeting of the Society of Automotive Engineers. Mr. Reese, Chairman, Reese Products, Inc., is the inventor of the Reese Strait-Line Hitch.

method of installing to the car frame or body.)

2. REMOVABLE BALL MOUNT. This is the part that has the hole for the trailer hitch ball and also suitable mechanisms for attaching the spring bar members.

3. SPRING BAR ASSEMBLY, or other flexible device, that is attached to the ball mount in an articulated manner, the other end being supported by a flexible member such as a chain which is attached to a trailer frame bracket.

or welded to the frame of the tow vehicle.

How the Weight-Equalizing Hitch Works.

We will assume that we have the tow bar attached very securely to the frame of the car . . . so rigidly, in fact, that any movement that the car makes will be reflected in the movement of the tow bar. In other words, if the car sets level the tow bar will be in a relatively level position. If the car sets low in the back and up in front, that is the whole body is slanted to the back, the tow bar naturally would be tilted to the back.

We will now assume that a pair of wheelbarrow handles are rigidly attached to the ball mount that has been inserted in the tow bar on the car and the car has no load, so it is naturally setting level. The wheelbarrow handles will extend back of the car parallel with the ground. We will now place about 400 pounds [~ 181.5 kg] of weight in the extreme rear of the trunk. The back of the car will drop 4 to 6 inches [~ 10 – 15 cm], and due to the lever action over the rear axle, the front of the car would rise 1 to 3 inches [~ 2.5 – 7.5 cm]. With the car thus loaded, the ends of the wheelbarrow handles would be slanted down, either resting on the ground or at least very little above it. Now, if a man were strong enough, he could bring the back of the car up to level by lifting up the end of the handles. Not only does this remove weight from the rear axle and wheels, but due to the lever length of the overhang of the rear of the car, plus the length of the wheelbarrow handles, it transfers much of the weight to the front wheels of the car. It can be more easily understood if the lift on the handles were exaggerated to the point that the rear wheels of the car were lifted free of the ground. The entire weight of the car and trunk load would be supported by the front wheels of the car and the one supporting the wheelbarrow handles.

Having itemized the principal parts, the author presents a general idea of how a weight-equalizing hitch distributes the hitch weight of a travel trailer over all the wheels of tow vehicle and trailer combined. He does this by using the analogy of a wheelbarrow.

Let us now substitute a trailer connected to a hitch ball for the weight in the trunk, and assume that the hitch weight of the trailer is the same 400 pounds [~181.5 kg] that was removed from the trunk of the car. The wheelbarrow handles will again assume the tipped down position and again could be lifted up so that the car would be level; also the trailer will set level, if the hitch has been properly installed.

Then we will substitute tapered spring bars for the rigidly attached wheelbarrow handles. These spring bars are attached to the ball mount in such a manner that they have no movement in a vertical plane, but are free to swing from side-to-side in a horizontal plane.

We will substitute chains and frame brackets for the strong man holding the wheelbarrow handles. [The chains are attached to the ends of the spring bars, and then to the brackets mounted on the tongue of the trailer.] In other words, we lift up the end of the spring bars with chain attached and hook the proper link onto the frame bracket. Since the spring bars are approximately 30″ [~76 cm] long and are pulling down on the trailer frame 30″ back of the hitch ball, a portion of the hitch weight is transferred through the trailer frame rearward to the trailer wheels.

Illustration 2 — Detailed Description

The A & J Peeler

One of the handiest implements to have around a kitchen is a patented peeler that is identified only by the letters A & J and the patent number, both stamped into the handle. It is much like a small paring knife in size and shape but novel in both design and construction. In spite of its novelty, it is made up of the three parts common to almost any kitchen knife: blade, tang, and handle.

The blade is designed to remove a uniformly thin peeling, as from an apple, without the need of any special care to keep the peeling thin or to prevent the blade from cutting too deeply into the fruit. For this purpose the blade is rounded, as if a piece of ⅜ in. [1 cm] tubing had been cut in halves lengthwise and the blade made out of one of the halves. The cutting edges, the tip of the blade, and

This description of a peeler, and the much shorter one that follows, are characteristic student exercises. The second is also characteristic of a great deal of writing found in routine scientific and industrial reports. This first one, on the other hand, is not so easily defined. It is quite common to find description as detailed as this in reports. It is not common, however, to find so prolonged and exhaustive a treatment of such a simple mechanism.

This description should therefore be thought of as essentially an exercise in—as we suggested earlier—seeing and communicat-

the shank of the blade will each be considered separately.

There are two cutting edges, but one of the novelties of this peeler is that they are not the outside edges of the blade. These cutting edges can easily be visualized by thinking again of the piece of tubing that has been cut in half lengthwise. In one of these halves, a broad slit has been cut from near the tip to near the shank of the blade, and the inner edges of this slit have been sharpened. When the convex side of this blade is moved along the surface of a fruit or vegetable, one or the other of these sharp edges peels the skin away. The peeling comes away through the opening in the blade formed by the slit. The sharpening has been done by grinding the convex side of the blade down to form a cutting edge along each side of the slit. As a result, the convex side is flattened along the slit. The blade is 3 in. [7.7 cm] long overall and ⅜ in. [1 cm] wide. The slit is 1⅞ in. [4.8 cm] long and ⅛ in. [0.3 cm] wide.

From the tip of the blade to the slit there is a distance of about ⅝ in. [1.6 cm], and this area has been rounded up toward the end into a shape resembling the tip of a tiny spoon. This shape is apparently intended to be helpful in digging out any small bad spots found in the fruit or vegetable after it is peeled.

The shank serves to fasten the blade to the tang. The shank can be visualized by once more thinking of the piece of tubing cut lengthwise. In effect, about a half-inch [1.27 cm] of the end of this cut piece of tubing has been rolled further inward around the long axis of the blade to form a cylindrical collar around the tang. The shank is held to the tang by a friction fit.

The tang is a slender steel rod which holds the blade to the handle. It is 3½ in. [9 cm] long and ⅛ in. [0.3 cm] in diameter. At the end opposite the blade it has a flange, or head—like the head of a nail; and ⅛ in. [0.3 cm] from this head two small projections or "ears" have been stamped into the tang. The head and projections both help to keep the tang in place in the handle.

The handle actually serves three purposes. In addition to its normal function as a handle, it provides a frame within which the blade and tang

ing. It is comparable to practicing scales on the piano or tackling a dummy on the football field. For you, it can serve as an exercise in the following ways. First, without trying to comprehend it in detail, examine its organization with respect to the generalized outline of the description of a mechanism presented above. Look for the statement of the principal parts, use of analogy, division into subparts, statement of purpose, and so on. Second, read the description slowly for its content. As you read, keep checking the word description against the illustrations on p. 116. And as you read the text and check the illustrations, keep asking yourself if the text is really communicating what the illustrations show. We have already pointed out one place in which the text is inaccurate.

You will probably find that reading this description with real comprehension requires intense concentration. Naturally, writing it must also have required intense concentration. Practicing this kind of concentration will help you acquire the ability to write accurately and clearly.

assembly can rotate, and also provides stops to limit this rotation to 90 deg. The purpose of this rotation will be explained below. The handle is constructed entirely of a strip of steel $3/64$ in. [0.1 cm] thick and $3/8$ in. [1 cm] wide which has been bent lengthwise into the necessary shape. The shape of the handle can be most easily visualized by starting with this strip of steel still in one long, flat piece as it presumably was after first being stamped out. At this stage of construction, a hole $5/32$ in. [0.35 cm] in diameter has been punched exactly in the center, and an open-ended slot has been stamped in each end. The inner ends of these slots are semi-circular. These slots are the same size, $5/16$ in. [0.8 cm] long and $1/4$ in. [0.6 cm] wide. The metal left along each side of the slots is only $1/16$ in. [0.2 cm] wide. These slender, pronglike sides ultimately become the stops mentioned previously.

This long, flat strip of steel can next be visualized as bent into a narrow U, about $3/4$ in. [~2 cm] across, with the punched hole at the bottom. And the following step in the visualization is to close the U by bending the ends of the sides toward one another. Actually, however, two bends are required.

The first might be imagined as the bending of $3/4$ in. [~2 cm] of each end inward, toward the other, at an angle of about 100 deg with their respective sides. When this bending has been completed, one end will overlap the other. The amount of this overlapping of the bent pieces should be adjusted in imagination by sliding them together, one over the other, beginning when the ends have just started to overlap. In this position, a rectangular opening (but with semi-circular ends) is formed by the combination of the slots in the ends. As the two pieces are being slid together, one over the other, it is helpful to imagine a nail with the same diameter as the shank of the blade inserted into this rectangular space. The nail should be perpendicular to the bent pieces. When the pieces are moved toward one another this rectangular space decreases in length until the nail is loosely gripped by the two rounded ends of the slots. Together (but one above the other) the two semi-circular ends have now formed an almost perfectly round hole. The nail withdrawn, this is the hole through which the shank itself will be inserted into the handle. The amount of overlap has now been properly adjusted.

The second of the two bends is now to be made, in
imagination. It is readily visualized by assuming
that a knife-edge is laid across the hole just
created, at right angles to the pronglike pieces
forming the sides of the upper slot. Each of these
two prongs is then bent upward, away from the
handle, to stand parallel to the axis of the blade
and tang assembly. A semi-circular hole is thus left
at the base of the prongs, its plane at right angles
to the blade. This bend might be compared to
laying a paper clip on a table, placing a ruler at
right angles across about ⅛ in. [0.3 cm] of the end
of the paper clip, and bending the principal length
of the paper clip upright. A ⅛ in. [0.3 cm] semi-circular "hole" would be left at the bottom; the
upright sides of the paper clip would correspond to
the upright prongs of the peeler handle. Each
prong is ¼ in. [0.6 cm] long.

After the prongs of the uppermost side are bent,
the prongs on the other side are to be bent in the
same way, again leaving a semi-circular hole.
When the inner faces of the prongs of the two sides
are then brought together, or mated, an almost
perfectly round hole is formed at their base, to accept the shank of the blade. The mated prongs are
tack-welded together. Each of the tack-welded
pairs serves as a stop to limit the rotation of the
blade and tang assembly. The welding gives
rigidity to the entire handle.

One other characteristic of the handle is that
beginning just back of the end nearest the blade,
where the handle is ⅞ in. [2.2 cm] wide, each side
is curved in about ³⁄₁₆ in. [0.5 cm] to provide a
comfortable grip.

As remarked earlier, the handle constitutes a
frame which permits a restricted rotation of the
blade and tang assembly. The hole just described,
at the base of the prongs, or stops, serves as a
bearing for the shank; the smaller hole, at the
other end of the handle, serves as a bearing for the
tang. That portion of the shank that is faired out to
become the blade stands between the two stops, as
does the base of the blade itself. The rotation of
the blade and tang assembly is limited by the
striking of the base of the blade against one or the
other of the stops. It is always the concave side of
the blade that makes contact.

The blade and tang assembly are fixed in position within the frame by the head and projections on the tang described earlier. The head is on the outside of the handle, the projections on the inside.

For removing the rind, or outer layer, from fruit or vegetables this peeler can be used with a rapidity that would be impossible with a conventional paring knife. Whichever cutting edge is being employed, it is prevented from slicing in too deeply because the opposite side of the blade is passing over the surface ahead of the cutting edge and limiting its angle of attack. The principle of operation is somewhat like that of a reel-type lawn mower, in which the angle of the blade is controlled by the wheels in front and the roller behind. As is now evident, the freedom of the blade to rotate permits an automatic adjustment to irregularities in the surface. The stops inhibit any tendency of the blade to "roll," and permit the application of some twisting force on the blade if that is necessary to help the edge bite in, as at the beginning of a cut.

Of course this peeler cannot be used for slicing, at least in any ordinary sense, but only for paring, or for such cutting as making carrot curls.

The handle, which has a decorative "cross-hatch" pattern stamped lightly into its outside surface, appears to be nickel plated. The blade and tang have been given no special finish.

Illustration 3 — General Description

The A & J Peeler

One of the handiest implements to have around the kitchen is a patented peeler that is identified only by the letters A & J and the patent number, both stamped into the handle. As shown in Figure 1 it is much like any small paring knife in size and shape but novel in both design and construction. In spite of its novelty, it is made up of the three parts common to almost any kitchen knife: blade, tang, and handle.

The blade is designed to remove a uniformly thin peeling, as from an apple, without the need of any special care to keep the peeling thin or to prevent the blade from cutting too deeply into the fruit. For this purpose the blade is rounded, as if a piece of $\frac{3}{8}$-in. [1-cm] tubing had been cut in halves lengthwise and the blade made out of one of the halves. The cutting edges, the tip of the blade, and the shank of the blade will each be considered separately.

There are two cutting edges, but one of the novelties of this peeler is that they are not the outside edges of the blade. Instead, they are the inner edges of the slit

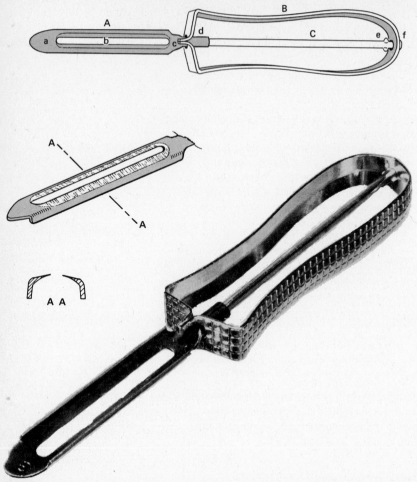

Figure 1   The A & J peeler. (Top) A: blade; B: handle; C: tang; a: rounded tip; b: slit; c, handle stop attachment; d: shank; e, "ears"; f: outer head. (Middle) Cross section at AA showing angle at which edges are ground. (Bottom) Photograph showing rounded end of blade and prongs or stops on handle.

(Figure 1-b) that has been cut lengthwise in the blade. When the convex side of the blade is moved along the surface of a fruit or vegetable, one or the other of these sharp edges peels the rind away. The peeling comes away through the slit. The sharpening has been done by grinding the convex side of the blade down to form a cutting edge along each side of the slit.

As shown in Figure 1, the end of the blade has been rounded up into a shape resembling the tip of a very small spoon. This shape is apparently intended to be helpful in digging out any small bad spots in the fruit or vegetable after it is peeled.

The shank (Figure 1-d), which serves to fasten the blade to the tang, has been formed, in effect, by rolling the end of the blade around the tang and securing it with a friction fit.

The tang (Figure 1-C) is simply a slender steel rod. It is free to rotate within the handle, and is held in place by a head at the end outside the handle (Figure 1-f) and two "ears" stamped into it just inside the handle (Figure 1-e).

The handle actually serves three purposes. In addition to its normal function as a handle, it provides a frame within which the blade and tang assembly can rotate, and also provides stops to limit this rotation to 90 deg. The purpose of this rotation will be explained below. The handle is constructed entirely of a strip of steel ³⁄₆₄ in. [0.1 cm] thick and ³⁄₈ in. [1 cm] wide. As can be seen in Figure 1, holes have been punched in both ends to receive the shank and tang, respectively. At the blade end, the edges of the handle remaining after the hole was stamped out have been bent parallel to the blade and tack-welded together. The prongs thus formed limit the rotation of the blade to 90 deg by intercepting the movement of the forward portion of the shank, where it is faired out into the blade.

For removing the rind, or outer layer, from fruit or vegetables this peeler can be used with a rapidity that would be impossible with a conventional paring knife. Whichever cutting edge is being employed, it is prevented from slicing in too deeply because the opposite side of the blade is passing over the surface ahead of the cutting edge and limiting the angle of attack. The principle of operation is something like that of a reel-type lawn mower, in which the angle of the blade is controlled by the wheels in front and the roller behind. As is now evident, the freedom of the blade to rotate permits an automatic adjustment to irregularities in the surface. The stops inhibit any tendency of the blade to "roll," and permit the application of some twisting force on the blade if that is necessary to help the edge bite in, as at the beginning of a cut.

Of course this peeler cannot be used for slicing, at least in any ordinary sense, but only for paring, or such cutting as making carrot curls.

The handle, which has a decorative "cross-hatch" pattern stamped lightly into its outside surface, appears to be nickel plated. The blade and tang have been given no special finish.

Pentax Super Program[4]

CAMERA TYPE

35-mm multimode SLR with programmed auto, aperture-priority auto, shutter-priority auto, metered manual, TTL autoflash, programmed autoflash modes

NORMAL LENSES

50-mm SMC Pentax A f/1.2, f/1.4, f/1.7, 28 → 135-mm f/4 (tested)

SHUTTER

Electronically governed, vertically traveling, metal focal-plane Seiko MFC E3 with stepless speeds

Descriptions of mechanisms like this one, which is a portion of a laboratory report on the Pentax Super Program camera, are very common; they differ in at least two respects from the type we have emphasized in this chapter: (1) since they are intended for readers who are well-informed on general relationships, an overall view is usually provided by a photo or drawing rather than by words, and (2) because

[4] Reprinted by permission from *Popular Photography Magazine*, January 1984, Copyright Ziff-Davis Publishing Company

from 15 to 1/2,000 sec on automatic, stepped speeds from 15 to 1/2,000 on manual, plus B and X-1/125 sec

VIEWFINDER

Fixed, eye-level type, noninterchangeable screen has central rangefinder spot in microprism collar on groundglass/Fresnel field; two readout windows at bottom outside picture area show liquid-crystal displays of automatic and manual shutter speeds, automatically set lens aperture, flashing exposure-factor "EF" warning, shutter-speed-selection error warning, weak battery, programmed auto "P" indication, shutter cocked, flash recycled, +/− symbols for use in manual-exposure mode

EXPOSURE METER

Through-lens, full-aperture reading; uses two gallium photodiodes (GPD) for ambient readings plus third cell in mirror box for off-film-plane measurements with special dedicated electronic flash units; EV range 1 to 19 with ISO 100 film and f/1.4 lens, ISO range 6-3,200; ±2 EV exposure override, light pressure on shutter-release button turns on metering system for approximately 30 seconds; system uses two S-76 batteries or equivalent

FLASH SYNCHRONIZATION

Dedicated hot shoe and PC outlet, X at 1/125 sec, shutter speed automatically set when used with dedicated electronic-flash units

LOADING

Conventional, swing-open removable back

FILM TRANSPORT

Single-stroke with 135° wind angle after 30° standoff

FRAME COUNTER

Additive, automatic-reset

OTHER FEATURES

Viewfinder illumination lamp, audio/visual self-timer with 12-second delay, external LCD display of shutter information, removable hand-grip, memo holder, shutter lock, manual shutter-speed-select buttons, depth-of-field preview lever, provision for winder attachment and databack, "Magic Needle" take-up spool

they usually appear in catalogs or other publications where space is at a premium, they are highly condensed and rarely make use of analogy or other figurative language. As you can see here, they are likely to consist primarily of a classified listing of distinguishing characteristics. The photograph which accompanied this description is not reproduced here.

WEIGHT

1314.6 g (46.37 oz.) with 28 → 135-mm lens

DIMENSIONS

L., 134.8 mm (5.31 in.)., H., 86.6 mm (3.41 in.).,
D., 163.8 mm (6.45 in.) with 28 → 135-mm lens

ACCESSORIES

Full line of lenses from extrawide-angle to
telephoto, zooms, micro and macro accessories,
dedicated electronic-flash units, winders, more

PRICE

Body only, $325; 28 → 135-mm lens, $540

DISTRIBUTOR

Pentax Corp., 35 Inverness Dr. East, Englewood,
Co. 80112

Suggestions for Writing

Selecting a Subject

The things to think about in selecting a mechanism to describe are (1) your personal interests, (2) the information you already possess or can readily obtain, (3) the suitability of the mechanism for a description conforming to the length desired, (4) the suitability of the mechanism for a description conforming to the kind of description desired, and (5) the selection of an imaginary reader.

The first two things are simple and obvious: if you yourself are choosing a mechanism to describe, you will naturally want one you are interested in and know something about, although for a beginning exercise, such as the description of the peeler quoted earlier in this chapter, the mechanism may be so simple that the choice may not matter very much. The last three things mentioned above, on the other hand, are all dependent on one another. For example, a mechanism that would be well suited for a highly generalized description limited to a few hundred words in length might prove impossibly complex for a detailed description of the same length. Or a mechanism you could easily describe for a reader having a good deal of technical knowledge might become extremely difficult and long for a reader who was unfamiliar with scientific concepts and terminology.

Among the five elements listed, your instructor will probably specify some, such as the length of the description, the degree to which it should be generalized or detailed, and the sort of imaginary reader to whom it should be directed. Any that are left unspecified you can think over and arrange to suit your own preferences.

The following is a list of possible subjects. The subjects designated by Arabic numerals are very broad areas, which have been chosen at random; under each of these subjects are three topics, beginning with one suitable for a detailed description and terminating with one suitable only for a generalized description. These topics will no doubt suggest others, some of which you might prefer.

1. Aircraft
   A nonfeathering propeller on a light plane
   The landing gear on a light plane
   A sailplane
2. Boats
   The running light mounted on the stern of a small motorboat
   The standing rigging of a small sloop
   A small catamaran
3. Automobiles
   A wheel
   A shock absorber
   An airconditioner
4. Farm machinery
   A manual post-hole digger
   A hammer mill
   A hay baler
5. Tree care
   A bow saw
   A manually operated sprayer
   A chain saw
6. Electricity
   A small transformer
   A simple speaker or earphone
   A cordless telephone
7. Carpentry
   A handsaw
   A plane
   A radial-arm saw
8. Metal working
   Tinsnips
   A power hacksaw
   An electric-arc welder
9. Hydraulics
   A garden hose
   A hydraulic ram
   A hydraulic lift in a service station
10. Optics
    A small hand lens
    Eyeglasses
    Binoculars
11. Earth-moving equipment
    A spade
    A scraper-blade attachment for a tractor
    A backhoe attachment for a tractor
12. Laboratory
    A ringstand

    A balance
    A centrifuge
13. Kitchen
    A spatula
    A coffee maker
    A food processer
14. Living room
    A simple magazine rack
    A folding snack table
    A venetian blind
15. Physiology
    A bird's foot
    A human hand
    An animal skeleton
16. Plant life
    A tulip bulb
    A maple seed
    A pine cone

Practicing Organization

When rearranged into the proper sequence, the numbered sentences below constitute a description of a lead pencil. There were six paragraphs in the original. Guided by the principles relevant to the description of a mechanism, rearrange these sentences into a logical sequence and group them into six paragraphs. For convenience, you might assign a Roman numeral to each paragraph.

A Wooden Lead Pencil

1. Typically, a wooden lead pencil is about 0.8 cm in diameter and about 19 cm long.
2. The shape of a wooden lead pencil resembles that of a piece of dowel, or the shaft of an arrow, with the exception that many pencils are hexagonal, rather than round, in cross section, and that occasionally still other cross-sectional shapes are found.
3. The purpose of the wooden case is to make the pencil comfortable to hold and to provide rigidity.
4. The case is composed of two parts: that is, it is divided lengthwise into identical halves.
5. A wooden lead pencil is a device for drawing or writing on a suitable medium with a piece of graphite permanently encased in a slender wooden shaft or holder.
6. Before it can be used, a wooden pencil must be sharpened: wood must be shaved off the end opposite the eraser end, and the graphite itself must be shaved down to a point.
7. The purpose of the eraser is to remove unwanted marks from the medium being written upon.
8. Each half of the case contains a semicircular groove centered along the length of

the flat side, to receive the graphite, and each has a slightly diminished cross-sectional area at one end over which to slip the metal ferrule.

9. In appearance, the ferrule is a simple piece of metal tubing about 1.5 cm long, and 0.8 cm in diameter, often painted to harmonize with the case and the eraser.

10. When assembled, the two halves of the case are glued together, with the graphite held in the hole created by the matching semicircular grooves.

11. It is usually in the form of a cylinder about 0.3 cm in diameter, and is the same length as the case.

12. As the graphite wears away in use, the pencil must be resharpened.

13. This diminished cross-sectional area is usually round, regardless of the shape of the rest of the case.

14. It is typically a rubber cylinder about 1.5 cm in length which will fit tightly inside the ferrule; and it is often made of colored rubber.

15. It is made up of four parts: the graphite, the wooden case, an eraser, and a ferrule by which the eraser is attached.

16. The purpose of the ferrule is to attach the eraser to the case.

17. The eraser and the case are inserted into opposite ends of the ferrule; they are secured in place by indentions which have been stamped into the ferrule after assembly.

18. The case is normally painted.

19. For pencils having a novel cross section, the graphite may be molded in some form other than that of a cylinder.

20. The function of the graphite is, of course, to make marks on the paper or other medium.

Writing a Brief Description

Write a short description of a table lamp, a lamp to which you have ready access so you can examine it carefully. Do not use any illustrations as part of your description. Instead, write on the assumption that you will read your description aloud in class while someone else attempts to sketch the lamp on the board, guided only by what you have written. On a separate sheet of paper, however, include a sketch of the lamp for your instructor.

7

Description of a Process

Introduction

A process is a series of actions, and fundamentally the description of a process is the description of action. The action may be either one of two types. One type is that in which attention is focused on the performance of a human being, or possibly a group of human beings. A simple example is planing a board by hand; in a description of this process, emphasis would fall naturally upon the human skills required. The other type involves action in which a human operator either is not directly concerned at all, or is inconspicuous. An instance is the functioning of an electric relay. Large-scale processes, when considered as a whole, are also usually of this second type, even though human operators may take a conspicuous part in some of the steps. The manufacture of paper is an example.

This chapter is divided into two main parts, according to these two types of processes. Before taking up the first type of process, however, we will consider three problems that arise in describing almost any process, regardless of type: (1) the adaptation of the description to the reader, (2) the overall organization, and (3) the use of illustrations.

Adapting the description to the reader depends, as always, upon an analysis of the reader's needs. As in the description of a mechanism, if the reader wishes to use the description as a practical guide, it becomes necessary for the writer to give careful attention to every detail. If the reader is interested only in acquiring a general knowledge of the principles involved and has no intention of trying to perform the process or to direct its performance, the writer should avoid many of the details and emphasize the broad outlines of the process.

The fundamental organization of a process description is simple, consisting merely of an introduction followed by a description of each of the steps in the process in the order in which they occur. But this simplicity is usually marred by the necessity of discussing the equipment and the materials used. In building a wooden boat, for instance, the equipment would include hand and power saws, miter boxes, and planes; the materials would include lumber, screws, paint, and others. It is not always necessary to mention every item of equipment or every bit of material, but no helpful reference or explanation should be omitted through negligence. Sometimes it is necessary to explain certain special conditions under which the process must be carried out, like that for developing photographic film in a darkroom.

There are basically two ways of incorporating the discussion of equipment and materials into the description as a whole. One is to lump it all together in a section near the beginning; the other is to introduce each piece of equipment and each bit of material as it happens to come up in the explanation of the steps in the process. The advantage of confining the description of equipment and materials to a single section near the beginning is that such discussion does not then interrupt the steps in the action itself. This method is usually practical if the equipment and materials are not numerous. If they happen to be so numerous or so complex that the reader might have difficulty in remembering them, the other method of taking them up as they appear in the process is preferable. The second method is by far the more common for process description, but the first method is usually the better choice if you are giving directions.

In summary, we can say that a process description is organized as follows (except that the discussion of equipment and materials may be distributed throughout the description instead of being confined to one section):

Introduction
Equipment and materials
Step-by-step description of the action
Conclusion (if necessary)

The use of illustrations, the last of the three general problems, needs little comment. To the degree that they are likely to be of use to the reader, as many illustrations as can be managed conveniently should be introduced. It is difficult to represent action graphically, but sometimes a sketch of how a tool is held or of how two moving elements in a device fit together can add greatly to the clarity of the text. The general problem of the use of illustrations is much the same as in the description of a mechanism.

PART 1: PROCESSES IN WHICH AN OPERATOR TAKES A CONSPICUOUS PART

In this part of the chapter we consider three subjects: the introduction to a process description, the step-by-step description of the action, and the conclusion. We also take note of the special problems encountered in giving directions.

The Introduction to the Description

The introduction to the description of a process is a comprehensive answer to the question, "What are you doing?" (The rest of the description is largely an answer to the question, "How do you do it?") An answer to the question, "What are you doing?" can be given by answering still other questions, principally the following:

1. What is this process?
2. Who performs this process?
3. Why is this process performed?
4. What are the chief steps in this process?
5. From what point of view is this process going to be considered in this discussion?
6. Why is this process being described?

It is not always necessary to answer all six questions, and it is not necessary to answer them in the order in which they happen to be listed. It will be helpful to consider each question in turn to get some notion of what needs to be done.

What Is This Process?   Very early in the report readers must be told enough about what the process is so that they can grasp the general idea. The way in which this explanation is given depends upon how much the readers are presumed to know about the process, as well as upon the nature of the process itself. As in the description of a mechanism, we have come up against the whole problem of definition of the subject of the description. Again we must refer to the chapter on definition for a full treatment of the problem; here we give some particular attention to the use of comparison and generalized description.

A report written for sophomore engineering students on how to solder electrical connections might start by saying merely, "It is the purpose of this report to explain how to solder electrical connections." This simple statement of the subject would be sufficient. If, however, a report on the same subject were being prepared for a reader who had no real understanding of even the word "solder," an entirely different approach would be needed. Let's consider a reader who is very different from the sophomore engineer. Suppose a description of how to solder electrical connections was being prepared for a group of adults taking an evening course in simple home repairs. Some of these people would probably have no knowledge at all of the process of soldering; consequently, the fundamental concept would have to be carefully explained. For these readers it would be wise to write a formal definition accompanied by a comparison to soldered articles that most of them had probably seen, and to similar processes that they would know about. Such a report might begin in the following manner:

> It is the purpose of this report to explain how to solder electrical connections. Soldering is the joining of metal surfaces by a melted metal or metallic alloy. This process may be compared roughly with the gluing together of two pieces of wood. Instead of wood, the solderer joins pieces of metal, and instead of glue he or she uses a melted alloy of lead and tin that, like the glue, hardens and forms a bond. Soldering is a very widely used technique; one evidence of its use, which probably almost everyone has noticed, is the streak of hardened solder along the joint, or seam, of a tin can of food.

The third and fourth sentences above constitute a comparison with a process with which the reader would probably be familiar. The last sentence in the example is a reference to a familiar device in which the process has been employed.

The preceding introduction might continue:

> The process of soldering consists essentially of heating the joint to a degree sufficient to melt solder held against it, allowing the melted solder to flow over the joint, and, after the source of heat has been removed, holding the joint immovable until the solder has hardened.

This example gives a general idea of the whole process. You will probably have noticed also that it looks much like a definition of the process, and at the same time like a statement of the chief steps (a subject to which we will come in a few moments). As a matter of fact, it is an acceptable definition; and although the list of steps is actually incomplete, the missing steps could easily be added. It is evident, then, that it would be possible to define a process and indicate its purpose, to give a generalized description of it, and to list the chief steps, all in one sentence. Would such compression be advisable? Sometimes; it depends upon the reader. For the readers identified above it probably would not be, since we are assuming some of them know nothing of the process. The more leisurely manner in the example above, including both parts, would provide such readers a little more time to get used to the idea. In short, the question "What is this process?" is simply a problem of definition, and therefore the use of comparison and of generalized description is often particularly helpful.

Who Performs This Process?   With one major exception, there is not a great deal to say about this matter of explaining who performs the process, except to emphasize the fact that it is sometimes a most helpful statement to make. For example, if a description of the process of developing color film was written for the general public, it might be rather misleading unless the writer explained that most amateur photographers do not care to attempt this complicated process, the bulk of such work being done commercially. Very often the statement about who performs the process will appear as a natural or necessary element in some other part of the introduction. Often no statement is required.

The major exception is that the reader should never be left in doubt as to whether he or she is expected to perform the process or whether the process is assumed to be performed by someone else. The careless use of "you," for example, can make it appear that the reader is expected to take action even though the writer did not intend the reader to become involved. On the other hand, careless use of language may produce the opposite effect by causing a reader to take no action at points where action on the part of the reader is expected. The simplest way to reduce these unwanted consequences is to give all directions to the reader in the form of commands (the imperative mood) and to avoid giving commands (by using the active or passive indicative) in the presentation of other types of process explanation. We will discuss the relative advantages of the imperative and the indicative later in this chapter.

Why Is This Process Performed? It is, of course, absolutely necessary that the reader know why the process is performed — what its purpose is. Sometimes simply explaining what the process is, or defining it, makes the purpose clear. Often the purpose of the process is a matter of common knowledge. There would be no point in explaining *why* one paddles a canoe, although not very many people may know *how* to paddle a canoe efficiently. Sometimes, however, the purpose of a process may not be clear from a statement of what it is or how it is performed. Then it is necessary to be quite explicit in stating its complete purpose. To take a simple instance — one might explain clearly and accurately how to water tomato plants, how much and in what manner, and still do readers a disservice by not informing them that if the supply of moisture is not sufficiently regular, there will be a tendency for circular cracks to appear around the stem end of the ripening tomatoes.

What Are the Chief Steps in This Process? The listing of the chief steps in the process is an important part of the introduction. It is important because it helps the reader understand the process before the details of its execution are presented. Even more important is its function in telling the reader what to expect in the material that follows. It is a transitional device. It prepares the reader for what lies ahead. Naturally, it serves the purpose of a transitional forecast best when it appears at the end of the introduction.

The list of steps may appear as a formal list, with a number or letter standing beside each step. If this method seems too mechanical, the steps may be stated in ordinary sentence form, with or without numbers or letters. Care should be taken with punctuation to avoid any possibility of ambiguity or overlapping of steps. The statement of the major steps in the process of soldering an electrical connection might be written as follows:

> The chief steps in this process are (1) securing the materials and equipment, (2) preparing the soldering iron (or copper), (3) preparing the joint to be soldered, (4) applying the solder, and (5) taping the joint.

Observe that itemized parts of the sentence are grammatically parallel, as they should be. The steps should be discussed in the order in which they are listed.

From What Point of View Is This Process to Be Discussed? Why Is This Process Being Described? These two questions, which are the last two, can conveniently be discussed together. Neither one is properly concerned with the question with which we started this section on the introduction: "What are you doing?" Nevertheless each represents an important aspect of the introduction. Each is concerned in its own way with the purpose of the report.

The latter question ("Why is this process being described?") calls for a specific statement of purpose — the purpose of including the description of this process in the report of which it is a part. In other words, readers will want to know why you are asking them to take time to read your description of the process. More often than not, your reason for including the description will be perfectly obvious. If so, there is no need to mention it. Sometimes, however, it may not at once be clear. Perhaps, for example, a later part of a discussion will be incomprehensible without a preliminary understanding of a certain process, and readers will need to have this fact explained. Be careful to keep in mind the distinction between the purpose of the process itself and the purpose you have in writing about it. These are very different matters.

Perhaps we should add that if your instructor asks you to write a paper devoted to describing a process, we don't think you will need to explain why you are writing the paper.

The first of the two questions above is likewise related to the matter of purpose, but here the interest is not in why the process is being described; rather it is in why it is being described in a particular way or from a given point of view. One illustration of this fact is contained in the different ways that were suggested earlier for the writing of the introduction to the report on soldering. There would be no difficulty in seeing at once that the report written for the evening class in home repairs was designed to explain the simple process of soldering so fully that a completely unini-tiated reader could successfully use the explanation as a guide. However, it is often wise to state the point of view explicitly, as in the following example:

> The explanation of how to correct the instability of this oscillator will be given in terms of physical changes in the circuit rather than as a mathematical analysis.

One concludes from this statement that the point of view in the report is going to be practical, the treatment simple. The point of view will perhaps be that of a radio technician rather than that of an electrical engineer.

So much for the introduction to a description of a process. In this discussion we have pointed out what facts the reader of a process description should be aware of upon finishing the introduction. Sometimes almost all the problems mentioned will be met by the writer in a single introduction; sometimes only a few. But probably they should all be considered. Much depends upon who the readers of the report will be and upon the general circumstances that cause the report to be written.

Of course the writing of introductions may involve many problems not men-tioned here at all. In this section we have discussed only those elements that are likely to be involved in the "machinery" of starting off a process description. For a discussion of other aspects of the writing of introductions see Chapter 10.

The Chief Steps

Organization   With the possible exception of the discussion of equipment and materials, the introduction to a description of a process is followed directly by a description of the chief steps in the process. Two problems appear in organizing the description of the chief steps. One problem is how to organize the steps; the other is how to organize the material within each individual step.

The organization of the steps can be dismissed at once. It is chronological, the order of the performance of the steps. Although there are processes in which two or more steps are, or can be, performed simultaneously, you can usually manage fairly well by explaining the situation plainly and then taking one step at a time.

The organization *within* the description of the individual steps requires more comment. For both content and organization of the description of each individual step one idea is so useful that it cannot easily be overemphasized. This idea is that each individual step constitutes a process in itself. The individual step should therefore be properly introduced, and, if necessary, divided into substeps. Its description is essentially a miniature of the description of the process as a whole. Furthermore, if a given individual step can be broken down into substeps, each substep is treated according to the same general principles applied to the whole process.

Of course it would be easy to go too far with this idea. What we just said should be taken with a little salt. In the introduction to the whole report, for instance, it is often desirable to say something about who performs the process, about the point of view from which the process will be described, and about why the description is being written. Usually, when you introduce an individual step, nothing of this sort need be said. Definition, statement of purpose, and division into parts, on the other hand, require the same attention in introducing the individual step that they do in introducing the whole report. The great importance of making the purpose of each step clear may be seen from another point of view in the discussion of a block diagram in Appendix C. Read the second paragraph under "Information Arrangement."

What is to be said in describing the action itself constitutes an entirely new problem. It is perhaps surprising to reflect that of all that has been said so far in this chapter about how to describe a process, which was originally defined as an action or series of actions, nothing has as yet been said about how to describe the action itself. Everything has been concerned with how to get the action in focus, together with all its necessary relationships. The only point in the whole report at which action is really described is in the individual step. And if there are substeps, the description of the action drops down to them.

The Description of the Action   In describing the action, the writer must say everything the readers need to know to understand, perhaps even to visualize, the process. The omission of a slight detail may be enough to spoil everything. Moreover, care should be taken not only in connection with the details of *what* is done, but also of *how* it is done. For example, in telling readers about heating a soldering iron, it would surely be wise to tell them that if the tip of the iron begins to show rainbow colors, it is getting *too* hot. And in an explanation of how to calibrate a wide-range mercury thermometer in an oil bath, it would be advisable to point out that the oil should not be allowed to get too hot because the thermometer may then blow its top off. Keep the readers in control of the action.

A further illustration of the importance of details and of analyzing the needs of readers can be taken from the following true incident. An army sergeant was instructing several cooks on the procedure for cleaning and storing some gasoline cookstoves they had used for the first time during a field training exercise. Although these cooks were experienced in food preparation, all their previous training had been in modern kitchens equipped with electric ranges. They did know that all the gasoline had to be removed from the portable stoves before storing them, but they didn't know how to remove the last few teaspooons remaining in the tanks after the bulk of the gasoline had been poured out. The sergeant, however, knew that if a strand of absorbent material were dropped into the tank with one end of the strand exposed to the outside air the liquid gasoline would travel up the strand and slowly evaporate at the exposed end. The sergeant, having at the moment other matters that required his immediate attention elsewhere and thus needing to give directions as briefly as possible, told his cooks, "Cut some strands from a dry floor mop and use them for wicks to get the remaining gas out of those stoves." The cooks dutifully inserted some dry mop strands into each of the tanks — and proceeded to set fire to the exposed ends! Fortunately the sergeant returned in time to snatch the burning wicks from the stoves before any explosions occurred. Perhaps you will feel that these cooks were not very alert, and perhaps they weren't. On the other hand, the sergeant was speaking *to them.*

We started out in this section by saying that the content of the description of a process is governed by the reader's need to comprehend every step in the action. There is little more that can be said about the description of the action in the various steps of the process, with one important exception: that is, the style.

Style   A general discussion of style in technical reports is given in Chapter 3, and what is said there applies to the description of a process. One problem peculiar to the description of a process is not taken up in that chapter, however. This problem is the choice of the mood and voice of the predicate, and of the noun or pronoun used as the subject. A good many possibilities exist, but (neglecting the noun or pronoun for the moment) three are of special importance: the active voice and indicative mood, the passive voice and indicative mood, and the active voice and imperative mood. We will illustrate each of these and then comment on them.

Active Voice, Indicative Mood
The next step is the application of the solder to the joint. This step requires the use of only the heated iron (or copper), and a length of the rosin-core solder. The solderer takes the iron in one hand and the solder in the other, and holds the iron steadily against the wire joint for a moment to heat the wire. Then he or she presses the solder lightly against the joint, letting enough of it melt and flow over the wire to form a coating about the entire joint.

Passive Voice, Indicative Mood
The next step is the application of the solder to the joint. This step requires the use of only the heated iron, and a length of the rosin-core solder. The iron is held steadily against the wire joint for a moment to heat the wire. Then the solder is pressed lightly against the joint until enough of it has melted and flowed over the wire to form a coating about the entire joint.

Active Voice, Imperative Mood
The next step is the application of the solder to the joint. This step requires the use of only the heated iron, and a length of the rosin-core solder. Take the iron in one hand and the solder in the other, and hold the iron steadily against the wire joint for a moment to heat the wire. Now press the solder lightly against the joint. Let enough of it melt and flow over the wire to form a coating about the entire joint.

The essential differences among these three ways can be expressed as the differences in the following three statements: (1) The solderer holds the iron. (2) The iron is held. (3) Hold the iron.

Which one of the three ways is best? It depends upon several factors.

The advantage of the first way, the active voice and indicative mood, is that it gives the reader the greatest possible assistance in visualizing the action. It is the most dramatic. It comes as close as it is possible to come in words to the actual observation of someone performing the action. The presence of the person carrying out the process is kept steadily in the mind of the reader. This technique is without question a very effective one, and its possibilities should not be overlooked. Probably its best use occurs when the following three conditions prevail: (1) the process being described is one that is performed by one person; (2) the description of the process is intended as general information rather than as a guide for immediate action; and (3) the description is directed to readers who know little about the process. If a guide for immediate action is desired, the terse imperative mood may be preferable — although this is a debatable point. And if the readers of the report already know about the process in general, they will have little need of aid in visualization.

The disadvantage of using the active voice is that it is likely to become monotonous unless handled with considerable skill. The monotony arises from the repetition of such terms as "the solderer," "the operator," or whatever the person performing the action is called, even though pronouns can be used to vary the pattern a little. Finally, for some curious reason, perhaps because the active voice is not the customary way of describing a process, the writer may feel reluctant and slightly embarrassed to continue saying "The operator does this, the operator does that," and so on. There is no logical reason for giving in to this feeling. Nevertheless, the use of the active voice has in recent years become somewhat impractical because of distaste for the use of masculine pronouns (he, his, him) in situations in which the referent may be either male or female. It becomes impractical to write numerous sentences like "After the solderer has tinned the iron, he or she is ready to begin soldering the joint." For a discussion of this problem, see p. 35.

The advantage of the passive voice is that there is no problem about handling this hypothetical operator. The disadvantage is that the positiveness and aid to visualization of the active voice are missing. For a process performed by one person, or perhaps even a few persons, a combination of the active and the passive voices is possibly a good compromise. We do not care to be dogmatic about this.

The advantages of the third way, the active voice and the imperative mood, are that it is concise, easy to write, and is a reasonably satisfactory guide for immediate action, as long as the process is not too complex. It is, however, not really a description at all; it is a set of directions. And, because it is a set of directions, there is likely to be a slighting of emphasis upon purpose, and a consequent weakness of the report as an explanation of the process. The imperative mood promotes action better than it promotes understanding.

There are numerous possibilities in addition to the three just illustrated. In fact, all the practical possibilities can be listed as follows:

Active Voice, Indicative Mood
The solderer (or "I," "we," "you," or "one") takes the iron. . . .

Active Voice, Subjunctive Mood
The solderer (or "I," "we," "you," or "one") should (or "must," or "ought to") take the iron. . . .

Passive Voice, Indicative Mood
The iron is taken. . . .

Passive Voice, Subjunctive Mood
The iron should (or "must," or "ought to") be taken. . . .

Active Voice, Imperative Mood
Take the iron. . . .

Almost all these forms may be found in use occasionally. We will comment on special problems related to a few of them. (1) We don't advise the use of "one," but still less do we advise the use of "you" as a substitute for "one" (for example, in "You take the iron. . . ." or worse, "You take your iron. . . ."). On the other hand, there can be no objection to "you" when its referent is the reader (for example, "You should take the iron. . . ."). But even if there are no objections to the latter use of "you," there is not much to be said in favor of it, and we do not advise its frequent use. (You will have noticed that we use it often in this text, but we are not describing a technical process. The style of this book is more colloquial than that of technical reports.) (2) The subjunctive mood should be used sparingly. It is a fine form in which to give advice — as we just did. It is also a useful form for emphasis, as in giving a warning about a special precaution that should be taken (for extreme hazards, the imperative is best). You should just keep in mind the difference between *describing* and giving advice. (3) It is all right to use different forms within the same process description, but discretion is necessary. It is probably best to use only one of the forms throughout if you can make it sound natural and easy. Please note that we did not say, "if you can *do* it easily." Good writing is easily read; it is not usually easily written. All in all, the three forms illustrated at paragraph length above (active indicative, passive indicative, and active imperative) are by far the most useful, with the active imperative running a poor third. These remarks refer only to the type of process in which there is a conspicuous operator.

The Conclusion

The last of the major parts of the description of a process is naturally the conclusion. It is not always necessary to write a formal conclusion. Whether one is desirable depends, of course, on whether it will help the reader. Sometimes the reader needs help in matters like the following:

1. Fixing the chief steps in mind (listing them again might help)
2. Recalling special points about equipment or materials
3. Analyzing the advantages and disadvantages of the process
4. Noting how this process is related to other processes, or other work that is being done, or reported on

Analysis of the entire communication situation is necessary to determine whether a conclusion is desirable.

A Note on Giving Directions

A particularly important relationship between reader and writer exists when the writer is giving directions.

First, what seems completely obvious to the writer will not always appear so obvious to the reader. The directions must be sufficiently clear and complete so that the reader is never left to guess what to do or when to do it. Consistent use of the imperative mood ("Remove the cover . . ., depress the outer cylinder . . ., lock the release lever by sliding the pawl to the left . . .") is the best means of letting the reader know what actions to take and when to take them.

Second, in giving directions, the writer assumes a degree of responsibility for the reader's welfare. Since you are the one giving the directions, you should know the process well enough to recognize which steps are most critical, where possible hazards exist, and where the reader is most likely to encounter difficulty with the process itself or to jump to wrong conclusions as to what is to be done. At points where loss of time or money can occur if the process is not performed properly, the reader should be made aware of the fact. Stronger and more conspicuous cautions or warning should be given at points where damage or injury can occur if the process is not followed properly. For an example of a report in which directions for carrying out an experiment are given, see "Identity of Artificial Color on Oranges," p. 150.

PART 2: PROCESSES IN WHICH AN OPERATOR DOES NOT TAKE A CONSPICUOUS PART

We turn now to that kind of process in which the human agent is less conspicuous. Such processes may be of great magnitude, like the building of a large dam, or relatively simple, like the functioning of a tire pump. They are distinguished by the fact that little emphasis falls directly upon the performance of a human being or beings. How does a tire pump work? An answer to this question would be the description of a process; but in that description there would be little need to mention the quality of the performance of the operator.

The fact is that the kind of process description requiring little attention to the operator turns up in technical writing more frequently than does the other kind. An answer is more likely to be needed for the question, "How does this work?" than for the question, "How do you do this?" So our subject here is an important one.

All that need be considered here is how the description of a process in which the operator does not take a conspicuous part differs from one in which the operator is important. The essential differences are three:

1. Emphasis is altogether on the action — on what happens — and not on the operator and how the operator performs certain actions.
2. The presentation is usually (not always) in the active indicative, the passive indicative, or a combination of the two. The imperative mood never appears.

3. The terms "equipment" and "material" take on a somewhat different meaning and significance.

Point 1 is fairly obvious. Once a train of events has been set in motion, as in a chemical process, interest in the operator who set the events in motion fades. From then on, interest lies in what occurs next. In a process of great magnitude like the manufacture of rubber, where hundreds of operators are engaged and where it is obviously impossible to keep an eye on an individual operator, or even a group, the emphasis must be on the action itself. The reader is simply not interested in the *who* involved. Similarly, in a description of how a machine functions, no person may be involved; the "operator" is the machine itself.

In view of this emphasis, it is easy to see that either the active indicative or passive indicative (or both) will likely be used and that the imperative cannot be used. The passages quoted below[1] illustrate these styles, the first principally in the passive indicative, the second in the active indicative.

> Following the work of Faraday, Ferdinand Carré developed and patented the first practical continuous refrigerating machine in France in 1860. Carré's idea was to use the affinity of water for ammonia by absorbing in water the gas from the evaporator, then using a suction pump to transfer the liquid to another vessel where the application of heat caused the liberation of ammonia gas at a higher pressure and temperature.

> Carré's machine is illustrated by the flow diagram in Fig. VI.[2] In this ammonia-water system, high-pressure liquid ammonia from the condenser is allowed to expand through an expansion valve and the low-pressure liquid then vaporizes in the surrounding refrigerated space. In these two steps the ammonia absorption system is exactly like the compression system. However, the gas from the evaporator, instead of being passed through a compressor, is absorbed in a weak solution of ammonia in water ("weak aqua"). The resulting strong solution ("strong aqua") is then pumped to the generator, which is maintained at high pressure. Here the strong aqua is heated and the ammonia gas driven off. The weak aqua which results flows back to the absorber through a pressure-reducing valve, the highly compressed ammonia gas from the generator is condensed, and the cycle is repeated.

Except for two in the last sentence, the verbs in the preceding paragraph are in the passive indicative. Now consider another account[3] of substantially the same process, this time in the active indicative:

> The process is shown diagrammatically and much simplified in Fig. X.[4] Beginning with the ammonia-hydrogen loop, the ammonia gas enters the evaporator from the condenser through the liquid trap which confines the hydrogen to its own conduit. In the evaporator it takes up heat from the surrounding space and vaporizes, its gaseous molecules mixing with those of the hydrogen. The addition of the heavier

[1] From draft materials prepared by Mr. John L. Galt of the General Electric company for presentation in a contest sponsored by the American Institute of Electrical Engineers. Reprinted by permission.
[2] Figure not shown.
[3] From the final draft of Mr. Galt's report.
[4] Figure not shown.

ammonia molecules increases the specific gravity of the vapor, and it sinks down the tube leading to the absorber.

In the absorber, the ammonia dissolves in the countercurrent stream of weak aqua, while the practically insoluble hydrogen, lightened of its burden of heavy ammonia molecules, ascends to the evaporator to perform again its task of mixing with and decreasing the partial vapor pressure of the ammonia.

Taking up the ammonia-water loop, the strong aqua in the absorber flows by gravity to the generator, where the application of heat drives the ammonia out of solution. A vertical tube, the inside diameter of which is equal to that of the bubbles of gaseous ammonia generated, projects below the surface of the boiling liquid. This "liquid lift" empties into the separator, where the ammonia vapor is separated from the weak aqua. The weak aqua then returns by gravity to the absorber to pick up another load of ammonia.

Finally, the ammonia loop, which has been traced as far as the separator, next involves the "condenser," an air-cooled heat exchanger which removes the latent heat from the ammonia gas, converting it into a cool liquid. Here it passes through the liquid trap that marks its re-entry into the evaporator to serve its purpose of cooling the refrigerated space.

In the account of closed-cycle refrigeration given above, there is an operator, of course, but after lighting the gas flame that starts the process, the operator takes a back seat, for the process completes itself without any further assistance.

Which of the two versions of the process is the better? They are both good. Pay your money and take your choice. And don't fail to ponder the value of the consistency of point of view illustrated in both versions.

The third of the differences listed above has to do with a change in the meaning of the terms "equipment" and "material," as these terms are used in the list of the major parts of a description of a process. Where an operator is conspicuously involved, their meaning is clear; the operator *uses* equipment and materials in carrying out the process. But in the description of a process like those quoted above, there is no operator, and — curiously enough — what in the other kind of process would be simply equipment and materials may now be said to be performing the process! For instance, in a description of how a tire pump works, there would be no operator, and instead of being merely the equipment, the pump might be granted the active voice — as in the statement, "The plunger compresses the air in the cylinder. . . ."

Once this fact is understood, you will have no difficulty. And now we can go on to point out an important fact about process descriptions in actual industrial and research reports. More often than not in such reports, the description of the device or devices involved (as discussed in Chapter 6) and the description of the process (as discussed in the present chapter) are inextricably intermingled, and often other elements (analysis, classification, and the like) are involved as well. As we have said several times, process description is one of the special techniques of technical writing; it is not a type of report. The complexity found in actual reports does not invalidate the principles discussed in these two chapters, although it naturally makes them more difficult to apply. But that is not the fault of the principles; it is, indeed, only through the principles that the complexity can be ruled and order created.

Aside from the three differences just discussed, the description of a process in which an operator is conspicuous and of one in which there is no operator rests upon the same principles.

Summary

1. The principal elements in the organization of the description of a process are the introduction, the discussion of equipment and materials, the step-by-step description of the action, and if necessary a conclusion.
2. The discussion of equipment and materials may be concentrated near the beginning or distributed throughout the description.
3. The introduction is in effect an answer to questions as to what the process is, who performs it, why it is performed, what the chief steps are, from what point of view it will be described, and why it is being described.
4. Processes are usually described in the active voice and indicative mood ("The solderer holds the iron"), the passive voice and indicative mood ("The iron is held"), or the active voice and imperative mood ("Hold the iron").
5. A human operator is conspicuous in some processes but not in others. The essential differences between these two types are that in a process without a conspicuous human operator, emphasis is on the action, the imperative mood is never used, and the terms "equipment and materials" take on a somewhat different meaning.

Illustrative Material

The following pages contain three examples of process description. The first is a description of a relatively simple process in which an operator takes an important part. The second is a fairly detailed description of the process of setting up the Apple Imagewriter printer designed to be used with an Apple computer. The third example is set up in the familiar form of directions for students on how to perform a laboratory experiment. An additional example, illustrating the use of itemized directions in a report designed for professional scientists, is found on p. 232. It is important to recognize that in these last two examples the use of itemized directions unaccompanied by explanations of purpose is justified by the character of the particular situation. The students read their directions as part of a total communication situation involving assigned reading and lectures on theory. The professional scientists read theirs in a total communication situation involving access to other reports on the same subject, as well as long personal familiarity with the laboratory techniques required.

The concept of the "communication situation" is discussed on pp. 26–30.

Removing an Unwanted Branch from a Tree

Homeowners are often confronted with a need to saw an unwanted branch off a tree, perhaps a

Answers to all the questions involved in introducing the

branch that is beginning to scrape the side of the house or one that is interfering with a view of a flower garden. Too often, such branches are merely sawed off several inches from the trunk of the tree, leaving a sizable stub. This stub may not look bad, but it nevertheless constitutes a hazard to the life of the tree. The hazard arises because the tree cannot heal the cut at the end of the stub. The stub will rot, meanwhile providing an entry into the tree for many varieties of disease. The following discussion will explain how a home-owner can remove a branch in such a way that the cut can heal properly.

For this job, only three pieces of equipment are needed: a ladder (in most instances), a saw, and tree paint. Saws used for tree pruning are coarse-toothed, designed for rapid cutting. A bow saw is probably the commonest choice. The special paint is stocked by garden supply stores.

There are five steps involved in the removal of a branch: placing the ladder, making an undercut, cutting off the main part of the branch, cutting off the stub, and painting the cut.

Placing the ladder certainly requires no particular skill, but it does demand some forethought, as many people have discovered to their sorrow. The ladder should be set on firm ground, well braced against the tree, in such a position that the sawing will be as easy as possible.

The next step is to make the undercut. An undercut is a cut made on the under side of the branch, going only part way through the branch (not more than halfway), and located typically 10 to 15 cm from the trunk. The purpose of the undercut is to avoid splitting of the wood in the trunk of the tree when the main cut is made later. If a heavy branch is sawed off without a preliminary under-cut, the branch will usually break off when the sawcut is almost but not quite completed. The stub will split, just below the cut; bark will be torn off the trunk, and the splitting of the wood itself will extend down into the trunk, leaving a long wound, which is unsightly and hard to deal with. The only precaution necessary when making an undercut, apart from locating it at a convenient distance from the trunk, is to make it deep enough to be effective but not so deep that the branch will break off. It is hard to do anything wrong here.

description of a process are found here. The first sentence indicates what the process is and why it is performed. The following sentences explain why the process is being described (that is, because many home-owners are not aware of the con-sequences of pruning improp-erly). The last sentence in the first paragraph states the point of view from which the process will be described, and at the same time makes clear that it is the homeowner who will be performing the process. The second paragraph comments on equipment and materials; the third paragraph states the major steps. Description of the steps follows. The third step is divided into substeps.

Most of this description is in the passive voice. The passive subjunctive is used, however, at those points at which special precautions are being discussed, as in placing the ladder and getting the cut branch safely to the ground.

The third step, cutting off the main part of the branch, actually involves two substeps: making the saw cut, and getting the branch out of the tree.

Making the sawcut itself is simple. The cut is located farther away from the trunk than the undercut, anywhere from 15 to 30 cm beyond the undercut, and of course is started on the upper side of the branch. The thicker the branch, the farther from the undercut the main cut is made.

Great care must be taken, however, in planning what will happen when the branch comes free and drops. The weight of a large-sized branch is great, and its destructive power when falling is astonishing. If anything valuable beneath the branch can't be moved and might be damaged, it may be necessary to cut off small sections of the branch, one at a time. And sometimes a branch may be so entangled with other branches that it will not fall straight down, but will swing one way or another. Occasionally, a branch must be roped to higher branches to prevent damage when it falls, but if the problem is that difficult it may be wise to let a professional tree surgeon handle it.

The fourth step in the process is cutting off the stub that was left after the branch was removed. As nearly as possible, this cut is made flush with the trunk of the tree. The reason for this requirement is that for the cut to heal, bark must grow over it. If a protuberance is left, the bark will take longer to cover it. If the protuberance is too great, the bark will never succeed in covering it at all, and disease is almost sure to enter. When the sawcut has been almost completed, it is helpful to hold the stub with one hand while sawing gently with the other. In this way the cut can be completed without the stub's falling just before the last strip of bark is sawn through and tearing bark off the trunk.

The last step is to paint the cut on the trunk of the tree with a special antiseptic paint that will protect the wound from disease. As noted earlier, this paint is obtainable from garden supply stores. It is too thick to brush on, and is applied with a small stick or a putty knife.

Learning to remove an unwanted branch from a tree is not difficult, and knowing that the job has been done in such a way that the tree will not be harmed can be a source of satisfaction to a homeowner.

Note that the passive indicative (The cut is located . . .) is used here rather than the imperative (Make the cut . . .). In such instances as this one where the reader can be assumed to be generally familiar with the process of sawing, the passive indicative *can* be used, but we repeat our earlier encouragement to use the imperative in giving directions.

Here is a clearly written and economically worded description of the process of setting up the Apple Imagewriter printer. From the *Imagewriter User's Manual*[5] (Part I: Reference), this process description is directed to the reader who wishes to use the printer with an Apple computer; it is, therefore, appropriately written in the second person, imperative mood. As in most sets of directions or instructions, the separate actions the user is directed to take are numbered. These numbered instructions are complemented with photographic illustrations of relevant parts of the printer.

Setting Up Your Printer

This chapter shows you how to get your printer up and running. You'll learn about loading the paper and ribbon, running the printer through a built-in test, and connecting it to your computer.

Note that this short, two-sentence introduction clearly answers the six questions an introduction to a process should answer except for those to which the answer is obvious.

LOADING THE PAPER

You can put ordinary typing paper, business forms, and letterhead into your Apple Imagewriter. You may find it more convenient to use special *pin-feed paper* or *roll paper* made for computer-driven printers.

The Apple Imagewriter prints on ordinary bond paper, up to 24-pound weight. (Standard pin-feed computer paper usually comes in 15- and 20-pound weights.) Heavier paper may not feed properly; lighter paper, such as onionskin, may give a poor print image unless it is backed by a sheet of 15-pound bond.

You can also use multiple forms in your printer, provided their total thickness (including carbon sheets) does not exceed 0.28 millimeter (0.011 inch), or about the thickness of four sheets of 15-lb. bond. You'll have to adjust the printer when printing on more than one sheet (see Adjusting for Paper Thickness).

You can adjust your Apple Imagewriter to accept pin-feed or roll paper of from 3 to 10 inches in overall width. This includes standard 9-½-inch-wide pin-feed paper, which produces 8-½-by-11-inch finished pages after you tear off the perforation strips. You can also use several standard sizes of pin-feed label stock, for jobs such as printing mailing labels. If you like, you can position the paper so that the printer prints right up to the perforation line on either side (but not both sides at once).

[5] Cupertino, CA: Apple Computer, Inc., 1983.

Warning

Never put paper with staples or paper clips in your Apple Imagewriter.

The next two sections describe how to load pin-feed paper into your Apple Imagewriter. If you plan to use roll paper or single sheets of paper without pin-feed holes, skip ahead to the section entitled Loading Plain Paper.

SETTING THE PIN-FEEDER WIDTH

The first time you use pin-feed paper in your Apple Imagewriter you will need to set the spacing of the *sprockets*. Here is how to do it:

1. Tear off a single sheet of the pin-feed paper you are going to use, to serve as a gauge.
2. Make sure the printer is off.
3. Remove both the carrier cover and the paper cover.
4. Set the release lever to the friction (forward) position.
5. The sprockets are the two black plastic parts inside the machine at the rear. Lift up the two tabs that point toward the center, so that the hinged paper clamps over the two sprockets swing outward. Find the white sprocket release levers on the side of each sprocket. When the white levers are pushed toward the rear, the sprockets are free to slide along their square shaft. When the levers are forward, the sprockets are locked in place.
6. Push the white sprocket release levers to the

Carrier Cover   On/Off Switch   Paper Cover
          Release Lever

Figure 1   Removing the Covers

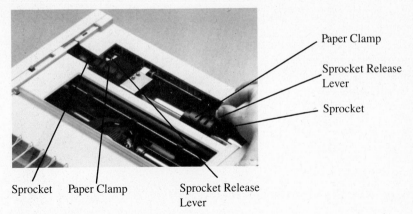

Paper Clamp

Sprocket Release
Lever

Sprocket

Sprocket   Paper Clamp   Sprocket Release
                                Lever

Figure 2   Finding the Sprockets

rear (so that the sprockets are free to slide)
and place the single sheet of pin-feed paper
you tore off to serve as a gauge in step 1 over
the sprockets. Position the sprockets so that
the pins fit through the holes in the paper.
Snap the paper clamps back into place.

7. Feed the forward end of the paper under the
rubber platen and into the printer, as you
would in a typewriter, by turning the platen
knob. Lift the roller shaft in the front of the
machine and guide the paper under it. Then
set the release lever to the pin-feed (rear)
position.

8. You can now move the paper sideways,
carrying the two sprockets with it, until it is
centered with respect to the printing area.
The two red rings at each end of the roller

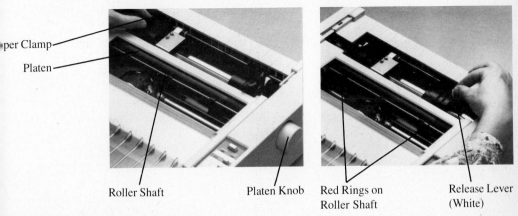

per Clamp

Platen

Roller Shaft   Platen Knob   Red Rings on   Release Lever
                                  Roller Shaft   (White)

Figure 3   Adjusting the Sprockets

shaft tell you where the margins of an
eight-inch printed line occur. When the paper
is in place, pull the release lever forward to
the friction position to hold it in place.

9. Open the hinged cover on each sprocket and
push the white lever forward. Make sure the
sprocket pins fit easily in the holes in the
paper, without pulling inward or outward.
10. Push the release lever back to the pin-feed
position.

LOADING PIN-FEED PAPER

Follow these steps to load pin-feed paper into your
Apple Imagewriter:

1. If the end of the paper you are going to use is
not straight, tear it off at the nearest perforation.
2. Make sure the printer is off.
3. Remove both the carrier cover and the paper
cover.
4. If necessary, set the pin-feeder width as
described earlier.
5. Set the release lever to the friction (forward)
position.
6. Grasp the roller shaft at both ends, and pull it
forward.
7. Lift the paper clamps on the two black
sprockets at the rear inside the machine,
folding them outward to expose the sprockets.
8. Place the paper over the sprockets so that the
pins go through the paper holes. Snap the

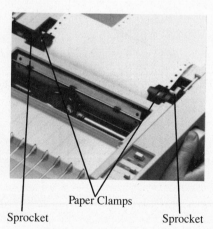

Roller Shaft   Paper Clamps

Sprocket   Sprocket

Figure 4 Loading Pin-Feed Paper

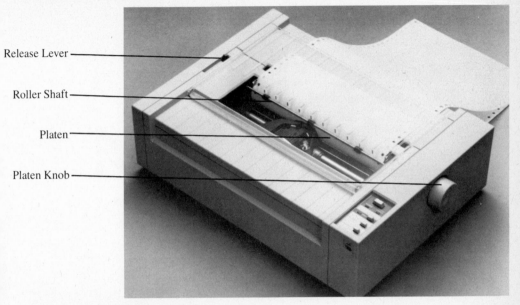

Figure 5 Pin-Feed Paper Loaded

    paper clamps back down to hold the paper in
place.
9. Pull the pin-feed paper into the printer by
turning the platen knob clockwise (top away
from you).
10. When the end of the paper comes up the
front of the platen, under the type head, snap
the roller shaft shut on it.
11. Close the paper clamps on the sprockets.
12. Push the release lever back to the pin-feed
position.
13. Replace the paper cover and the carrier cover.
14. Line up the first perforation with the top of
the type head.

LOADING PLAIN PAPER

To load paper without pin-feed holes—either a
single sheet or a roll—do the following:

1. Turn the power on and then off. This moves the
type head as far left as it will go. Leave the
power off.
2. Open the clear plastic lid on the top of the
machine.
3. Set the release lever to the pin-feed (rear)
position.

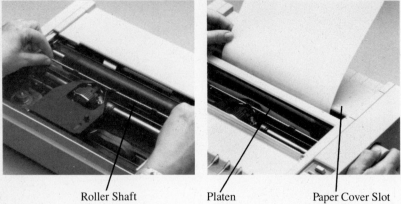

Roller Shaft   Platen   Paper Cover Slot

Figure 6 Loading Plain Paper

4. Grasp the roller shaft by both ends, and pull it forward.
5. If you are using a roll of paper and the end of the paper is not straight, cut across it with scissors. Push the end of the paper down into the slot in the paper cover, feeding it around the platen as you would in a typewriter.
6. When the paper comes up in front of the platen, move it sideways to the location you want. The red rings on the roller shaft indicate the limits of the printer's line of type. Snap the roller shaft shut.
7. Set the release lever to the friction (forward) position.

SETTING THE TOP OF THE PAGE

When you press the FORM FEED button in the control cluster, the Apple Imagewriter feeds paper through to the top of the next page.

Pin-feed paper is perforated between pages, so pressing the FORM FEED button brings the paper to a handy place to tear it off.

By the Way
Your printer was set at the factory for a page length of 11 inches (66 lines). If you want to change the page length, see Chapter 4, Controlling Your Printer.

To position the top of form after you have loaded a new supply of pin-feed paper, make sure the power is off. Then simply turn the platen knob to bring the paper to the desired position. (Usually

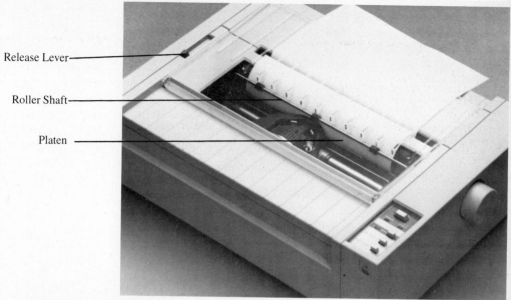

Release Lever

Roller Shaft

Platen

Figure 7 Plain Paper Loaded

the top of the page should be level with the top of the type head.) Unless you turn the platen knob again or send control sequences to the printer to change the settings (instead of moving paper up by means of the LINE FEED button), the FORM FEED button will feed paper to the top of the next page each time you press it.

ADJUSTING FOR PAPER THICKNESS

Your Apple Imagewriter will make higher-quality copies and give you longer service if you take care to adjust it for the thickness of paper you are using.

Remove the carrier cover and look inside the printer. On the right side, near the platen knob, you will find a white plastic lever. When this lever is pushed all the way back (toward the platen), the printer is correctly set to print on a single thickness of ordinary paper. When the lever is pulled all the way forward (toward the control panel), the printer is set for a four-sheet multiple form. As you move the lever back and forth you can see a slight movement of the horizontal metal carrier bar on which the type head slides, compensating for the paper thickness. You can also feel that it clicks in four positions, corresponding to one to four sheets of ordinary paper.

Paper Thickness Lever
(Orange Lever Beneath Cover)
Figure 8   Adjusting for Paper Thickness

INSTALLING AND REMOVING THE RIBBON

The ribbon cassette supplied with your printer is a special type, made specifically for this machine. You can get additional ribbons from your Apple dealer.

Changing ribbons is fast and easy. To install the ribbon that was packed with your printer, do the following:

1. Make sure the power is off.
2. Remove the carrier cover.
3. Take up slack in the ribbon by turning the knob on the cassette once or twice in the direction of the arrow.

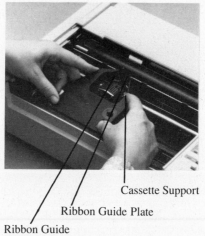

Cassette Support

Ribbon Guide Plate

Ribbon Guide
Figure 9   Changing the Ribbon

Figure 10   The Ribbon Loaded

4. Carefully slip the exposed portion of the ribbon between the black plastic ribbon guide and the thin metal ribbon guide plate (the part that nearly touches the paper). At the same time, guide the cassette downwards onto the ribbon deck. The two black plastic cassette supporters (the parts that stick up) fit into the notches on the sides of the cassette.

5. The cassette should easily snap in place with a single motion. If it refuses to go down completely, or if the ribbon is caught in the space between the ribbon guide and guide plate, turn the knob on the cassette slowly in the direction of the arrow as you seat it. When properly installed, the cassette should lie flat on the ribbon deck.

To remove a used ribbon cassette, turn off the power and remove the carrier cover. Gently spread the two black plastic cassette supporters and lift up the cassette.

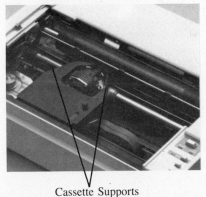

Cassette Supports
Figure 11   Removing the Ribbon

Warning
When pulling the cassette away, be careful not to
get the ribbon caught between 'he ribbon guide
and the guide plate.

TESTING YOUR PRINTER

Inside your Apple Imagewriter is a microprocessor
with a permanent program that can print a test
alphabet on command. The printer does not need
to be connected to a computer to run this program. It is a handy way to put the machine
through its paces before connecting it to your
system.

To run the built-in test, follow these steps:

1. If you have not already done so, load paper and
   ribbon into the printer.
2. Plug the printer into an electric power outlet
   and turn the printer on with the ON/OFF
   switch. The POWER light should come on.

Warning
All electrical connections to your computer and
peripheral equipment **must be grounded!** The
electrical ground prevents damage from a power
surge to your computer or monitor or you. If your
electrical outlet doesn't have a third hole (the
round one), have a qualified electrician determine
if the outlet box itself is grounded. If it is not, have
the electrician install the correct wiring before
using your computer and printer. If it is grounded,
you can plug your computer into the outlet using
an adapter available in any hardware store. This
kind of adapter has a "pigtail," or metal tab, that
can be secured with the screw that holds the cover
plate to the outlet.

3. If the SELECT light also comes on, press the
   SELECT button once to turn it off.
4. Press the LINE FEED button a few times to
   make sure the paper feeds smoothly through
   the machine.
5. Turn the power off.
6. Press the FORM FEED button and continue
   holding it in while you turn the power back on.
   Then release the FORM FEED button. The
   Apple Imagewriter will print its complete set of
   characters repeatedly until you press the
   ON/OFF switch to turn the power off. It takes
   96 lines to print the entire pattern.

Examine the print-out carefully. All the characters
should be complete (no dots missing) and neatly
aligned. The lines should appear equally black
from end to end. Spacing between characters and
between lines should be even. If this is not the
case, check for correct ribbon insertion, paper
loading, and thickness settings. If the problem per-
sists, please contact your Apple dealer.

CONNECTING YOUR PRINTER

Now that you know your new printer works, it is
time to connect it to your Apple computer. Details
of the electrical hookup are different for every
Apple computer. Turn to the *Apple Imagewriter User's
Manual, Part II* for instructions on how to perform
the hookup. After you have successfully connected
the printer to your computer, return to this book
and read the remainder of this chapter.

So now your printer is all set up for operation. In
the next chapter you'll learn how to maintain it in
top condition.

The third example of the description of a process is set up in the familiar form
of directions for a laboratory project.[6] The reader for whom it is intended is a student
in a beginning course in chemistry.

Because this description is brief, and was prepared for use by students in a
laboratory, the introduction is virtually limited to an answer to the question: "Why
is this process performed?" There is no need to explain why the description was
written, or to answer the other questions likely to come up in the writing of an
introduction to a description of a process. The equipment is itemized; the steps are
numbered. It's a fine, clear presentation, and fundamentally a good way to describe
a process — provided the circumstances of the reader permit so limited a discussion.
What "limited" means here is interestingly illustrated by a comparison of the
content of the "Procedure" section with the discussion of the fourth step (cutting the
stub) in the article about cutting a branch from a tree. In the latter, much attention is
given to the purpose of the step — in contrast to the blunt directions in the "Procedures" section of this description of the experiment on color in oranges. The reason
for this difference no doubt lies in the fact that the author of the description for the
students assumed they were being given lectures and reading assignments on the
theoretical (explanatory) aspects of what they were doing.

We have omitted a brief "Discussion" section and a bibliography that appeared in the original.

---

[6] FDA Publication No. 54. Science Project Series. Food and Drug Administration, U.S. Department of
Health, Education, and Welfare. Adapted from *Journal of the Association of Official Agricultural Chemists, 43*
(1960).

Identity of Artificial Color on Oranges
presented by the Educational Services Staff, Food and Drug Administration

Sometimes oranges get ripe as far as taste is concerned before their skins turn completely orange. The development of the orange color in the peel is sometimes hastened artificially by exposing the oranges to ethylene gas in a special chamber designed for the purpose. The ethylene gas reacts with the pigmenting material in the peel to develop the color very much as it would naturally develop if the orange were left on the tree longer. Nothing is added to these oranges, and no special labeling is required.

But not all oranges respond satisfactorily to the ethylene treatment, and growers prefer to use artificial color. You may have noticed that some oranges are individually stamped "Color Added" or "Artifically Colored," while others do not bear this stamp. The uncolored oranges may be streaked with green, although they are ripe to the taste.

The color is permitted for use in making the oranges more attractive, provided they are in fact ripe and otherwise of acceptable quality, and provided the consumer is notified by the stamp that artificial color has been used. The Federal Food, Drug, and Cosmetic Act (FD&C Act) prohibits the use of the color, however, when it would serve to conceal inferiority or damage, or to make the product appear better than it really is.

The color most often used on oranges is called Citrus Red No. 2. It was developed specifically for the purpose after it was shown that a previously permitted orange color could produce injury to test animals when it was consumed in large quantities.

Like all other colors used in foods, drugs, and cosmetics, the colors used on oranges must be specifically authorized for this use, and the amount used must be within the safe tolerance limit set by the Food and Drug Administration. In addition, a sample from every batch of the color produced by the manufacturer must be submitted to the Food and Drug Administration for tests. If the color is found to be pure and suitable for use, it is "Certified."

The color previously permitted on oranges, but now banned as not having been proved safe, is called Oil Red XO. It is possible — although unlikely — that some growers or shippers might use this nonpermitted color by mistake.

Perhaps it is more likely that Citrus Red No. 2 would be used, but the oranges not stamped. Such undeclared use of color would be a violation of the Federal Food, Drug, and Cosmetic Act and of State laws.

The *spectrophotometric absorption curve* of a color is preferred as a means of identification. However, in the absence of a suitable spectrophotometer, the probable identity of a color can be determined by other means, such as chromatographic properties.

In the following experiment, you can use a paper chromatographic technique to test colored oranges to determine that the proper color has been used, and to determine whether undeclared color was used on oranges not stamped "Color Added."

EXPERIMENT

Problem:
To determine by ascending paper chromatography which colors are added to artificially-colored oranges, and whether colors have been used on oranges without proper labeling.

Equipment Needed:
2 funnels, 125 mm.
6 glass rods
2 pipettes, 25 ml.
2 Erlenmeyer flasks, 125 ml.
2 beakers, 100 ml.
1 micro pipet; a capillary or melting point tube drawn to a fine point
1 steam bath
1 funnel support
2 sheets Whatman No. 3 MM filter paper, 7″ × 9″
1 chromatographic tank, inside dimensions 8″ × 9″ × 4″. Any glass or stainless
    steel tank of the approximate size will do. Small aquariums are useful and in-
    expensive.

Reagents:
cotton
oranges, at least 3 stamped "Color Added" and 3 not so stamped
chloroform, 300 ml.
light mineral oil, 5 grams
ethyl ether, 95 ml.
acetone, 200 ml.
distilled or deionized water, 100 ml.
Citrus Red No. 2,1-(2,5-dimethoxphenylazo)-2 naphthol, about 0.1 gram
Oil Red XO, 1-xylylazo-2-naphthol, about 0.1 gram

ADVANCE PREPARATION

Place the Whatman filter paper sheet so that the 9″ dimension is vertical and the 7″
dimension is horizontal. Using a soft lead pencil, draw a horizontal line across the
sheet 1 inch from the bottom edge. Then mark off on the line three 1-½″ segments
about ½″ apart. Label the 1-½″ segments as follows:

First Sheet
1. Citrus Red No. 2
2. "Color Added" oranges
3. Oil Red XO

Second Sheet
1. Citrus Red No. 2
2. Oranges not stamped "Color Added"
3. Oil Red XO

Make a solution of mineral oil in ethyl ether by dissolving 5 grams of the oil in 95
ml. of the ether. Stir until well mixed. Transfer to a 100 ml. graduated cylinder.
Immerse one rolled sheet of the paper in the mineral oil solution for a few minutes.
Remove and dry it by suspending it in air. Treat the second sheet in the same way.
(Note — Ether is highly inflammable and should be kept away from flames or
electric heating elements.)

Make a mixture of 130 ml. of acetone and 70 ml. of distilled water. Stir until mixed
and store until ready for use in a glass-stoppered bottle.

Place a small piece of cotton in the bottom of a 125 ml. filtering funnel supported
on a stand. Position three glass rods in the funnel in such a manner that they will
support an orange so it will not touch the sides of the funnel.

Prepare standard solutions of the two dyes as follows: dissolve 10 mg. of Citrus Red No. 2 in 50 ml. of chloroform. Store in a glass-stoppered, labeled bottle, away from light.

Dissolve 10 mg. of Oil Red XO in 50 ml. of chloroform. Store in a glass-stoppered, labeled bottle, away from light.

PROCEDURE

1. Set water to boil if steam bath is not available (for step 4).
2. Place a 125 ml. Erlenmeyer flask beneath the funnel and support an orange on the glass rods. Wash the color off a "Color Added" orange by spraying it with 25 ml. of chloroform in the form of a fine stream from a pipet. (Note: Use rubber tube on end of pipet. Avoid getting chloroform in your mouth.) Surface oils, waxes, and natural pigments, as well as the artificial color will be washed off.
3. Repeat step 2 with two more of the "Color Added" oranges, combining the washings in the same flask.
4. Transfer a portion of the solution in the flask to a 100 ml. beaker. Allow the solution in the beaker to evaporate by placing the beaker on a steam bath in the hood, or over a suitable container of hot water, if a steam bath is not available. When the solvent has evaporated, add another small portion of the solution. Continue the evaporation in this manner until all of the solution has been transferred and all of the solvent has evaporated.
5. Using another 125 ml. flask and a clean funnel, glass rods, pipet, and beaker, repeat steps 2, 3, and 4 above, using oranges not stamped as artificially colored.
   While waiting for the cloroform in the two beakers to evaporate, proceed with step 6.
6. Pour the prepared solution of acetone and water into the bottom of the chromatographic tank. Transfer a 50 microliter portion (need not be measured exactly) of each of the prepared solutions of Citrus Red No. 2 and Oil Red XO to the marked sheets of filter paper prepared as indicated above, by dipping a pointed capillary tube or melting point tube into the solutions and drawing the liquid in a band along the appropriate 1-½" lines on the filter papers.
   Permit the chloroform to evaporate from the paper for about 5 seconds, and retrace the line with the tip of the pipet twice more. Permit drying between applications. Finally, suspend the papers in air to dry.
7. Allow the beakers to cool after removing them from the steam bath. Dissolve the residue in each beaker in 3 ml. of chloroform. Transfer a 50 microliter portion of the "Color Added" extract to the first sheet of prepared filter paper in the same manner as you did with the authentic dye solutions; repeat on the second sheet with the extract from the presumably uncolored oranges.
8. Lower the two sheets of filter paper in the chromatographic tank so that the 9" dimension is vertical and the 7" dimension is horizontal, and suspend from a rod in such a way that the bottom edge of the paper (with the 1-½" marked segments) is immersed about ¼" in the solvent. Do not allow the papers to touch each other or the sides of the tank. Cover the tank and allow the papers to remain undisturbed for about 1 hour.
9. Remove the papers from the tank and suspend in air to dry. For each sheet of paper, measure the distance the colors have traveled. By comparing the distance traveled by the known color solutions with that of the solutions washed from the oranges, you can determine which one of these colors was used on the "Color Added" oranges, and whether either of them was used on the other oranges without the required declaration. Usually Citrus Red No. 2 will travel a distance

of about 2-¼″ from the point of origin, and Oil Red XO will travel about ¾″ from the origin. The natural coloring materials present will remain at the origin.

SPECIAL CAUTIONS

Preparation of ether-mineral oil solution for treatment of the chromatogram papers should be carried out in the hood.

Evaporation of chloroform solutions should also be carried out in the hood, over a steam bath if available.

Suggestions for Writing

The preliminary remarks in the "Suggestions for Writing" in the preceding chapter, "Descriptions of a Mechanism," apply equally well to the selection of a process to describe, so we will not repeat them here. Perhaps we should remind you, however, of the importance of deciding whether you are describing a process or giving directions about performing a process.

Many of the same general areas of subject matter that were considered as sources of topics for the description of a mechanism are appropriate here. As before, the first of the three topics listed under each general subject will be simple, and suitable for a detailed treatment, and the second and third will be progressively greater in scope. And, again as before, we assume that these topics will suggest others to you, some of which you may prefer.

Topics for Description

1. Aircraft
   Refinishing the fabric of a fabric-covered airplane
   Landing a single-engine, propeller-driven, light airplane
   Landing a twin-engine, propeller-driven airplane
2. Boats
   Tacking a small sloop
   Installing a steering-wheel, including cables and fittings, on an outboard motorboat
   Constructing a wooden or fiberglass boat of 16 to 20 feet in length in your backyard
3. Automobiles
   Changing a tire
   Installing a separate radiator to cool the oil of an automatic transmission
   Rebuilding a worn engine
4. Agriculture
   Attaching a plow to a Ford tractor
   Making an earth dam to create a pond or tank
   Planting, cultivating or weeding, and harvesting a crop (specify the crop)
5. Tree care
   Making a graft
   Taking down a large tree in a congested area in a city
   Taking proper care of a farm wood-lot

6. Electricity
   Making a simple toy motor
   Installing electric heating in the living room of a house
   Installing a cassette player in an automobile
7. Carpentry
   Making a small article like a tie rack or a footstool
   Adjusting a radial arm saw
   Constructing the framework of a summer cottage
8. Metal working
   Making a silver-solder joint
   Making a greensand mold
   Making a utility trailer
9. Optics
   Checking the quality of binoculars on sale in a store
   Grinding a lens for a telescope
   Reconditioning a camera
10. Laboratory
    Making a thermocouple
    Carrying out a laboratory procedure such as determining the flash and fire
    points of an oil
    Making an annealing oven
11. Masonry
    Making a concrete bird bath
    Making a fieldstone and mortar wall
    Making a fireplace
12. Gardening
    Measuring the pH of a soil
    Raising a crop of tomatoes from seed
    Constructing a small greenhouse with provision for control of temperature,
    humidity, and light
13. Fishing
    Fly casting
    Fly tieing
    Equipping a boat for deep-sea sport fishing
14. Cooking
    Making a velouté sauce
    Making a simple beef stock
    Making and storing crêpes

Practice in Organizing

The itemized list below contains enough information for the writing of a brief
description of how to start a plant in a peat pellet. Using this list, which is in random
order, write a brief description of the process. Avoid the imperative mood.

Starting Plants with Peat Pellets
  1. A peat pellet is a disk, typically about 2½ to 3 cm in diameter by 0.7 cm in
     height, formed of compressed peat moss and plant food. When moistened with

water, this disk quickly expands into a "pot" six or eight times its compressed height. The expanded pot is actually a loosely integrated mass of peat fibers and plant food that, on the outside, has the shape of a small pot.

2. When a plant started from seed in a peat pellet pot has achieved sufficient growth, the entire pot can be planted in the ground or in a suitably large container filled with soil.

3. The peat pellet should be set out into the ground when a plant growing in it has developed its first two true leaves.

4. Peat pellets can be purchased from most garden supply stores.

5. Before the pellets are moistened, they should be placed in a suitable tray.

6. Two or three seeds should be put into the peat-pellet pot; when the seeds sprout, the less vigorous seedlings should be pulled out.

7. For the trays needed when using peat pellets, small aluminum pans, plastic trays, or milk cartons with one side cut out can be used.

8. After the pellets have been placed in a tray and moistened, and have fully expanded, a planting hole should be made in the center, with a finger or tool, to the correct depth for the kind of seed being planted, as indicated on the seed packet.

9. Before the peat-pellet pots with growing plants in them are taken from the tray and placed in soil, they should be watered thoroughly.

10. The use of peat pellets permits the gardener to start plants indoors while it is still too cold to start them in the garden.

11. After the seeds have been planted, they should be covered with finely shredded peat moss and moistened with water.

12. The trays in which the pellets are to be placed should be cleaned with hot soapy water and then rinsed well.

13. After all the seeds have been planted and covered with peat moss, the tray and the pots should be enclosed in a transparent plastic bag, to retain the moisture.

14. If the tray is not thoroughly clean, fungus or other disease organisms may attack the plants.

Writing A Set of Directions

Using the same information as in the previous exercise, write a set of directions for performing the process. Use the imperative mood.

8

Classification and Partition

Introduction

If you were to list, just as they occur to you, all the terms you could think of that name kinds of engines, you might write down a list something like the following: steam, internal-combustion, in-line, aircraft, radial, diesel, gasoline, marine, automobile, two-cycle, four-cycle, rocket, jet, eight-cylinder, six-cylinder, and so on. Such a list, quite apart from its incompleteness, obviously makes little sense as it stands; it has no order or system. If you were then to experiment with the list further in an effort to bring order and meaning to it, you would probably rearrange the items in the list into groups, each group in accord with a certain way of thinking about engines. In other words, you would list kinds according to a point of view. Thus, the term "internal combustion" might suggest a grouping according to where the power-producing combustion occurs and would give you two kinds of engines: internal-combustion engines and external-combustion engines (steam engines suggest the latter type). Other terms of the list would naturally suggest other ways of grouping engines: according to cylinder arrangement, use, number of cylinders, and so on. You would, in fact, be on the way to making a classification of engines, for classification is the orderly, systematic arrangement of related things in accordance with a governing principle or basis. The classifier notes the structural and functional relationships among things that constitute a class.

In recording these relationships, the classifier employs certain conventional terms. Acquaintance with these convenient terms will make the rest of what we have to say easy to follow.

Genus and Species   A genus is a class; a species is a subdivision within a class. If "engineering subjects in college" is the genus, then mathematics is a species; if mathematics is the genus, then algebra, geometry, and calculus are species; if calculus is the genus, then differential, integral, and infinitesimal are species. These two terms, "genus" and "species," are very commonly used, but many others can be used if a more complex classification is needed. Recent classifications of animal life, for instance, give as many as 21 categories, from subspecies through species, subgenus, genus, subtribe, tribe, subfamily, family, superfamily, infraorder, suborder, order, superorder, cohort, infraclass, subclass, class, superclass, subphylum, phylum, and finally kingdom, the broadest group of all. Elaborate classifications like this are designed to tell all that is known about the structural and functional relationships among the individuals of the classifications.

Classification   The term "classification" has a loose popular meaning and a more precise technical one. Popularly, classification is almost any act of noting relationships. Technically, classification is the act of locating a specimen of all the different kinds of objects that possess a given characteristic or characteristics. Initially, of course, classification must begin with the recognition that different things possess similar characteristics. Suppose that one day you happened to see a strange creature swimming around in the water, a creature with the body of a horse, feet like a duck, and a tail like a whale (we're thinking of some local statuary). You'd probably only stare; but if you presently saw a second creature just like the first, except that it had a tail like a salmon, you'd possibly say, "There's another of *those things!*" And if, soon after, you saw a third, slightly different from the first two, you might be moved to think up a name (like Equipiscofuligulinae) for the whole family, and to spend many years thereafter hunting for new species and giving them names. You would be classifying.

Logical Division   When you got around to sending off some papers to the learned journals on the discovery described above, you would find yourself engaged in logical division. By this time, you would have found all, or at least all you could, of the existing species, and so would have completed your classification. You would write, "The genus Equipiscofuligulinae is made up of 17 species. . . ." Thus, in dividing into 17 parts the collection that had previously been made, you would be doing what is technically called "logical division." In short, classification and logical division are the same gun seen from opposite ends of the barrel.

In report writing it is usually logical division, not classification, that you will be concerned with; nevertheless, the term "classification" is the one you are likely to use even where "logical division" is technically correct. You might write, "Tractor engines can in general be classified as full diesel, modified diesel, and gas." This is logical division simply because it is a division into three groups of information already known. But it is likely that a scientist or engineer would say "classified" rather than "logically divided"— to judge from our own acquaintance with technical literature. For that reason, and for convenience in general, we will use the term "classification" to mean either logical division or classification. And we will be chiefly concerned with logical division, since the report writer is almost always concerned with arranging a collection of facts or ideas, in order to discuss them in turn, rather than with hunting down new species. After all, the hunting will necessarily have been done before the writing starts.

If we're going to use the term "classification," you may wonder why we bothered to distinguish between it and logical division at all. The reason is twofold; first, just to get down all the facts; and second, to avoid confusion when we go on to the next term on our list.

Partition   Partitioning is the act of dividing a unit into its components. The parts do not necessarily have anything in common beyond the fact that they belong to the same unit. A hammer may be partitioned into head and handle. *Hammers* may be logically divided according to the physical characteristics of their heads as claw, ball peen, and so forth. Classification, or logical division, always deals with several (at least two) units. Partition deals with the parts of only one unit. A hammer is a single unit. A hammer head without a handle is not a hammer. The head and the handle are parts of a single unit. You have probably become familiar with a variety of partitioning in a chemistry course when you determined the components of a chemical compound. Partitioning is further discussed later on in this chapter, and is illustrated in the material quoted on pp. 169–171.

Basis   Suppose we go back for a moment to the strange creatures with a body like a horse, feet like a duck, and a tail like a fish or a whale. And suppose we imagine two people discussing them after only a very brief glance as the creatures swam by. The first person, A, catches only a glimpse of the body, and immediately says, "Why, these animals are clearly related to the horse." But B, who is a little slower, sees only the tail, and protests, "Not at all! They're obviously fish!" Now, what these people need is a basis of classification. Classified on the basis of their main body structure, the creatures are horses. Classified according to their tail structure, they are fish or whales.

Or we might go back to an illustration shown at the beginning of this chapter. How should we classify engines? On the basis of power, use, the kind of fuel they burn, or where the combustion occurs? The manufacturer who is looking for a gas engine to use in a power lawn mower scheduled for production would not thank you for a classification of small engines according to the place where the combustion occurs. What basis would be helpful? Power? Weight? Cost?

These terms, then, are the ones to remember: genus and species, classification, logical division, partition, and basis. The rest of this chapter is devoted to a consideration of the times when the techniques of classification and partition may be found useful, what rules govern the use of classification, what rules govern the use of partition, and what particular writing procedures may be involved.

When Is Classification a Useful Expository Technique?

The foregoing discussion has suggested why classification is a useful technique of exposition: it permits a clear, systematic presentation of facts. When to use this technique depends on whether a writer is dealing with classifiable subject matter and whether his or her writing can be made more effective by means of the technique.

To get an idea of when this technique is useful, let's consider a specific writing problem. Let's suppose that a report is needed on kinds of vat dyes. And let's suppose further that the readers of this report will need to be given an understanding of the properties of these dyes so they can use the dyes effectively. It would be possible to discuss each of the 40 or 50 dyes in turn, of course, giving all the pertinent information about dye No. 1, dye No. 2, and so forth. But it is hard for any reader to keep in mind 40 or 50 individual sets of characteristics. A better possibility is to classify all these dyes into groups having characteristics in common. As a matter of fact, all the dyes can be divided into two groups on the basis of their derivation, and can be further divided into several groups on the basis of their dyeing behavior. It would be much easier for readers of the report about these dyes to remember the characteristics of a group, and to relate an individual dye to a group, than it would be to memorize the behavior of all the dyes individually.

Classification, then, is useful when you have a number of like things to discuss, among which there are points of similarity and difference which it is important for the reader to understand. Obviously, however, the relationship among the things classified must be a significant one. Consider these items:

Lead pencil
Lead, tetraethyl
Lead, Kindly Light

There is a point of similarity among the items in this list, but it is difficult to imagine that it could be very significant!

Suggestions or "Rules" to Follow in Presenting a Classification

In using classification as an effective way of presenting related facts, it is helpful to follow a number of "rules," all of which are simply commonsense suggestions for clarity and meaningfulness. There are seven rules altogether:

1. *Make clear what is being classified.* Making clear what is being classified requires a definition of the subject if there is a question as to whether the reader will be familiar with it. For instance, "colloids" would need definition for some readers before a classification is begun. Although a formal definition of a classifiable subject is rarely necessary in reports — the nature of the discussion will already have made it clear — remember that grouping the related members of a class will

mean little to a reader who does not know what you are talking about in the first place.

2. *Choose (and state) a significant, useful basis, or guiding principle, for the classification.* The basis of a classification governs the groupings of members of a class. If we were to classify roses according to color, each species in our listing would necessarily name a color. Color would be our basis. Thus, we would list *red* roses, *pink* roses, *white* roses, and so on until we had named every color found in rose blossoms.

It is possible to classify most subjects according to a number of different bases, some of them informatively significant, and some of them unimportant or of limited importance. Let's consider another example. A classification of drafting pencils according to the color they are painted would be of no value at all, except perhaps to the aesthete who prefers a yellow to a blue pencil. Disregarding personal tastes about color, a person would choose a pencil with lead of a desired hardness or softness. In short, a significant, informative classification could be made according to a basis of hardness or softness of lead, but not according to the color of the encasing wood. The basis should point to a fundamental distinction among the members of a class.

A word or two about a commonly chosen basis for classification: *use*. Everyone is familiar with numerous, practical classifications of objects according to the use to which they are put. A common example may be seen in the terms "sewing machine oil" and "motor oil." What we want to call your attention to is this: Classifications according to a *use* are of limited value except for those who understand what the qualities are that make an object particularly suitable for a special use. In other words, the *real* basis for such a classification is not use at all; the real basis is the qualities or properties that make the various uses possible. To a person with any technical knowledge at all, the terms "sewing machine oil" and "motor oil" automatically suggest a possible real basis of the classification —viscosity. Before employing use as a basis for any classification, ask yourself two things: (1) Is *use* really what I want my reader to understand? Or (2) Is the quality or property that distinguishes an object for a special use what I want my reader to understand? We do not mean to suggest that classifications should never be made according to a basis of use; as a matter of fact, they may be very helpful. But do not confuse a use with a quality or distinguishing characteristic.

Finally, we advise stating the basis, clearly and definitely, as a preliminary to naming members of a class. It is true, of course, that this rule need not always be followed; sometimes the basis is clearly implicit; for example, as in the color classification of roses mentioned earlier. But, in general, it is a good plan to put the basis in words for the reader to see. The actual statement helps guarantee that your reader will understand you and also helps you stick to the basis chosen.

3. *Limit yourself to one basis at a time in listing members of a class.* Limiting yourself to one basis at a time is simply common sense. Failure to do so results in a mixed classification. This error results from carelessness, either in thinking or in choice of words. The student who wrote, for instance, that engineers could be classified according to the kind of work they do as mechanical, civil, electrical, petroleum,

chemical, and research was simply careless in thinking. A little thought would have suggested to this student that research is not limited to any special branch of engineering. This error is obvious, especially once it is pointed out, but the other kind mentioned—improper choice of terms—is less obvious. Just remember that the *names* of the members of the class should themselves make clear their logical relationship to the basis that suggests them. An author, in illogically listing fuels as "solid, gaseous, and automotive," may actually have been thinking correctly of "solid, gaseous, and liquid," but no matter what this author was thinking, the term "automotive" was illogical. Still another practice to avoid is the listing of a specific variety instead of a proper species name, as listing fuels as gas, liquid, and *coal* (instead of "solid).

4. *Name all species according to a given basis.* In making this suggestion that every species be listed, we are simply advising you not to be guilty of an oversight. As we pointed out earlier in this chapter, a writer would scarcely consider using classification as a method of presenting facts unless he or she has the facts to present. Just as important is the need for telling your readers what the limitations are upon the classification you are presenting, so they will not expect more than it is your intention to give. A complete classification, or one without any limitations placed upon it, theoretically requires the listing of every known species; but sometimes species exist which it is not practical to list. A classification of steels according to method of manufacture, for instance, would not need to contain mention of obsolete methods. Limiting a classification means making clear what is being classified and for what purpose. Thus, a classification of steels might begin: "Steels commonly in use today in the United States are made by. . . ." The rest of the statement would name the methods of production. In this statement three limitations are made: steels made by uncommon methods of production are neglected, steels made by methods of the past are omitted, and finally steels made in other parts of the world are ignored (though these latter might, of course, be made by methods originating in the United States).

5. *Make sure that each species is separate and distinct—that there is no overlapping.* The species of a classification must be mutually exclusive. This is clearly necessary, for the whole purpose of the classification is to list the *individual* members of a group or class; it would be misleading if each species listed were not separate and distinct from all the others. Usually, when the error of overlapping species is made, the writer lists the same thing under a different name or, without realizing it, shifts the basis. Classification of reports as research, information, investigation, recommendation, and so on illustrates this error, for it is perfectly obvious that no one of these necessarily excludes the others; that is, a research report may most certainly be an investigation report or a recommendation report. To guard against this error, examine the listing of species you have made and ask yourself whether species A can substitute for species B or C, or for any part of B or C. If so, you may be sure that you have overlapping species.

6. *Help your reader understand the distinction between species.* When classification is being used as an expository technique, make your reader understand each individual species. Ensuring understanding may require that you discuss each spe-

cies, giving a definition, description, or illustration of each — perhaps all three. In a discussion of steels it might be desirable, according to a basis of the number of alloy elements, to list binary, ternary, and quaternary alloys. It would then be natural to explain, unless it was certain the intended readers already understood, what each of these terms means, what alloy elements are used, and what special qualities each one contributed to the steel. What we are talking about here is not peculiar to classification writing; it is the same old story of developing your facts and ideas sufficiently so that your reader can thoroughly understand you.

7. *Make certain that in a subclassification you discuss characteristics peculiar to that one subclassification only.* Suppose you had classified grinding wheels according to the nature of the bonding agent used in them as vitrified, silicate, and elastic; then, in discussing elastic wheels, had pointed out that they are further distinguished by being made in several shapes, including saucer and ring. Now you give this a little more thought and realize that shape is not a distinguishing characteristic of elastic grinding wheels alone, but of all species, regardless of the bonding agent. It is clear, therefore, that while shape may be a suitable and useful basis for classifying grinding wheels, it is an unsuitable one for subclassifying elastic wheels. What is wanted for a thorough exploration of the subject of elastic •wheels is a characteristic, or basis, peculiar to them and to none of the others. Thus, you might subclassify elastic-bonded grinding wheels according to the specific elastic bonding material used, such as shellac, celluloid, and vulcanite. These would constitute subspecies, which could not possibly appear under the heading of vitrified or silicate-bonded grinding wheels. In other words, you would have pointed out something significant about this particular kind of grinding wheel and not something characteristic of *any* grinding wheel.

Whenever you find yourself employing a basis for a subdivision that is applicable to the subject as a whole, you can either use it for the latter purpose, if you think it worthwhile, or incorporate the information into your prefatory discussion of the subject proper. An introduction to a discussion of grinding wheels, for instance, might contain the information that all kinds of grinding wheels, however made, come in various shapes.

In the process of subdividing a subject, a point is often reached at which no further subdivision is possible. At this point, one is dealing with varieties of a species. Electronic devices for playing video recordings continue to be popular. We might classify home video machines according to the type of recording medium used as either disc players or video cassette recorders (VCR's). In further discussion of the tape players we could point out that they are of two kinds, depending on whether they use the VHS or the Beta variety of tape cassette. Then we could say that a given tape player is a VCR that uses VHS tapes and that another model is a VCR that uses Beta tapes. We could continue the subdivisions depending on how many events the VCR's could be programmed to record or whether they could be operated by remote control. Ultimately we would reach the end of the possibilities along a particular line of inquiry. Note that in discussing the VCR's we used a principle peculiar to the tape machines alone — disc players presently have no recording capability and thus cannot be programmed for any number of events to be recorded.

A Note on Partition

Earlier in this chapter we defined the term "partition"; now we want to comment briefly on the use of partition in exposition. Classification, as we have seen, is a method of analysis (and exposition) that deals with plural subjects. You can classify houses, for instance, by considering them from the point of view of architectural style, principal material of construction, number of rooms, and so on. But you cannot classify *a house* except in the sense of putting it into its proper place in a classification that deals with *houses.* You can analyze a particular house, however, by naming and discussing its parts: foundation, floors, walls, and so on. This analytical treatment of a single thing (idea, mechanism, situation, substance, function) is called "partition," or simply "analysis." As you know, it is a familiar and useful way of dealing with a subject.

The seven classification rules we have discussed also apply to partitioning. Let's review the especially pertinent ones:

1. *Rule 2:* Any breakdown of a subject for discussion should be made in accordance with a consistent point of view, or basis, and this basis must be adhered to throughout any single phase of the discussion. Furthermore, this point of view must be clear to the reader; if it is not unmistakably implicit in the listing of parts, it must be formally stated. You might, for instance, partition an engine from the point of view of functional parts — carburetor, cylinder block, pistons, and so on — or from the point of view of the metals used in making it, such as steel, copper, aluminum. The importance of consistency in conducting such a breakdown is too obvious to need discussion.
2. *Rule 3:* Each part in the division must be distinctly a separate part; in other words, the parts must be mutually exclusive.
3. *Rule 4:* The partitioning must be complete, or its limitations must be clearly explained. It would be misleading to conduct an analytical breakdown of an engine and fail to name all its functional parts. For special purposes, however, incompleteness can be justified by a limiting phrase, such as "the chief parts employed. . . ."
4. *Rule 7:* Ideally, a subpartitioning of a part should be conducted according to a principle or a point of view exclusively pertinent to the part. It would be inefficient, for instance, to conduct an initial breakdown of a subject according to functional parts and then turn to a subpartitioning of one part according to metallic composition if all parts had the same composition. Besides being inefficient, since a general statement about composition could be made about all parts at once, such a subpartitioning would be misleading if the reader were to get the idea that the metallic composition of the particular part under discussion *distinguished* it from other parts of the engine.

You don't need us to urge you to break down a subject for purposes of discussion. You would do it anyway, since it is a natural, almost inevitable, method of procedure. After all, a writer is forced into subdividing subject matter for discussion because of the impossibility of discussing a number of things simultaneously. What we do want to emphasize is that you follow logical and effective principles in carrying out such divisions.

Conclusion

Here is a restatement in practical terms of the fundamental ideas to keep in mind when you present information in the form of classification. These ideas will be stated as they would apply to the writing of an article of classification. Remember that classification is a writing technique, not a type of report, and like all writing techniques it must always be adapted to the context in which it happens to appear.

1. Devote your introduction to general discussion, including definition when necessary, of the genus that is to be classified. Anything you can say that will illuminate the subject as a whole is in order. It may be advisable to point out the particular value of classifying the subject, the limitations of the classification, a variety of possible bases for classifying the subject besides the one (or ones) that will be employed, and your own specific purpose. Be sure to state the basis unequivocally.
2. List the species, either informally or formally, and then devote whatever amount of discussion you think is needed to clarify and differentiate the listed species. Subdivision of individual species, according to stated principles for division, may be carried out in the discussion.
3. Write a suitable conclusion (see Chapter 12).

Illustrative Material

The following four items illustrate the technique of classification. All are clear and helpful; in each there is, nevertheless, an example of a common problem encountered in classifying.

The first item, which was written by a student, is clear and explicit in setting up the basis of classification. Toward the end, however, the author ran into the common problem of having to make a rather arbitrary decision about which genus a certain species belonged in. See what you think of the way this problem is handled.

The second item, concerned with the grading of eggs, makes clear that a basis of classification is being presented, but it falls into the common difficulty of not defining the basis precisely.

The third item, on cockroaches, illustrates a problem in nomenclature: some of the species' names have to do with geography, the others with color.

The fourth item, a brief excerpt from Robert M. Pirsig's *Zen and the Art of Motorcycle Maintenance* is an example of partition.

Types of Sailboats

One way to classify modern sailboats is according to the number of hulls. On this basis, there are three general classes: single-hulled, twin-hulled, and triple-hulled. The single-hulled, the largest class and the most familiar, needs no comment.

This classification begins with an explicit statement of the basis: the number of hulls. The three species are then formally stated. Subsequently, each species is

The popularity of both multihulled classes is relatively recent, however, having developed only since World War II. Of the two, the triple-hulled, or trimaran, is the more novel. A trimaran is comprised of a relatively narrow central hull flanked on each side by a smaller hull to provide stability. This class ranges in size from small day-sailers to ocean-going cruisers. The twin-hulled class is made up of two groups: catamarans and outriggers. The distinction between them depends upon whether the two hulls are the same size or not. In catamarans, which make up much the larger of the two groups, the hulls are the same size and are either identical in shape or are mirror images of one another. That is, if you examined the individual hulls of some catamarans you would find that one side had more of a bulge just aft of the bow than the other side. This bulge might remind you of the camber in the upper surface of an airplane wing. When the two hulls are attached to one another by a "bridge" or deck to form the finished boat, the cambered sides face one another. The theory is that when the boat is sailing heeled over in a crosswind, or on a "reach," with the windward hull partly out of the water, the cambered side of the other hull creates a force in the water that helps prevent the boat from drifting downwind. This theory is controversial. Like the trimaran, the catamaran ranges in size from small day-sailers to ocean cruisers. The outrigger, the second group in the twin-hulled class, is comprised of a relatively narrow main hull which is given stability by a slender pontoon fastened parallel to it on one side only. Since this pontoon could not very reasonably be called a hull, it is possible that the outrigger should be classified as a single-hull sailboat. On the other hand, boats in the single-hull class are not dependent for stability on anything not a part of the hull—a point on which the outrigger fails to qualify. The outrigger is a very ancient design. It is currently not widely used, and is probably to be found only in the form of day-sailers.

discussed in turn. However, the twin-hulled species is taken up last, instead of in second place where it was first listed. There seems little reason for this change except possibly that the discussion of this class is the most complex of the three sections of the classification.

The twin-hulled class is subdivided. The basis of subdivision is explicitly stated in the words, "The distinction between them depends upon whether the two hulls are the same size or not." Then a sub-subdivision follows: hulls that are of the same size are subdivided into those that are identical and those that are not identical but instead are mirror images. In this third-order division, the basis is not stated as explicitly as before. Is it clear enough?

An ambiguity in the classification is recognized in the sentence beginning, "Since this pontoon. . . ." In strict logic, the outrigger does seem to belong in the single-hull class because as the author points out, the pontoon can scarcely be called a hull. Yet, as the author also points out, some kind of distinction should be made between outriggers and other single-hulled boats. The problem could be solved by creating one more species, for a total of four, as follows: single-hulled, single-hulled with pontoon support, twin-hulled, triple-hulled. Would this added complexity be worth the logical clarification? Apparently the author thought not, perhaps influenced by the fact that outriggers form a very small class indeed.

Eggs[1]

The four consumer grades for eggs, U.S. Grade AA, A, B, and C, refer to specific interior qualities as

Eggs are here classified according to two different bases: grade and

---

[1] From "Shoppers' Guide to U.S. Grades for Food," Home and Garden Bulletin No. 58, U.S. Department of Agriculture.

defined by the "United States Standards for Quality of Individual Shell Eggs."

GRADES

Grade AA and Grade A eggs are of top quality. They have a large proportion of thick white which stands up well around a firm high yolk and they are delicate in flavor. These high-quality eggs are good for all uses, but you will find that their upstanding appearance and fine flavor make them especially appropriate for poaching, frying, and cooking in the shell.

Grade B and Grade C eggs are good eggs, though they differ from higher quality eggs in several ways. Most of the white is thin and spreads over a wide area when broken. The yolk is rather flat, and may break easily.

Eggs of the two lower qualities have dozens of uses in which appearance and delicate flavor are less important. They are good to use in baking, for scrambling, in thickening sauces and salad dressings, and combining with other goods such as tomatoes, cheese, or onions.

WEIGHTS

Six U.S. Weight Classes cover the full range of egg sizes. Only 4 of these 6 classes are likely to be found on the retail market—Extra Large, Large, Medium, and Small. The other two are Jumbo and Peewee. Each of these size names refers to a specific weight class, based on the total weight of a dozen eggs. The weights per dozen eggs in each class, in ounces, are as follows: Jumbo–30, Extra Large–27, Large–24, Medium–21, Small–18, Peewee–15.

The grade letters (U.S. Grade A, etc.) indicate quality only. The weight class is stated separately and it indicates the weight of the dozen in ounces. Grade A eggs have the same quality whether they are small or large. The only difference is weight. Grade A eggs are not necessarily large; large eggs are not necessarily Grade A.

Cockroaches: How to Control Them

Cockroaches are pests throughout the United States. They carry filth on their legs and bodies and may spread disease by polluting food. They destroy food and damage fabrics and bookbindings.

weight. But what does "grade" mean? This article is not so explicit in regard to basis as the preceding article was. The meaning of "weight" as a basis is obvious, but all we know about grade, at least at first, is that it is related to quality. What, then, is quality? Comparison of the first two paragraphs reveals that quality has to do with flavor, and with the physical characteristics of the white and the yolk. What these characteristics are it is difficult to tell. The comments on the white might have something to do with viscosity. But what is a "high yolk"? In our experience, when a yolk begins to get "high" everybody immediately and positively loses interest in eating it. (Merriam-Webster's *Seventh New Collegiate Dictionary* says *"high adj 2b: beginning to taint.")* No doubt what the writer of the pamphlet actually had in mind was the fact that a fresh, good yolk will form a mound if deposited gently on a flat surface, rather than make a watery puddle. It's a perfectly good sort of test to apply, similar in principle to the familiar slump test for concrete, and probably the word "high" is commonly used in this way by people who grade eggs. This pamphlet was not directed to experts, however, but to the average shopper. The pamphlet would have been clearer had the writer been more precise about the meaning of "quality," or had he or she simply declared that grade is determined by how fresh and good the egg is.

In this clearly and simply written leaflet on cockroaches and how to control them, published by the United States Department of

Cockroaches have an offensive odor that may ruin food. Unless dishes over which the insects run are thoroughly washed, they may give off the odor when warmed.

There are about 55 kinds of cockroaches in the United States, but only seven kinds are troublesome in buildings. Most of the others live outdoors. They may enter houses by coming in on firewood or by flying to lights, but most of them cannot develop indoors. They either leave or die.

Those that do develop indoors are fairly easy to control in most homes.

DESCRIPTION

Cockroaches have a broad, flattened shape, and six long legs. They are dark brown, reddish brown, light brown, or black. The adults of most species have wings. The young look like the adults, except that they are smaller and do not have wings.

Cockroaches hide during the day in sheltered, dark places. They come out and forage at night. If disturbed, they run rapidly for shelter and disappear through openings to their hiding places.

The seven kinds that are troublesome in buildings can be distinguished from each other by their appearance and by the places where they are found. The table and illustrations will help you learn to distinguish them.

DEVELOPMENT

German and brown-banded cockroaches mature in 2 to 5 months; other kinds mature in about a year.

Agriculture (1980), the species are differentiated on the basis of appearance and places they are found. The names of four of the seven species, however, suggest nothing about appearance or places found. The tabular classification does identify them and how they differ clearly enough. Only three pages of the nine-page leaflet are quoted here.

American cockroach (approximately 60% natural size; some are larger.)

Cockroaches grow slowly when food, temperature, and moisture are unfavorable. Conditions are not usually ideal for rapid growth in buildings and homes where good sanitation is practiced.

The female lays her eggs in a leathery capsule, which she forms at the end of her body. The German cockroach carries the egg capsule about a month, and drops it just before the eggs hatch; about 36 cockroaches hatch from the capsule. The other kinds carry the egg capsules a day or two, then glue them to some object in a protected place; about 12 cockroaches hatch from each capsule after an incubation period of 1 to 3 months.

PREVENTING ENTRY

Cockroaches may enter the house from outdoors, in infested containers from other buildings, or from adjoining homes or apartments. To keep them out, fill all cracks passing through floors or walls, and cracks leading to spaces behind base-boards and door frames, with putty or plastic wood. Pay special attention to water and steam pipes entering rooms.

Cockroaches develop in large numbers in dirt and filth. Thorough cleaning reduces the likelihood of heavy infestation. When you bring baskets, bags, beverage cartons, or boxes of food and laundry into the house, look for cockroaches that may be hiding in them, and kill any that you find.

Oriental cockroach: *Left,* female; *right,* male. (approximately 90% natural size.)

Cockroaches Troublesome in Buildings

Name	Description	Where Found
American cockroach[1]	Reddish brown to dark brown. Adults 1½ to 2 inches long.	Develop in damp basements and sewers, forage mostly on first floors of buildings.
Australian cockroach[2]	Reddish brown to dark brown. Yellow markings on the thorax; yellow streaks at base of wing covers. Adults 1 inch long.	Develop in warm, damp places, in or out of doors; forage mostly on first floors of buildings.
Brown cockroach[3]	Reddish brown to dark brown. Adults 1¼ to 1½ inches long.	Develop in warm, humid environments. May occur in all areas of buildings.
Brown-banded cockroach[4] (also called tropical cockroach)	Light brown. Mottled, reddish-brown wings on female; lighter wings on male. Adults ½ inch long.	Develop and live all over the building.
German cockroach[5] (also called croton bug and water bug)	Light brown. Black stripes running lengthwise on back. Adults ⅝ inch long. Most common of the 7 kinds.	Develop and live all over the building, particularly in kitchens and bathrooms.
Oriental cockroach[6] (also called black beetle and shad roach)	Black or dark brown. Traces of wings on females; short wings on males. Female adults 1 to 1¼ inches long, male adults a little shorter. More sluggish than other species.	Develop in damp basements and sewers; forage mostly on first floors of buildings.
Smokybrown cockroach[7]	Dark brown to black. Adults 1¼ to 1½ inches long.	Develop in warm, humid environments. May occur in all areas of buildings.

[1] Periplaneta americana.
[2] Periplaneta australasiae.
[3] Periplaneta brunnea.
[4] Supella longipalpa.
[5] Blattella germanica.
[6] Blatta orientalis.
[7] Periplaneta fuliginosa.

Zen and The Art of Motorcycle Maintenance[2] (excerpt)
Robert M. Pirsig

A motorcycle may be divided for purposes of classical rational analysis by means of its component assemblies and by means of its functions.

Note that Robert Pirsig is not here dealing with *motorcycles;* rather he is breaking down or partition-

[2] Copyright © 1974 by Robert M. Pirsig. By permission of William Morrow and Company. Quoted from the 1979 Bantam Books edition.

If divided by means of its component assemblies, its most basic division is into a power assembly and a running assembly.

The power assembly may be divided into the engine and the power-delivery system. The engines will be taken up first.

The engine consists of a housing containing a power train, a fuel-air system, an ignition system, a feedback system and a lubrication system.

The power train consists of cylinders, pistons, connecting rods, a crankshaft and a flywheel.

The fuel-air system components, which are part of the engine, consist of a gas tank and filter, an air cleaner, a carburetor, valves and exhaust pipes.

The ignition system consists of an alternator, a rectifier, a battery, a high-voltage coil and spark plugs.

The feedback system consists of a cam chain, a camshaft, tappets and a distributor.

The lubrication system consists of an oil pump and channels throughout the housing for distribution of the oil.

The power-delivery system accompanying the engine consists of a clutch, a transmission and a chain.

The supporting assembly accompanying the power assembly consists of a frame, including foot pegs, seat and fenders; a steering assembly; front and rear shock absorbers; wheels; control levers and cables; lights and horn; and speed and mileage indicators.

That's a motorcycle divided according to its components. To know what the components are for, a division according to functions is necessary:

A motorcycle may be divided into normal running functions and special, operator-controlled functions.

Normal running functions may be divided into functions during the intake cycle, functions during the compression cycle, functions during the power cycle and functions during the exhaust cycle.

And so on. I could go on about which functions occur in their proper sequence during each of the four cycles, then go on to the operator-controlled functions and that would be a very summary

ing a motorcycle in terms of its components and its functions. In other words, he makes succinctly clear what the basis of his analysis is, and in subpartitioning each element he simply names the parts which comprise an element.

description of the underlying form of a motorcycle. It would be extremely short and rudimentary, as descriptions of this sort go. Almost any one of the components mentioned can be expanded on indefinitely. I've read an entire engineering volume on contact points alone, which are just a small but vital part of the distributor. There are other types of engines than the single-cylinder Otto engine described here: two-cycle engines, multiple-cylinder engines, diesel engines, Wankel engines — but this example is enough.

Suggestions for Writing

1. We can't claim that this first exercise is entirely serious, but it illustrates with great clarity how the selection of a basis of classification affects the way something is looked at. Under each genus below (identified by a capital letter) are two terms designating certain "roles" in our society. The two roles under "A," for example, are that of a member of a legislative committee on taxes and that of the owner of a car ferry. Of course the same person might have both roles by chance, but that is a complication we don't need to worry about. All we need think about is whether the job of operating a car-ferry service would lead one to classify cars on a different basis from that most naturally chosen by someone else with the duties of a member of a tax committee. For each of the two roles listed under each genus, indicate an appropriate basis of classification for the species making up this genus.

   A. Automobiles
      (a) a member of a legislative committee on taxes
      (b) the owner of a car ferry
   B. Musical instruments
      (a) a composer
      (b) an apartment house owner
   C. Cattle
      (a) the owner of a dairy farm
      (b) the manager of a meat-packing plant
   D. Dogs
      (a) an official of the American Kennel Association
      (b) a postal service letter carrier
   E. Public lectures
      (a) a newspaper reporter
      (b) a janitor

2. Write an article of classification of one of the following terms. Specify the imaginary reader. Be sure to pick a subject you know a good deal about.

   Lawn mowers
   Clothes-washing machines

Adjustable wrenches
Glues
Paint applicators
Power-saw blades for the home-craftsman type of saws
Lighted commercial signs
Fireworks
Detergents
Golf clubs
Automobile tires
Photographic films
Paper
Big game rifles
Tape decks
Bridges
Motor oil
Road surfaces
Student organizations on campus
Intramural sports
Professional courses in _____
      (deal with courses in your field)

3. State a basis for partitioning each of the following terms, and make the partition.

A hacksaw
A straight chair
The electrical wiring in a small house
A tree

9

Interpretation

Introduction

Interpretation, as we'll use the word in this chapter, is the art of establishing a meaningful pattern of relationships among a group of facts. It differs from formal analysis (see Chapter 8) in that it does not attempt to be exhaustive and is freer of conventional form. It is nevertheless rigorously logical; and formal analysis naturally enters into interpretation rather frequently.

Interpretation, in the sense just indicated, is one of the most important elements of science and engineering. Practical decisions such as where to drill an oil well, or what lightning protection system to use on a stretch of electric power transmission line, are the result of interpretation of a body of facts. So are Newton's laws of motion. Interpretation is a creative activity, requiring both knowledge and imagination. Sometimes the results of interpretation can be at best only tentative, as in long-range weather forecasting. Sometimes the results are fairly certain, as in determining the cause of the failure of a particular gas engine. And there is an extreme in which the results are absolute, as in a mathematical equation where all the factors are exact quantities.

From one point of view the study of interpretation is simply the study of logic. From another point of view, however, the study of interpretation is a study of the art of communication, of communicating to other people what you have found out through the application of logic to a certain group of facts. This kind of communication is, of course, common in technical writing. Indeed, one form of it is so common that it is often regarded as a type of report called the "recommendation" report (see Chapter 14). Later in the present chapter we will give some attention to the sort of problem characteristic of recommendation reports, although our primary concern will not be with any particular report-writing situation. Rather, it will be with the fundamental technique of how to organize a group of facts and communicate conclusions drawn from them.

Interpretative writing is often interesting to do because it is an integral part of the whole process of figuring something out. The subject matter itself is likely to be interesting to you. You've started with a challenge. Probably you're asking yourself, "Why did that happen?" Or, "What should I do?" The most abstract and, as we said, "rigorously logical" report often represents a personal experience of intellectual excitement and satisfaction. It is difficult to illustrate this fact with examples of technical writing because, as explained in Chapter 3, the personal element is usually subordinated in technical reports. In the writing done by a good journalist, on the other hand, there may be strong emphasis upon the personal element. Logic suggests, therefore, that if we could find a skillfully written example of "technical journalism," we might be able to observe good technical interpretation in combination with the personal experience from which it was drawn. Printed below are parts of an article in which just such a combination is found. We are going to show it to you without further comment except to point out that, as will quickly become obvious, the style of the article is not designed for "professional" readers, but for the mass audience of readers who have just enough knowledge of flying to be interested in some of the finer points. The author says, incidentally, that the accident described actually happened, but not to him.

So You'd Like to Fly a Fighter Plane[1]
Mike Dillon

The F-51 skids a little as we turn from base to final. As the runway comes into line, level the wings and put down full flaps. Lower the nose a bit to maintain one hundred and thirty mph. Make one last cockpit check. Mixture "RUN." Prop set for twenty-seven hundred rpm. Gear down and locked. Fuel selector to right tank.

All set. Take a deep breath. Try to relax. It's just an airplane. Add a little power. Don't let that airspeed drop. The Mustang slides down final and crosses over the fence high. *Damn,* that runway looks short! Nothing to do now but ease the power back and hold her off. Stick back slow. Easy . . . easy . . . don't let her balloon. Add a little power to catch it. With half the runway gone the 51 touches, hops and then settles solidly on all three wheels. We swerve drunkenly as the long, wide nose blocks our vision. Runway is going fast. She's not going to stop! Go around, it's the only thing. Throttle forward. Easy now — remember this big machine has torque. Full throttle and the seven thousand pound fighter begins to accelerate. Stick back hard to get off the ground quick. Can't be much runway left. Right rudder! More! Still more! The plane angles off to the left and into the grass. If it will just fly we can still make it. The trees are a thousand feet away. The big bird bumps and lurches as it hurdles across the turf.

Suddenly it's flying! But the left wing starts to drop. We already have full right rudder. Stick hard right to raise the wing. *Nothing!* It just keeps rolling over. Your head slams hard against the side of the canopy as that left wing plows into the ground. All is violence as the huge Hamilton Standard prop blades churn the earth sending dirt and aluminum flying. The battered F-51 thrashes, wildly, then cart-wheels into the air. Back down it comes buckling the right wing and smashing the tail sideways.

Your sense of vision is lost, replaced by a blur. Only your hearing remains unim-paired. Just so you can listen to those horrible tearing, buckling sounds. Now it is the broken left wing that digs in, snapping the fighter around on its nose, right wing and tail. Mortally wounded the bird pauses vertically on its tail for a moment as if it were making one last reach for the sky. Finally, it topples over on its back smashing the Plexiglas canopy — flat.

As suddenly as it began, the violence stops. In the shattered cockpit you hang from the straps, painfully aware that you have just destroyed a great airplane and that you have allowed a great airplane to almost destroy its pilot. What happened? What went wrong?

The answers to these questions is what this report is all about. The accident described actually happened (not to your author).

For those who may think this sort of thing is rare, let us guide you for a few minutes as we nose around a little on this one, relatively small, southwest airport.

Here, inside the hangar, we find the carcass of another 51 that a doctor tore up on take-off in Albuquerque.

Off the runway is the almost buried remains of still another F-51 that a pilot spun in on his second Mustang flight.

Over thataway near the field boundary is the wreckage of an SBD whose owner spun in on his first flight. Note: All of these on one airport.

[Because of limitations of space, we have here deleted a portion of the article in which the author gives a case history of two accidents involving fighter planes. These histories are based on a study made by Dr. R. G. Snyder. The author of the article then asserts that, including the 25 accidents reported by Dr. Snyder, he is himself aware of a total of 55. You might be interested in his figures: one P-38, two Hawker Seafurys, three P-63's, seven P-40's, one SBD, five F8F Bearcats, thirty-four F-51 Mustangs, and two Corsairs. In these accidents, 25 people were killed. The article continues as follows.]

There is a consistent pattern to most of these accidents. We believe the root cause can be traced to the pilot's not accepting the fact that the surplus ex-WW-2 fighter is not just another airplane.

Four factors set such fighters apart. High wing loading, torque, asymmetrical thrust, and, in the case of the F-51, the laminar flow wing airfoil.

High wing loading is fairly easy to compensate for: You fly the plane faster. Thus, Common Sense Rule 1: Keep your speed up. Read the handbook; know the recom-mended speeds.

Torque, it seems, is much harder for first-time would-be fighter-type pilots to handle. We figure the reason for this is that many pilots just don't know what the word torque means. So what is torque? When you raise the nose of a Cessna 182 to climb, it will turn left if right rudder is not used to hold the plane straight. This left turning tendency, though often called such, is *not* torque. Torque is not a yawing force, but is, instead, a twisting, rolling force. It is the action-reaction between the propeller and airplane. The engine is twisting the prop clockwise (as viewed looking forward from the cockpit), therefore the prop is trying to twist the airplane counter-clockwise (AVLFFTC). This tendency for the airplane to roll in opposite direction of prop rotation is directly proportional to the amount of power the engine is producing, the rate with which the prop is being accelerated, and the ratio of prop weight to airplane weight. Bear in mind that an F-51 prop itself weighs more than four hundred pounds. When the pilot rams the throttle forward from idle the engine applies a tremendous force, 1000-plus-hp, to twist that heavy prop faster to the right. That same 1000-plus-hp is also trying to roll your entire machine to your left. At low speeds this rolling force exceeds the amount of control available from the ailerons. To demonstrate this we'll slow the Mustang to 80-mph with full flaps and gear and just enough power to maintain level flight. Now, throttle all the way forward to the stop — As we push the throttle we also feed in full right stick and rudder. The F-51 rolls smoothly and smartly to the left onto its side and stops in a vertical slip, losing two hundrd feet of altitude. Hm-m-m. Had we been at fifty feet instead of five thousand, we'd be like maybe dead.

Herein lies Common Sense Rule 2. Avoid abrupt power increases at low speeds in a Mustang. You have 1470-hp out in front. You don't need to use each one of those "horses."

O.K., you say. So that's torque. What *does* make my 182 turn left while climbing? Your author smiles and replies, "Glad you asked, ole buddy boy, 'cause that's our next point — P effect or, if you will, asymmetrical thrust. When any prop-driven air-plane is at a positive angle of attack the prop blades at work on the right hand (starboard) side of the engine (assuming right-hand rotation) have a greater effective angle of pitch and therefore "pull" harder than when they get around on the left. . . ."

This asymmetrical thrust (or pull) that is so easily controlled in your 182 gives to the fighter a characteristic normally thought of as applying only to multi-engined planes. It has a VMC. That is to say, single engine fighters have a speed below which the pilot cannot maintain control at high power due to asymmetrical thrust.

For instance, in the case of the P-40 the VMC at *cruise* power is about 75-mph. The author determined this empirically — not to mention, accidentally while trying to stall his own ex-fighter plane (a P-40) at cruise power. To shorten a long story I found that if the Warhawk's nose was raised smoothly an extreme angle of climb was necessary to keep the air-speed dropping. By the time the speed had dropped to 80-mph the plane was about seventy degrees nose up. As the speed fell to 75-mph the fighter, without buffeting, and against full right rudder and aileron, yawed and rolled to the left. The nose dropped to about the horizon and the plane sort of stopped inverted. This was a bit confusing at first but several more tries showed the same results. The plane had not stalled. There simply was not enough rudder or aileron control at that low speed to control the torque and that "uneven" pull of the prop blades. C.S. Rule #3, therefore, becomes a composite of Rules #1 & #2. Keep your speeds up to those recommended in the handbook, beware of high power settings at low speeds with high angles of attack. An F-51 in this configura-tion has much the same control problem as a twin Beech with its left engine feathered. Herein lies an interesting matter of judgment. Quite a few bold pilots think themselves capable of handling a Mustang, but few of these same chaps would be anxious to solo a twin Beech with no prior instruction — yet the Beech is no harder to fly.

The F-51 has one extra quirk — its laminar-flow wing. This low-drag, high-speed airfoil is the main reason for the Mustang's success as a long-range fighter. The prospective atomic-age F-51 pilot should realize that this airplane handles best at speeds above three hundred mph. For example, at 250-mph it takes only 4.5 G's to stall your Mustang. This, coupled with the fact that it usually requires as much as ten thousand feet to recover from a power-on spin, has claimed the life of more than one pilot.

Common Sense Rule #4: When your 51 starts to buffet in a tight turn, ease off a bit — that plane is trying to tell you something. Beware any lack of communication!

To return to our beginning. Why did the first pilot we introduced you to wrap his F-51 up in a ball? First, his decision to go-round was in error. The 51 decelerates very rapidly once on the ground. Said pilot could have stopped on the runway. Second, if the go-round was necessary, the pilot should have retracted his flaps immediately and retrimmed. With full-flaps and nose-up trim, the plane would break ground at a very low speed. Without the rudder being trimmed to compensate for torque and asymmetrical thrust, the pilot probably thought he was using full-right rudder when, in fact, he wasn't. The airplane was at or near its minimum control speed and full rudder travel would have been necessary even to have had a chance of retaining control.

We should explain that we have deleted the final paragraph of the article, but that in it the author says he did not mean to "knock" fighters; that they are a great pleasure to fly if the pilot has adequate training and a suitable temperament.

You probably found this article clear and easy to follow (perhaps with the exception of the comments on asymmetrical thrust, and on this point clarity suffered because we were unable to retain a graphic illustration of the meaning of this term that appeared in the original article). Why is this article so easy to follow, on the whole? One reason is that the author wrote clear, effective sentences. Another is that it is organized with professional skill. There are three specific points at which this skill in organization can be seen.

First, the author states exactly what the reader can expect to find out. And he states it not only once, but twice. The first time is at the end of the initial description of the accident. Remember that this is an article designed for a popular magazine, and that the writer is keenly aware that he had to arouse interest quickly or lose the reader. So he begins with a few paragraphs of exciting action, and only at the conclusion of these paragraphs does he state what the reader can expect to learn from the article as a whole. The second time he indicates what the reader can expect to learn is when he turns from a description of accidents that have happened to a relatively abstract interpretation of why they happened. In this second statement, which appears directly after our italicized interpolation, the writer also becomes somewhat more specific. Here are the two statements again, one after the other. Both indicate that the author's purpose is to discuss the cause of the accidents.

(1) What happened? What went wrong? The answers to these questions is what this report is all about.

(2) There is a consistent pattern to most of these accidents. We believe the root cause can be traced to the pilot's not accepting the fact that the surplus ex-WW-2 fighter is *not* just another airplane.

As you will remember, this second statement is followed in the article by an extended interpretation in which an answer is given to the question of what happened, or what caused the accidents. Thus the reader has been told twice, in different terms, what to expect.

A second point in which the organization makes the article easy to follow is that before the relatively abstract interpretation is begun, the author carefully describes how he acquired evidence about the seriousness of the problem he is examining. As noted earlier, this part of the article has been deleted and summarized in the italicized interpolation. The author also explains, later, how he obtained another kind of evidence — that is, how he learned, through personal experience, about such things as torque and asymmetrical thrust. From a technical point of view, however, this material is quite sketchy, a treatment forced upon the author by the fact that many of his intended readers are not technically trained.

The third point about the organization of the article is that the author indicates exactly how the interpretative part of it will be organized. He does so, as you will recall, in the following paragraph:

Four factors set such fighters apart. High wing loading, torque, asymmetrical thrust, and, in the case of the F-51, the laminar flow wing airfoil.

Each of the four factors is then taken up in turn, in precisely the order stated. The reader is therefore never in doubt as to how the interpretation is organized.

The three points of organization just considered in relation to this skillfully written article are the three principal points requiring attention in virtually any interpretation. Leaving the article now, we turn to an examination of each of these three points in greater detail, with reference to their part in any interpretation you yourself may have to write. Following this examination, the discussion in this chapter will be concluded with a brief review of the place of the scientific attitude in interpretation. The three points of organization are, again, these three questions:

1. What is to be found out?
2. How was evidence obtained?
3. How should the interpretation be organized?

What Is to Be Found Out?

The first job in writing a technical interpretation, as distinguished from a journalistic article, is to tell the reader exactly what problem you intend to discuss. Probably no single part of an interpretation is more important than the initial statement of objective. With rare exceptions, this statement will be made up of two parts: first, a statement of the problem in a concise form, and second, in an expanded form. The nature of the expanded statement depends on whether your objective is merely to explain your data or to present a decision about action to be taken.

Stating the Problem in Concise Form   Boiling down a complex problem to one short, simple statement isn't always easy. On the other hand, this process is often the most interesting part of the whole operation. Let's take a brief example. Suppose it became necessary to state the fundamental design problem in a tank-type vacuum cleaner. In other words, what is it we have to find out if we are to design an effective tank-type cleaner? Perhaps it would be most natural to begin by considering such matters as general shape and size, the configuration of the intake area, the location of a motor, provision of wheels and a handle. But these are, in a sense, details; after thinking it over we might want to begin by saying that fundamentally what we have to find out is the best overall design for a cylinder, open at both ends, in which dirt is filtered out of a stream of air drawn through by a motor-driven propeller. Having this concise statement of the problem, we would be in a good position to organize and interpret all the details we might care to discuss concerning such matters as filters and motors. Similarly, the author of the article about the fighter plane boiled down his problem to the question of how a WW-2 fighter plane differs from other airplanes. In his subsequent interpretation, he analyzed the factors that make it different. Most interpretations that are effectively handled proceed in more or less this way.

Stating the Problem in an Expanded Form   As in the examples of the vacuum cleaner and the fighter plane, chances are that any problem requiring interpretation will naturally break down into a number of parts, or subordinate problems. A second element in the whole process of stating what is to be found out is usually, therefore, the presentation of the subordinate problems. For the fighter plane, you will recall, these subordinate problems were the high wing loading, torque, asymmetrical thrust, and laminar-flow wing. Below is another example, taken from a classic work[2] on atomic energy.

> No one who lived through the period of design and construction of the Hanford plant is likely to forget the "canning" problem, i.e., the problem of sealing the uranium slugs in protective metal jackets. On periodic visits to Chicago the writer could roughly estimate the state of the canning problem by the atmosphere of gloom or joy to be found around the laboratory. It was definitely not a simple matter to find a sheath that would protect uranium from water corrosion, would keep fission products out of the water, and would not absorb too many neutrons. Yet the failure of a single can might conceivably require shutdown of an entire operating pile.

In this quotation the last part of the first sentence is a concise statement of the primary problem, and at the same time is a definition of a bit of technical jargon. In the next to the last sentence, the primary problem is divided into subordinate problems.

The principle of the breaking down of a problem into subordinate problems

[2] Henry D. Smyth, *Atomic Energy for Military Purposes,* rev. ed. (Princeton, N.J.: Princeton University Press, 1946), p. 146. Reprinted by permission of the author.

involves no special difficulty, and for our purposes nothing need be said about the principle itself. The way in which the subordinate problems are presented, however, depends on the writer's objective, and here there is a point that needs further consideration.

Interpretations can be written for either of two basically different purposes. One is simply to help the reader understand certain phenomena, or a certain body of data. The article on the fighter plane is an example. The other purpose is to present a recommendation concerning a course of action, or a decision. In the latter case the subordinate problems are written in a special form, which we can call "standards of judgment."

This term is new, but the idea itself is familiar. For instance, if you'll go back to the quotation from Smyth about the "canning" problem, you'll see that what we called subordinate problems could be rewritten as standards of judgment. One of the subordinate problems was to design a sheath that would protect the uranium from water corrosion. To transform this problem into the statement of a standard of judgment, we would rewrite it as follows: "The sheath must protect the uranium from water corrosion." The effect of this statement is to tell the reader that no design for a sheath would be acceptable if it did not meet the standard of preventing water corrosion of the uranium.

Let's take another simple example. Suppose your neighbor asks you to stop at the store on the way home to get a dozen eggs. You say you will; but because you have read the explanation of the grading of eggs in Chapter 8, you warily inquire, "What kind of eggs?" The only reply you receive is, "Nice big ones." On the way to the store you reason to yourself, " 'Nice' means high in quality, so I'll get grade AA. And the biggest size is the Jumbo, so I'll get that." By this reasoning you have established two standards of judgment. To be acceptable, the eggs must meet the standards of being AA in grade and Jumbo in size. (Often the word "specifications" is used to mean the same thing as our term "standards of judgment.")

Likely enough, however, when you get to the store you discover that it doesn't have any Grade AA Jumbo. What it does have are Grade A Large and Grade AA Medium, plus some others too low in grade and too small to merit consideration. So you undertake a quick mental interpretation of the data.

Before thinking about the results of your interpretation, let's ask once more — exactly what is the problem? If the problem is to obtain eggs for your neighbor's children to color on Easter morning, it looks at first glance as if the Large eggs would be an obvious choice; if the problem is to obtain eggs for your neighbor's breakfast, it looks as if the Grade AA would be an obvious choice. But what about the price — another problem? If you are really serious about these eggs, what you will want to do is to transform your specific problems into standards of judgment and to make your choice with their help. If the eggs are for breakfast, you might transform your problems into the following standards of judgment. The eggs must (1) be as high as possible in grade, and not less than A; (2) be as large as possible; and (3) be as cheap as possible, and in any case cost not more than x cents per pound (eggs of different sizes would have to be compared by weight).

In summary, the difference between a problem and a standard of judgment can be put like this. If you say, "A system is needed for grading eggs according to quality," you have expressed a problem. But if you say, "I have to obtain some eggs having a firm, well-mounded yolk," you have expressed a problem in the form of a standard of judgment. If your objective in interpreting a body of data is to make a choice, then you will naturally emphasize the standards of judgment by which your choice is guided. On the other hand, if your objective is merely to explain a body of data, as in the article on the fighter plane, then you may have little interest in standards of judgment.

Another consideration with reference to presentation of either subordinate problems or standards of judgment is the possible need for explanation of their meaning. Merely stating a problem may not be enough. For example, the problem of torque is explained at some length in the article about the fighter plane. Similarly, an explanation of how eggs are graded would be necessary if an uninformed reader were to understand the significance of "Grade A" in the statement of a standard of judgment.

Two final comments. The first comment is that there is sometimes a puzzling interdependence between or among subordinate problems or standards of judgment. When this is true, it may be hard to see how to discuss one of them without discussing the other or others simultaneously, which is impossible. Usually, this difficulty can be solved by pointing out the overlap to your reader and explaining how you are going to manage it. An effective and simple solution of this kind of problem can be seen in the treatment of "Common Sense Rule #3" in the fighter plane article (p. 176).

The other comment is that what has been said here concerning the statement of a problem will often need to be supplemented by more general principles involved in writing introductions. These principles include such elements as the statement of scope and the discussion of background material. See Chapter 10.

How Was Evidence Obtained?

An interpretation can be no better than the data on which it is based. Consequently, a second major part of an interpretation is the provision of any necessary explanation about how the data was obtained, or of a statement of its probable reliability. For example, in a large tank of crude naphthalene there may be considerable random separation of naphthalene and water, and a sample taken at a given point might prove to be 100 percent water. Any discussion of the contents of the tank would be useless unless carefully controlled sampling methods were used. And any reader should refuse to accept a statement about the contents of the tank that did not explain the method by which samples were taken, or at least the probable accuracy of the results. Another example of the same principle can be seen in a botanist's complaint that carelessness in reporting the conditions under which an unusual plant has been found growing may rob the find of much of its value.

The principle to remember is that the reader should have enough information about the data to make an independent evaluation of the validity of the conclusions.

How Should the Interpretation Be Organized?

Once the problem has been stated and, if necessary, the source and validity of the data have been commented on, the next step is to present and explain the data and state conclusions. At this point, particularly if you have a lot of data to work with, you may feel that the situation is little short of chaotic. What you need to do is to divide your material into units and deal with one unit at a time. Our immediate purpose is to explain how to do this. We will take up the questions of how to organize the interpretation, and of how to present supporting data.

Organizing the Major Factors   There are four major factors in the organization of an interpretation: (1) the statement of the problem in a concise form; (2) a statement about the source and reliability of the data; (3) the statement of the subordinate problems, or of the standards of judgment; and (4) the presentation and interpretation of the data. Obviously, the nature of the organization as a whole depends on whether the writer wants to recommend a certain course of action or only to offer an explanation of the data.

If what the writer intends is an explanation only, the organization consists virtually of items 1 through 4 above. In simple outline form, the organization looks like this:

   I. Statement of problem
       A. In concise form
       B. In the form of subordinate problems
  II. Statement about the source and reliability of data
 III. Subordinate problem number 1
       A. Restatement of the problem
       B. Explanation of the problem
       C. Presentation and interpretation of the data
       D. Solution of the problem (where appropriate)
  IV. Subordinate problem number 2
       [as above in III]
V., VI. [additional subordinate problems]

Usually, a summary or conclusion would terminate the discussion. We will comment on this additional element in a moment. As you see, the outline above is essentially the organization found in the article about the fighter plane.

The organization of an interpretation in which standards of judgment are employed is, typically, somewhat different because of the need to apply a given standard to several different possible choices of action. With reference to the problem of designing cans for the uranium slugs mentioned earlier, for instance, it is evident that a considerable variety of possible designs were considered. The effect of this need for consideration of several possible choices is seen in tabular form as

Choice A	Standard 1
Choice B	Standard 2
Choice C	Standard 3

What this simple table tells us is that we need an organization that permits discussion of *every possible choice* with reference to *every standard of judgment.*

A practical illustration will help to clarify this principle. Let us consider the problem faced by a salesperson in making a choice of a new car. Let us assume that 90 percent of this salesperson's driving is done within a large city, and that he or she has set up the following standards on which to base a choice, as well as having made a preliminary survey narrowing the purchase possibilities to only four brands of car.

Possible Choices	*Standards of Judgment*
Car A	1. There should be a large trunk, for samples.
Car B	
Car C	2. Operating costs should be low.
Car D	3. The price should be low.
	4. The performance should be good at low speeds.
	5. The appearance should be neat and conservative.

Only a glance at the list of standards is needed to see that some of them will require considerable explanation. (Exactly how large a trunk is needed, and is the shape important? What is meant by low operating costs? And so on until all standards have been made explicit.) With this need for explanation of the standards in mind, consider the following three ways of combining the major factors. The standards are those listed above; the different possible choices of car are represented by capital letters.

Version 1
Statement of the problem
List of the standards (with very little explanation)
Explanation of why only four cars are to be evaluated
Comment on source and reliability of data
Explanation of standard 1, and judgment (according to this standard) of each car in
     turn
Explanation of standard 2, and judgment (according to this standard) of each car in
     turn
(Explanation of, and judgment according to, standards 3, 4, and 5, same as above)
Summary of conclusions
Final choice

Version 2
Statement of the problem
Explanation of all standards
Explanation of why only four cars are to be evaluated
Comment on source and reliability of data
Judgment of each car in turn, according to standard 1
Judgment of each car in turn, according to standard 2
(Judgment according to standards 3, 4, and 5, same as above)

Summary of conclusions
Final choice

Version 3
Statement of the problem
Explanation of all standards
Explanation of why only four cars are to be evaluated
Comment on source and reliability of data
Judgment of car A according to all five standards in turn
Judgment of car B according to all five standards in turn
(Judgment of cars C and D, same as above)
Summary of conclusions
Final choice

All possible combinations of the major factors are now shown in the three versions above (for example, there might be a section or sections devoted to a general description of each of the cars, either near the beginning or later). However, the three versions do illustrate pretty clearly the kind of decision the interpreter has to make in organizing the major factors of a complex problem.

Which version is the best? For the car problem, we would choose the first version. But for other applications it would be impossible to say which one is best without detailed knowledge of the whole situation. The chief point of these remarks, anyhow, is that if you are aware of the various possibilities, then you can select the most suitable organization for whatever subject and problem you have.

It should be remembered that the outlines above are by no means complete, being confined to illustrating relationships among the major factors. Later we'll illustrate a more nearly complete outline of an interpretation, but before doing so we'll make two additional comments about the major factors. These comments have to do with the elimination of possible choices and with the handling of conclusions.

If it should happen that one of the four cars being considered in the foregoing problem appeared to be a more likely choice according to every standard but one, it might nevertheless be necessary to rule out the car on that one point. Suppose, for instance, that car C was excellent with respect to four of the five standards, but that it had an extremely small trunk. If the necessary samples couldn't be carried the car could be eliminated completely from any further consideration. This procedure speeds up and simplifies the whole interpretation. But a warning is needed here: Don't eliminate a possible choice on the basis of failure to meet a single standard if there is any chance that that possible choice (in our illustration, car C) would be the best one in spite of the one disadvantage.

A recurrent question about conclusions is whether they should be stated in the body of the interpretation and then restated at the end, or whether they should be stated only at the end. The answer is that almost invariably they should be stated at both points. If, when the four cars are judged by standard 1, car B is found to be superior to the others, a clear and rather formal statement of that conclusion helps prepare the reader to accept whatever final conclusion is offered at the end of the interpretation. Of course, where anything like an introductory summary is used, the conclusions appear at the beginning as well.

Another important fact about conclusions in an interpretation is that considerable discussion of them may be needed at the end. This need can be seen in the simple example of the purchase of eggs. If Grade A Large eggs are cheaper (by the pound) than Grade AA Medium, which would you recommend? People might disagree about this choice. Where a choice becomes as subjective as it does here, the employment of standards of judgment may not be sufficient for a satisfactory decision. At the end of your interpretation in such a case, you may have to point out the area in which the choice is essentially subjective. You can then either indicate your own choice or leave it to the reader, as circumstances suggest.

In bringing to a close these remarks on organization of the major factors, we'll add a somewhat more detailed outline. The outline below indicates one way of organizing a discussion about the choice of a car: it is by no means the only way, and it is more generalized than would be desirable in practice, but it has the virtue of filling out the introductory portion more completely than do the three short versions presented earlier, and thereby of removing some possibly misleading implications of the earlier versions. Reference to Chapter 19 will illustrate differences in overall organization that would be desired by certain companies.

I. Introduction
    A. Statement of the problem
        1. Discussion of the need for a recommendation
        2. Concise statement of the problem
        3. Concise statement of the standards of judgment
    B. Scope
        1. Statement of the cars to be considered
        2. Explanation of why the cars are restricted to the group named
    C. Comments on source and reliability of data
    D. Plan of development
        1. Comments on the presentation of data
        2. The overall plan
II. Judgment according to the first standard
    A. Explanation of the standard
    B. Judgment of car A
        1. Presentation of data
        2. Interpretation of data
    C. Judgment of car B
        1. Presentation of data
        2. Interpretation of data
    D, E. [Same as A,B,C]
III, IV, V. [Same as I,II for the remaining standards]
  VI. Summary of conclusions
 VII. Final choice

As we said at the beginning of this discussion of interpretation, our objective has been to make suggestions to you about how to communicate to other people what you have learned from examination of a group of facts. For discussion of the complex techniques of experimental and statistical determination of what the facts are in a given situation, the following books are helpful: K. C. Peng, *The Design and Analysis of Scientific Experiments* (Reading, Mass.: Addison-Wesley Publishing Company, 1967), and Geoffrey Keppel, *Design and Analysis: A Researcher's Handbook* (Englewood Cliffs, N.J.: Prentice-Hall, Inc., 1973). There are many books on research in specific disciplines.

Presenting Data   In addition to organizing the major factors just discussed, some decisions have to be made about the presentation of data. You're likely to commence writing an interpretative report with a thick pile of data at your elbow and questions like the following going around in your head: How much of this data should I put into the report? Where should I put it? What form should it be in? How much explanation should I include about what it means, and how much should I assume readers will see for themselves?

The question of how much data to include must be answered according to circumstances. A college instructor often asks students for all the raw data they took, and sometimes all the raw data is included in industrial and research reports. Unless it is quite clear that the raw data should be included, however, it is better to leave most of this material out. If it is put in, it should usually go into an appendix. Don't clutter up the text with it.

Whether or not all the raw data is put in, it must, of course, be sufficiently represented in the body of the interpretation to convince the reader that the conclusions reached are sound. And so we come to the question: In what form should the data be introduced into the body of the interpretation? The answer is — in any form at all. But remember, as the architects like to put it, that form follows function. If your purpose is to communicate, then whatever form will best convey your idea is the one to choose. Graphic aids provide a tremendous range of possibilities for the illustration or presentation of factual material, often in very dramatic form (see Chapter 20). In addition, there are such possibilities as presenting small samples of data, providing short lists of key figures or facts from the data, working out a typical or illustrative problem or calculation, summarizing trends in terms of range and percentage of change, and many others. Actually, however, the only special knowledge you need about the form in which data can be presented is an acquaintance with the basic concepts of graphic aids. Other forms in which to present data will arise naturally out of the situation you are discussing. (We refer here only to the writing problem. For suggestions about statistical and other technical methods of handling data, see the Peng and Keppel titles noted above.)

After deciding upon the general organization of subordinate problems and (possibly) standards of judgment, your principal tasks are to decide how much data to put in, where to put this material, what form to present it in, and how to reveal significant relationships without, on the one hand, confusing the reader with a mass of detail or, on the other hand, failing to offer sufficient supporting evidence.

Success in this last task of revealing significant relationships is made most certain by a very clear decision, made before the writing is begun, as to what relationships should be explained. A carefully worked-out outline is of great value. Start with the assumption that your reader is intelligent but uninformed; then caution yourself that you cannot and should not discuss every detail.

There are three specific "don'ts" here that are of particular importance:

1. Don't put into writing the kind of information that is easier to grasp in the form of graphs or tables.
2. Don't restate all the facts that have been put into tabular or graphic form. From our observation, this is a mistake students are especially likely to make.
3. Don't assume that, having made a table or a graph, *nothing* need be said about it. A little explanation of how to read the graph or table is often helpful. And almost invariably the significant relationships revealed by the table or graph should be pointed out.

These principles will be illustrated in the second report at the end of this chapter.

Attitude

The attitude the interpreter brings to his or her writing should be what is generally known as the scientific attitude. This fact is self-evident, and yet it is not always easy to adhere to in practice. Detachment and objectivity are particularly difficult in the evaluation of evidence on an idea that one has intuitively felt at the outset to be true, but which has come to look less certain as investigation progressed. We are all in some measure the creatures of our emotions. A counterbalance to the natural human desire for infallibility even in intuitions, however, is the deep emotional satisfaction of feeling above and in command of a given set of facts, with no obligation beyond saying that a given idea is true, or false, or uncertain. It is this emotional "set" that should be brought to problems of interpretation.

An illustration of the kind of attitude that should *not* be taken turned up in some student papers we once read. We had given a class a sheet of data on the records of a number of football coaches and asked the students to write an interpretation of the data. Personal loyalties evidently got mixed up in the analysis, for one student concluded solemnly that whatever the data might indicate, Coach X was definitely the best coach in the group because all the football players who had played under him said so!

Another problem of attitude that often arises in interpretation is that of adapting the manner of the interpretation to the individuality of a certain reader or readers. Human nature being what it is, novel or unexpected conclusions are almost certain to meet with opposition. A cool appraisal of probable opposition and an allowance for it in the manner of the presentation is not only wise and profitable, it is also kind. Kind, that is, as long as the conclusion being offered is an honest one.

Summary

1. Interpretation, which is the art of establishing a meaningful pattern of relationships among a group of facts, has as its first important step the statement of the problem being investigated.
2. Usually, the problem is then broken down into a number of subordinate prob-

lems, or of standards of judgment. The statement of a subordinate problem is transformed into the statement of a standard of judgment by being phrased in such a way as to serve as a guide for a decision.

3. In organizing an interpretation, the major factors to be considered are the primary problem, the subordinate problems or standards of judgment, a comment on the source and reliability of data, and the presentation and interpretation of the data.

4. Supporting data should be put into graphic or tabular form whenever possible; the writing should be devoted to pointing out significant relationships.

5. Where a choice is to be made among a number of possibilities, early elimination of some possibilities speeds up the whole process.

6. Conclusions should be stated as they are reached in the body of the interpretation, even if they are to be summarized elsewhere.

7. The attitude throughout should be impartial and objective, although not without a little human consideration of the individuality of the intended reader.

Illustrative Material

Three examples of interpretation are presented below. The first is the most nearly perfect example of an interpretation in miniature we have ever seen. Brief as it is, it reveals the way in which an interpretation is structured. This brevity was no doubt forced upon the author because this miniature article was published as an advertisement. Many changes have occurred in high-fidelity sound equipment since this advertisement was written, and we are reprinting it not for its content but as a fine example of how to organize an interpretation.

The second example is also well organized, but on a much larger scale. Added to the interest of its illustration of good structure is its subject matter, which concerns an analysis of opinions expressed by college graduates of their training.

The third selection, on using personal computers in college, is the first several pages of an article which appeared in *Computer* (April 1984).

Sound Talk[3]
Dr. W. T. Fiala, Chief Physicist

HIGH FREQUENCY HORNS

The high frequency horn is an important part of any high fidelity speaker system. It must properly load the driver element, provide smooth distribution from its lower frequency limit to beyond the range of the human ear, offer no interference to the frequency response of the driver, and be free from resonances that introduce a "character" to the reproduced sound.

The first paragraph presents the standards of judgment. In a general way, the first sentence is a statement of the problem — *that is,* how should a high-frequency horn perform? The remainder of the paragraph is a statement of standards.

---

[3] From an advertisement of Altec Lansing Corporation appearing in *High Fidelity,* November 1958, p. 131.

Horns available for high fidelity reproduction fall into four general types: diffraction horns, ring or circumference radiators, acoustic lenses and sectoral horns. Of these four, only one meets all the requirements for an acceptable high frequency horn. Diffraction horns provide no distribution control. At lower frequencies the distribution pattern is unusably wide. At higher frequencies it becomes progressively narrower, eventually becoming a narrow beam of sound. Good listening quality can only be found directly in front of the horn. Even there, since at lower frequencies the sound energy is widespread while it is concentrated as the beam becomes more directional, an un-natural accentuation of higher frequencies will be experienced.

The ring radiator, like the diffraction horn, makes no attempt to control high frequency distribution. It has the additional fault of phasing holes whenever the distance between the near and far sides of the radiator equal ½ the wave length of the frequency being reproduced.

The acoustic lens provides a smooth spherical distribution pattern at all frequencies. The lens elements used to achieve this distribution, how-ever, act as an acoustic filter and seriously limit high frequency reproduction, tending to introduce a "character" to the reproduced sound.

Sectoral horns, when built to a size consistent with their intended lower frequency limit, provide even distribution control. The smooth exponential development of their shape assures natural sound propagation of the full capabilities of the driving element. They are the only horns that fully meet all of the requirements for high fidelity reproduction.

We believe that ALTEC LANSING sectoral horns, built of sturdy nonresonant materials, are the finest available. Listen to them critically. Compare them with any other horn. You will find their superior distribution and frequency characteristics readily distinguishable: their "character-free" reproduction noticeably truer.

The second paragraph presents possible choices among horns.

The body of the discussion is organized in terms of the types of horn, in the order in which the horns were named in the second paragraph. The principle of elimination is used throughout. That is, each of the first three types is eliminated because of at least one alleged important weakness.

In conclusion, we'd like to point out that it would evidently be fairly easy to expand this highly condensed discussion into a long examination of the relative merits of these horns, and in any such long examination still retain the organization found here.

Note: We are here taking the unusual step of reprinting an advertisement because it provides a remarkably concise illustration of some of the important principles considered in the preceding discussion. Our reprinting of it does not reflect any opinion of our own about, or even any interest in, the question of what kind of horn is really best.

Research into the Amount, Importance, and Kinds of Writing Performed on the Job by Graduates of Seven University Departments that Send Students to Technical Writing Courses[4]

Paul V. Anderson

Because introductory courses in technical writing are usually designed to prepare students for their careers, both teachers and students can benefit from learning about the writing that college graduates perform on the job. Teachers can use this knowledge to help them make sound decisions about the design of their courses—what topics to cover, how to design assignments, and so on. Students can use it to predict the role that writing will play in their futures and, hence, to help them decide how much effort to devote to their technical writing course and what aspects of the course to concentrate on most fully.

In this paper, I report on a survey in which I asked graduates of a heterogeneous set of seven departments that send students to a technical writing course to tell me about the writing they do at work.* My purpose was to answer two questions:

After a brief justification of investigating the amount, importance, and kinds of writing done by graduates of seven university departments which send students to technical writing courses, Professor Anderson wastes no time in setting forth his purpose in conducting the investigation. He then proceeds to present a careful account of the methodology of his study, including an explanation of his statistical analysis procedure, followed by a detailed analysis of the data he collected, considered both from the point of view of the entire group of respondents to his questionnaire and from the point of view of the individual departments from which the respondents had graduated. The final section of his report presents a detailed account of the conclusions he reached as a result of the study, along with a careful interpretative statement of the implications of those conclusions. The report is supported by figures and tables of statistical data.

Professor Anderson's study is, we believe, a fine example of the kind of interpretation we have discussed in this chapter: he clearly explains what he wanted to find out, how he collected evidence, and what the meaning and significance of the evidence is.

* I thank Professors Steve Hinkle, John Skillings, and C. Gilbert Storms for their helpful suggestions concerning this project. I am also grateful for help provided by the following students: Rick Bayster, Lisa Beede, Jennifer Kirch, and (especially) Deb Schoenberg.

[4] We are grateful to Dr. Paul V. Anderson for permission to quote this report. Dr. Anderson is associated with the Technical Writing and Communication Program at Miami University, Oxford, Ohio 45056. This report is copyright 1980, 1984 by Paul V. Anderson.

—What generalizations can be made about the amount, importance, and kinds of writing performed by graduates of these seven departments?

—What important differences, if any, exist among the seven departments in terms of the amount, importance, and kinds of writing their graduates perform on the job?

SIMILAR SURVEYS OF WRITING AT WORK

By surveying alumni of departments that send students to their courses, teachers of business writing have produced a great deal of information that is useful to these teachers and their students (Bennett, 1971; Huegli and Tschirgi, 1974; Persing, Drew, Bachman, and Galbraith, 1977; Andrews and Koester, 1979; Stine and Skarzenski, 1979; Storms, 1984; Swenson, 1979). Similarly, technical writing teachers in two-year colleges have learned many helpful things by surveying graduates of technical and occupational programs at their schools (Skelton, 1971; Glenn and Green, 1979).

However, we have no comparable information from alumni of four-year institutions who majored in departments that send students to technical writing courses. Davis (1978) has surveyed persons listed in *Engineers of Distinction*, and Barnum and Fischer (1984) have surveyed alumni of Southern Technical University who graduated with the Bachelor of Engineering Technology. As useful as these studies are, by themselves they do not entirely satisfy the needs of students now enrolled in technical writing or of their teachers. Without further research, we cannot be sure that on-the-job writing experiences of engineers of distinction are typical of those that will be encountered by the vast majority of engineering students, who will not earn such recognition, or that the experiences of persons who earn *engineering technology* degrees will be similar to those of persons who earned degrees in *engineering*. Even more importantly, we might suspect that the typical experiences of alumni in *all* engineering-related fields may be significantly different from the experiences that typically will be encountered by the majority of students who study technical writing in college. At many universities, technical writing courses are populated by students from a wide variety of departments, including those in the applied sciences, the physical and natural sciences, and the service professions. It is possible that after they graduate the students in many of these departments will write in ways that are significantly different from the ways that either engineers of distinction or engineering technology graduates write. From the point of view of technical writing students and teachers, a similar problem limits the usefulness of a study by Faigley, Miller, Meyer, and Witte (1981). These researchers surveyed a sample of college graduates working in *all* fields, but did not separately analyze the responses from people who had graduated from the departments that send students to technical writing courses.

METHOD

To gather information about their on-the-job writing experiences from a group of alumni who work in the kinds of jobs that technical writing students may soon hold, I mailed a questionnaire to 2335 graduates of Miami University (Ohio) who met all of the following criteria:

—Earned a baccalaureate from one of seven departments that enroll students in Miami's introductory technical writing course, which is taught to juniors and seniors. The departments are Chemistry, Home Economics, Manufacturing Engineering, Office Administration, Paper Science and Engineering, Systems Analysis, and Zoology.

—Received his or her baccalaureate at least one year earlier.
—Had a U.S. address listed in the files of the university's Alumni Office.

The questionnaire went to every graduate of the Manufacturing Engineering, Office Administration, Paper Science and Engineering, and Systems Analysis Departments who met those criteria. It went to every other person on the list of Home Economics graduates and every fourth person on the lists of graduates from the Chemistry and Zoology departments who met the criteria.

Procedure

When mailed, the four-page questionnaire was accompanied by a letter that explained the purpose of the survey and asked the alumni to cooperate by returning their completed questionnaires in a prepaid envelope that was enclosed. Ten days later, a postcard was sent to the alumni, reminding them to return the questionnaire.

Data Analysis

Besides being summarized descriptively, the responses to many of the questions on the questionnaire were analyzed statistically, after first being coded in the following way:

—One question asked the alumni to indicate "The percentage of your time at work that you spend writing." The alumni answered by checking one of seven alternatives: 0%, 1–10%, 11–20%, 21–40%, 41–60%, 61–80%, and 81–100%. Their responses were coded on a six-point scale for which 0% = 0; 1–10% or 11–20% = 1; 21–40% = 2; and so on.

—Another question asked the alumni, "How important would the ability to write well be to someone who wanted to perform your present job?" The respondents answered by checking one of five alternatives: Negative Importance, Minimal Importance, Some Importance, Great Importance, and Critical Importance. Their responses were coded on a five-point scale for which "Negative Importance" = 0 and "Critical Importance" = 4.

—Twenty-four questions asked the alumni to tell how often various kinds of people read their writing, how often they prepare various kinds of communication, and how often they write for each of two purposes (at someone else's request or on their own initiative). The respondents answered by checking one of five alternatives: Never, Rarely, Sometimes, Often, Always. Their responses were coded on a five-point scale for which "Never" = 0 and "Always" = 4.

The coded responses were analyzed in a variety of ways, which are described at the appropriate points in the "Results" sections of this paper. For all statistical tests, the $\alpha = .05$ level of significance was used.

SAMPLE

In all, 1052 alumni returned the questionnaire, for an overall response rate of 37%. However, 211 of the questionnaires were unusable, in most cases for one of these reasons: (1) the respondent did not identify his or her department, (2) the respondent indicated that despite the Alumni Office records he or she actually graduated from some other department than one of the seven under study, or (3) the respondent had not gained any employment experience since receiving his or her baccalaureate, usually because the respondent was attending graduate school.

The 841 alumni who returned usable questionnaires were distributed in the following way among the seven departments.

Department	Number
Science	
Chemistry	52
Zoology	60
Applied Science	
Manufacturing Engineering	143
Paper Science and Engineering	93
Systems Analysis	230
Service Professions	
Home Economics	163
Office Administration	100
Total	841

When asked how many years they had been employed since obtaining their baccalaureates, the alumni provided the information shown in Figure 1.

The alumni were also asked how much additional education they had received since earning their baccalaureates. Their responses are summarized in Figure 2. The alumni who placed themselves in the "Some" category had engaged in a wide variety of educational activities, such as: earning a second baccalaureate or a teaching certificate, taking additional college courses, either graduate or undergraduate, without receiving a degree or certificate; completing flight training school or an American Dietetics Association Internship; and participating in in-house courses offered by their employers. The 39 respondents with doctorates earned the following degrees: Ph.D. (17), M.D. (11), D.D.S. (5), J.D. (3), D.O. (2), and D.V.M. (1).

Information supplied by the 841 alumni who returned usable questionnaires is discussed in the following two sections. In the first, the entire sample of 841 is considered as a single group. In the second, the respondents are classified according to the department from which they graduated so that the departments can be compared with one another.

RESULTS FOR ENTIRE SAMPLE CONSIDERED AS A SINGLE GROUP

By returning the questionnaire, the alumni provided a great deal of information about the amount, importance, and kinds of writing they perform at work.

Amount of Writing

Taken as a single group, the alumni spend a considerable amount of their time at work writing, as Figure 3 shows. Sixty-nine percent reported that they write more than 10% of their time at work, which is equivalent to saying that on the average they write more than one-half day in every 40-hour week. Furthermore, 48% reported writing more than 20% of their time (more than one day a week) and 15% reported writing more than 40% of their time (more than two days a week).

Importance of Writing

The alumni also indicated that writing is a very important part of their work. As Figure 4 shows, 93% said that the ability to write well (not just write, but write well) would be of at least "Some Importance" to someone who wanted to perform their jobs. (The questionnaire defined "Some Importance" to mean, "Ability to write well would help a significant amount.") Furthermore, 67% — well over half — stated that it would be of at least "Great Importance," and 16% said that it would be of "Critical Importance," which the questionnaire defined to mean, "Ability to write well would be indispensable."

Kinds of Writing

To gather information about the kinds of writing the alumni prepare on the job, the questionnaire asked them to indicate on a five-point scale how often:

— their writing is read by each of eleven groups of *readers.*
— they write each of eleven *forms* of communication.
— they write for each of two *reasons.*

Readers. Overall, the alumni's answers to the 11 questions about their readers indicate that the writing of the typical alumnus or alumna is read, not by any single group of readers, but by a wide variety of groups — a conclusion that is very important to students and teachers of technical writing. Table 1 shows how often the alumni's writing is read by four groups that are defined in terms of how much they know about the alumni's specialties: those who know more about it than the respondents do, those who know about the same amount, those who are familiar with it but know less, and those who are unfamiliar with it. An analysis of variance (randomized complete block) indicated that there is a significant difference among the alumni's answers to these four questions, $F (3, 2376) = 117.01$, $p = .0001$. Consequently, Duncan's multiple range test ($\alpha = .05$) for variable response was used to determine which group or groups of readers read the writing of the typical alumnus or alumna significantly more often than do the others; the results are shown in the righthand column of Table 1, where brackets enclose means that are *not* significantly different at the $\alpha = .05$ level.

The information shown in Table 1 is especially interesting for two reasons. First, it shows that on the job the alumni most often address readers who are different from the readers that college students most often address in their writing assignments. College students usually write to readers (their professors) who know much more than the students do about the students' own disciplines. In contrast, the alumni report that their writing is most often read by readers who know less than they do or only about as much as they do about their specialties. On the other hand, Table 1 also shows that at least 60% of the alumni at least "Sometimes" address all four groups of readers, including those who are unfamiliar with the alumni's specialties. Thus, the results shown in Table 1 indicate that in general the alumni write to a variety of groups of readers, rather than to a single group.

Table 2 shows how often the alumni's writing is read by people at three levels *within* their own organizations: a higher level than the alumni's, the alumni's own level, and a lower level. An analysis of variance (randomized complete block) indicated that the alumni's writing is read significantly more often by people in one or two of these groups than by people in the other (or others), $F (2, 1582) = 145.74$, $p = .0001$. Consequently, Duncan's test ($\alpha = .05$) was performed. As the righthand column in Table 2 indicates, the alumni's writing is read significantly more often by people above them than by people at their own level or below; in addition, their writing is read significantly more often by people at their own level than by people below. However, even people at a lower level read the writing of 70% of the respondents at least "Sometimes." These results provide further evidence that in general the alumni write to a variety of groups of readers.

Table 3 shows how often the alumni's writing is read by four groups of readers *outside* the alumni's own organizations: customers, vendors (who sell products to the alumni's organization), the general public, and legislators and other government officials. An analysis of variance (randomized complete block) showed that the alumni's writing is read more often by one or more of these outside groups than by the other (or others), $F (3, 2352) = 93.26$, $p = .0001$; the results of Duncan's test are shown in the righthand column of Table 3. Clearly, the alumni's writing is read much more often by customers than by any of the other three outside groups.

Finally, a Duncan's test ($\alpha = .05$) comparing the means displayed in Table 2 with means displayed in Table 3 showed that the alumni's writing is read significantly more often by all three groups of people within the alumni's organizations than by people in any of the four outside groups. That finding is explained by the very large percentage of alumni who reported that their writing is "Never" read by the four outside groups.

Forms. In response to another set of questions, the alumni indicated how often they prepare each of the eleven forms of written communication that are listed in Table 4. Admittedly, there is some overlap among the forms. For example, a person could write general instructions in a memorandum as well as in a special form called "general instructions." Nevertheless, this kind of question seems well worth asking because so many textbooks and courses in technical writing are organized around forms of communication.

An analysis of variance (randomized complete block) showed that the alumni write one or more of these forms significantly more often than the others, $F (10, 7919) = 461.86$, $p = .0001$. Duncan's test produced the results shown in the righthand column of Table 4. Although there is some overlap among the forms (for example, alumni could write general instructions in a memorandum), the results clearly indicate that the alumni write memoranda, letters, step-by-step instructions and general instructions significantly more often than they write formal reports and proposals, which are two forms given special attention in many technical writing courses. Furthermore, the alumni write all ten of the forms, even advertising, significantly more often than they write articles for professional journals. That finding is particularly interesting because some teachers ask their students to model their written projects upon journal articles rather than upon the other, more frequently written forms of communication. (Of course, the ranking of these forms differs for individual departments; these differences are discussed below.)

Reason for Writing. In answering the final pair of questions on the questionnaire, the alumni indicated how often they write at someone else's request and how often on their own initiative. The results are summarized in Table 5, which shows that *both* reasons were checked either "Often" or "Always" by a majority of the alumni. Furthermore, a paired t-test indicated that there is not a significant difference between the responses to these two questions: the alumni write about as often for both reasons, $t (792) = 1.48$, $p = .14$, $\overline{X}ser = 2.49$, $\overline{X}oi = 2.55$.

Summary

In summary, the responses from all the alumni considered as a single group indicate that they spend a considerable amount of time at work writing, and that their writing is important to them. The communications they prepare are read by a wide variety of readers, including those with various levels of knowledge of the alumni's specialties and at various levels within the alumni's organizations. Although many alumni also write communications that are read by people (particularly customers) outside the alumni's organizations, their readers are more often in their own organizations than outside. When they write, the alumni use a variety of forms, most notably memoranda, letters, step-by-step instructions, general instructions, and preprinted forms the writer must fill out. The forms they write least often are advertising and articles for publication in professional journals. Finally, the alumni of this heterogeneous set of seven departments write about as often on their own initiative as at someone else's request.

The implications of these findings for teachers of technical writing are discussed in the final section of this paper.

RESULTS OF A COMPARISON AMONG THE SEVEN DEPARTMENTS

Because the seven departments represented in this study are so different from one another — ranging all the way from Systems Analysis to Home Economics, from Zoology to Office Administration — it is reasonable for teachers and students to inquire whether the generalizations made about the group as a whole hold true for the individual departments. Students will want to know the answer to that question so that they can determine whether the results reported above apply to them, or only to students in other departments. Teachers will want to know the answer because it might help them settle their debate over the following issue: are the needs of students from such a variety of departments so different that the students should be segregated into homogeneous classes, each containing students from only one department, or are their needs similar enough that they can (and even should) be taught in heterogeneous classes, each containing students from several departments?

As the following discussion indicates, the writing performed on the job by alumni of each of the seven departments examined in this study is remarkably similar to on-the-job writing performed by alumni of the other six departments.

Amount of Writing
A comparison of the seven departments in terms of the amount of time their alumni spend writing produced one of the most surprising results of the survey. Although one might naturally expect that alumni of some departments would spend more time writing than would the alumni of others, that is not the case; an analysis of variance (one way) showed that alumni of all seven departments spend about the same amount, $F(6, 829) = 1.40, p = .211$.

Importance of Writing
In contrast, an analysis of variance (one way) did show that a significant difference exists among the amounts of importance that the alumni of the seven departments place on writing, $F(6, 793) = 4.97, p = .0001$. Table 6 shows the responses that the alumni of each department made to the question on the importance of writing; the righthand column displays the results of a Duncan's test ($\alpha = .05$).

One result shown in Table 6 seems especially noteworthy. Even the departments that find writing to be least important still consider it to be very important. For example, 84% of the Zoology alumni reported that writing is of at least "Some Importance" and 42% reported that it is of at least "Great Importance." Thus, despite the difference among the departments, alumni of all seven agree that writing is an important part of their work.

Kinds of Writing

A comparison of the seven departments in terms of their alumni's responses to the 24 questions asking about the kinds of writing they perform at work also supports the conclusion that what is true about all 841 respondents taken as a single group is, for the most part, true for the respondents from each of the individual departments.

Readers. Analyses of variance (one way) of the four questions asking about the reader's *level of knowledge* of the writer's specialty showed the following:

—The alumni of all departments responded in essentially the same way when asked how often their writing is read by people who know as much about their specialties as they do and by people who are completely unfamiliar with their specialties: F (6, 787) = 1.88, p = .082; F (6, 787) = 1.93, p = .073.

—The alumni of some departments responded in significantly different ways than did alumni of others when asked how often their writing is read by people who know more about their specialties than they do and by people who are familiar with the specialties but know less about them: F (6, 787) = 3.30, p < .003; F (6, 787) = 5.31, p = .0001.

Despite these differences, however, the alumni's responses to these four questions are remarkably similar from department to department; that similarity can be seen by comparing the results (shown in Table 7) of a series of analyses of variance (randomized complete block) and Duncan tests (α = .05). For each individual department (just as for all the alumni taken as a single group), the alumni's writing is read about equally often by those who know more than they do and by those who know about as much as they do about their specialties—and that their writing is read significantly more often by those two groups of readers than by readers who are completely unfamiliar with their specialties. Furthermore, a department-by-department examination of the responses to these four questions showed that at least 50% of the alumni from *every* department reported that their writing is read at least "Sometimes" by each of the four kinds of readers. Clearly, the alumni of all seven departments typically address a variety of kinds of readers, not just one kind.

That conclusion is given further support by a comparison of the responses given by alumni of the seven departments when they were asked how often their writing is read by people at *three levels within their own organizations.* Analyses of variance (one way) showed that there are significant differences among departments in terms of the frequencies with which the alumni's writing is read by people at a higher level than their own, by readers at their own level, and by readers at a lower level: F (6, 786) = 6.07, p = .0001, F (6, 786) = 3.83, p = .0009, F (6, 790) = 4.00, p = .0006. However, as Table 8 shows, the responses that the alumni of each department gave to these three questions indicated a fundamental similarity among them; this table displays the results of a series of analyses of variance (randomized complete block) and Duncan's tests (α = .05) performed on a department-by-department basis. For every department except Zoology, the alumni indicated that their readers are significantly more often at a level above the alumni's level than they are at the alumni's level or at a lower one. Similarly, in every department where the alumni's writing is read significantly more often by one of the latter two groups of readers than by the other, those readers are at the alumni's own level. Finally, a department-by-department examination of the responses to these four questions showed that at least 48% of the alumni from *every* department (and 69% from every department except Office Administration) reported that their writing is read at least "Sometimes" by readers at each of the three levels.

Three major conclusions are supported by analyses of the alumni's responses to the questions that asked them how often their writing is read by readers *outside their own organizations.* First, customers are an important group of outside readers for alumni of all seven departments. As Table 9 shows, the alumni from all seven departments address customers about as often, or significantly more often than they address any of the other three groups of readers. Also, an analysis of variance (one way) showed that the alumni from each of the seven departments address customers about as often as do the alumni from the other six departments, $F(6, 778) = 1.67, p = .126$.

The second major finding is that the frequency with which the alumni address any of the other three groups of outside readers (vendors, government officials, and the public) depends largely upon the department from which the alumni graduated, $F(6, 775) = 7.80, p = .0001; F(6, 786) = 5.01, p = .0001; F(6,787 = 34.71, p = .0001$. Some differences among the departments in this regard are striking, as Table 9 shows. For example, whereas the public reads the writing of Home Economics alumni significantly *more* often than do either vendors or government officials and just as often as do customers, the public reads the writing of Systems Analysis alumni significantly *less* often than do any of the other three groups.

The third major finding is that for every department except Office Administration, the alumni's writing is read significantly more often by all three groups of readers within the alumni's organization (higher level, same level, lower) than by any of the four groups of outside readers. Even for the one department that is an exception, Office Administration, the alumni's writing is read more often by people at a higher level and at the alumni's own level in their organizations than it is by any of the groups of outside readers; thus, the extent of the exception involving Office Administration graduates is merely that their writing is read about as often by customers as it is by people lower in their own organizations.

Thus, this department-by-department analysis shows that the conclusions drawn above for all 841 respondents considered as a single group applies equally to the respondents from each of the seven departments: *regardless of their department,* the alumni typically address a variety of kinds of readers, not just one kind, including not only readers with widely different levels of knowledge of the alumni's specialties but also readers at all three levels within the alumni's organizations. Furthermore, regardless of department (with only one exception, Office Administration), the alumni more often address readers at all three levels in their own organizations than readers in any of the four outside groups. The only important differences among the departments involves the extent to which their alumni address some kinds of readers outside their own organizations. Even there, however, a very important group, customers, are addressed about equally often by alumni of all seven departments.

Forms. As Table 10 shows, analyses of variance (one way) indicated that there are significant differences among the seven departments in terms of how often they write ten of the eleven forms of communication asked about in the survey (all except general instructions). Accordingly, the following procedure was used to determine whether these differences mean that the *ranking* of these forms is different in some remarkable way for some departments than it is for all 841 alumni taken as a single group.

First, a series of analyses of variance (randomized complete block) and Duncan's tests ($\alpha = .05$) were performed upon the responses that the alumni from each department gave to these eleven questions. Then, the results of these tests were compared with the results of the Duncan's tests performed on the responses to these same questions by all the 841 respondents considered as a single group (Table 4).

This comparison showed that if one arranges the eleven forms in order, beginning with the one most frequently written by the 841 respondents considered as a single group (memoranda) and ending with the one least frequently written by the 841 respondents considered as a single group (articles for professional journals), the resulting list very closely resembles the lists of the same sort that are constructed for each of the seven individual departments. This resemblance is illustrated in the following discussions of the five forms that the sample as a whole writes most often.

Rank By Group As A Whole	Form	Discussion
1	Memoranda	For all seven departments, no other form is written significantly more often ($p < .05$) than memoranda. Memoranda are written at least "Sometimes" by at least 61% of the respondents from every department and by 82% of the respondents from every department except Zoology.
2	Letters	For only one department do the alumni write anything other than memoranda significantly more often than they write letters. The exception is Systems Analysis, whose alumni write step-by-step instructions significantly more often. Letters are written at least "Sometimes" by at least 74% of the respondents from every department.
3	Step-by-Step Instructions	For all seven departments, no forms besides memoranda and letters are written significantly more often than step-by-step instructions. Such instructions are written at least "Sometimes" by at least 65% of the respondents from every department.
4	General Instructions	For only one department do the alumni write anything but memoranda, letters, and step-by-step instructions significantly more often than they write general instructions. The exception is Office Administration, whose alumni fill out preprinted forms more often than they write general instructions. Such instructions are written at least "Sometimes" by at least 59% of the respondents from every department.

| 5 | Preprinted Forms (To Be Completed by Respondent) | For all seven departments, no forms of communication except memoranda, letters, step-by-step instructions, and general instructions are written significantly more often than preprinted forms that are to be completed by the respondent. Such forms are filled out at least "Sometimes" by at least 57% of the respondents from every department. |

The same kind of agreement exists among the departments with respect to the two forms written *least* often by the group as a whole.

| 10 | Advertising | For all seven departments, no form other than articles for publication in professional journals is written significantly less often than advertising. Advertising is written "Rarely" or "Never" by at least 59% of the respondents from every department and by at least 86% of the respondents from every department except Home Economics and Office Administration. |
| 11 | Articles | For only one department is any form at all written significantly less often than articles for professional journals. The exception is Zoology, whose graduates write advertising significantly less often. Articles are written either "Rarely" or "Never" by at least 67% of the respondents from every department and by at least 88% of the respondents from every department except Chemistry and Zoology. |

In contrast, the individual departments do vary considerably in terms of the frequency with which their alumni write the four forms (proposals, formal reports, minutes, and scripts) that were ranked seventh, eighth, and ninth in the Duncan's test performed on the responses of all the alumni taken as a single group. For example, for the group as a whole, both minutes of meetings and conversations and also scripts for speeches and presentations are written significantly less often than any other forms except advertising and articles for professional journals (see Table 4). However, the Home Economics alumni indicated they write scripts significantly more often than formal reports, and the alumni from the Chemistry, Office Administration, and Zoology departments indicated that there is not a statistically significant difference between the frequency with which they write scripts and the frequency with which they write proposals and formal reports. Similarly, Office Administration alumni write minutes significantly more often than either proposals or formal reports, but alumni of the Manufacturing Engineering, Paper Science and Engineering, and Zoology departments write minutes and formal reports about equally as often.

In summary, there is a set of five forms that are among those most frequently written by the alumni from every one of the seven departments. These forms are: memoranda, letters, step-by-step instructions, general instructions, and preprinted forms that the writer must complete. In addition, two forms — advertising and articles for publication in professional journals — are among those least frequently written by the alumni from every one of the seven departments. Finally, the remaining four forms — proposals, formal reports, minutes, and scripts — are written relatively more often or relatively less often than other forms, depending upon the department from which the alumni graduated.

Reason for Writing. The last two questions asked the alumni how often they write for each of two reasons: on their own initiative and at someone else's request. As mentioned above, a paired t-test showed that when they are considered as a single group, the alumni write equally often for each of these two reasons. A series of paired t-tests were used to see if that generalization holds true for the alumni from each of the individual departments. As Table 11 shows, it does hold for the Chemistry, Office Administration, Paper Science and Engineering, and Zoology departments. However, the t-tests also showed that alumni of the Home Economics and Manufacturing Engineering departments write significantly more often on their own initiative than at someone else's request and that the alumni from the Systems Analysis department write significantly more often at someone else's request than on their own initiative. Nevertheless, at least 87% of the respondents from every one of the seven departments said that they write at someone else's request at least "Sometimes," and at least 84% of the respondents from every department said that they write on their own initiative at least "Sometimes." Clearly, then, alumni from all seven departments do a great deal of writing for each of the two reasons.

Conclusions and Implications

The information provided by the respondents to this survey lead to seven major conclusions, each with implications for students and teachers of technical writing.

1. Writing is an important part of the work performed by graduates of the types of departments represented in this study. These graduates spend a substantial amount of their time at work writing, and the ability to write well is important to the performance of their jobs.

Implication. Students majoring in science, in engineering and other applied sciences, and in service professions can benefit from studying technical writing while in college. If they aren't required to take such a course, students should consider enrolling in such a course as an elective. Once in the course, they should treat it as seriously as they treat courses in their major.

2. In terms of the writing their graduates perform on the job, all seven departments are remarkably similar. For example, the departments are not significantly different ($p < .05$) in terms of the *amount* of writing their alumni perform on the job. Similarly, although there are significant differences among the departments in terms of the *importance* their graduates attach to writing, the graduates of all seven departments find writing to be very important. Furthermore, all seven departments are very much alike in terms of the *readers* their graduates address and in terms of the *forms* of communication their graduates write most frequently.

Implication. Nothing found in this survey indicates any need to establish special sections of introductory courses so that students majoring in different departments are segregated from one another. Students from all departments (or, at least from all seven of the departments represented in this survey) need to be prepared to address the same kinds of readers in a variety of forms of writing.

3. The typical graduates of all seven departments write to a variety of kinds of readers — not only to readers who (like their professors) know more than they do about their specialty, but also to readers who know only as much as they do, to readers who know less, and to readers who are completely unfamiliar with their specialty. In addition, the graduates often write not only to readers at their own level within their organizations but also to readers at levels above and below theirs.

Implication. Introductory courses should provide students with instruction and practice in writing to a variety of kinds of readers. Teachers who have been accustomed to asking their students to write every assignment to readers with the same level of knowledge should reevaluate that practice. This practice is probably most widespread among teachers new to technical writing, some of whom attempt to deal with their unfamiliarity with their students' subject matter by telling the students to write all their assignments so that the assignments can be understood by "the intelligent but uninformed layperson," meaning the teachers themselves. While such a practice does seem justifiable for brand new teachers, even they should immediately start learning enough about their students' specialties so that they can ask the students to write some of their assignments to readers more knowledgeable than the uninformed layperson. In an equally untenable position are teachers who ask their students to write all their assignments to readers with any one of the other levels of knowledge.

Likewise, teachers should be sure that they include not only assignments that ask their students to write to readers at their own level or above but also assignments that ask the students to write to readers below them in their organizations. This latter kind of reader is often ignored in technical writing classes, except when (and if) instructions are taught.

Finally, the importance of preparing students to address a variety of kinds of readers suggests that classes that mix students from a variety of departments may be preferable to classes in which all the students are from a single department. In classes that mix students from a variety of majors, the students will probably be much better at helping one another with reactions and advice in situations where they are trying to address audiences who know less than the writer does about the writer's specialty. (I do not mean, however, that other students should be identified as the intended readers of the assignments, only that students from other departments may be best able to take the point of view of a reader less knowledgeable than the writer.) Students, it follows, should realize that the advice offered to them by classmates in other majors can be as valuable as advice offered by students in their own major.

4. The alumni write more frequently to audiences inside their organizations than to audiences outside.

Implication. Introductory technical writing courses should emphasize intra-organizational communications.

5. The alumni write in a variety of forms. Five of the eleven forms treated in the survey are written at least "Sometimes" by at least 50% of the respondents from every department.

Implication. Introductory technical writing courses should provide students with instruction and practice in writing a variety of forms, not just one or two. It does not seem reasonable to conduct courses in which the students prepare only one assignment (such as a formal report) or courses in which all the assignments in the course are parts of a single communication.

6. The five forms that are most often written by the alumni are: memoranda, letters, step-by-step instructions, general instructions, and preprinted forms that the writer must fill out. (These are the five forms written at least "Sometimes" by at least 50% of the respondents from every department.) On the other hand, the alumni write advertising and articles for publication in professional journals much less frequently than they write many other forms of communication.

Implication. Introductory technical writing courses should include instruction and practice in preparing each of the five most frequently written communications. Although it still seems reasonable to include other forms, such as the formal report, it is not reasonable to devote so much attention to those other forms that any of the five most frequently written forms are excluded from consideration. Similarly, although there may be good pedagogical reasons for asking students in courses in their own majors to prepare term papers and other communications that resemble articles for publication in professional journals, there is no justification in an introductory technical writing course for teaching the article rather than any of the other forms that students, once they graduate, will write more frequently. The same is true for advertising.

7. The alumni frequently write on their own initiative as well as on assignment.

Implication. In introductory courses, the rhetorical situations in which students are to imagine themselves as they write should include not only situations in which the students are writing because a superior or a client has asked them to, but also situations in which they are preparing unsolicited communications.

Of course, one might reasonably ask about the extent to which the results of my survey are generalizable to graduates of other departments at other universities. Overall, it appears that it would be reasonable to accept the generalizability of these results as a working hypothesis at least until similar studies have been conducted that involve other groups of college graduates. After all, even though all the respondents to my survey earned their undergraduate degrees at the same university, they studied in a wide variety of departments — from Chemistry to Office Administration, from Manufacturing Engineering to Home Economics. Furthermore, these seven departments belong to three distinct categories: sciences, applied sciences (including engineering), and professional studies. Therefore, the similarities among the responses from the alumni are particularly striking, and the conclusions drawn from them particularly compelling for teachers and students of technical writing.

References

1. Andrews, J. D., and Koester, R. J. Communication difficulties as perceived by the accounting profession and professors of accounting. *Journal of Business Communication,* 1979, *16*(2), 33–42.
2. Barnum, C., and Fischer, R. Engineering technologists as writers: results of a survey. *Technical Communication,* 1984, *31*(2), 9–11.
3. Bennett, J. C. The communication needs of business executives. *Journal of Business Communication,* 1971, *8*(3), 5–11.
4. Davis, R. M. How important is technical writing?—a survey of the opinions of

successful engineers. *Journal of Technical Writing and Communication,* 1978, *8,* 207–216.

5. Glenn, T., and Green, M. Re-evaluation and adaptation—revising a course to meet graduates' needs. *Proceedings of the 26th International Technical Communication Conference,* 1979.

6. Huegli, J. M., and Tschirgi, H. D. An investigation of communication skills application and effectiveness at the entry job level. *Journal of Business Communication.* 1974, *12*(1), 24–29.

7. Penrose, J. M. A survey of the perceived importance of business communication and other business-related abilities. *Journal of Business Communication,* 1976, *13*(2), 17–24.

8. Persing, B., Drew, M. I., Bachman, L., and Galbraith, E. The 1976 ABCA followup evaluation of the course content, classroom procedures, and quality of the basic course in college and university business communication. *ABCA Bulletin,* 1977, *39*(2), 18–24.

9. Rader, M. H., and Wunsch, A. P. A survey of communication practices of business school graduates by job category and undergraduate major. *Journal of Business Communication,* 1980, *17*(4), 33–41.

10. Skelton, T. A survey of on-the-job writing performed by graduates of community college technical and occupational programs. In T. M. Sawyer (Ed.), *Technical and Professional Communication.* Ann Arbor: Professional Communication Press, 1977.

11. Stine, D., and Skarzenski, D. Priorities for the business communication classroom: a survey of business and academe. *Journal of Business Communication,* 1979, *16*(3), 15–30.

12. Storms, C. Gilbert. What business graduates say about the writing they do at work: implications for the business communication course. *ABCA Bulletin,* 1983, 13–18.

13. Swenson, D. H. Relative importance of business communication skills for the next ten years. *Journal of Business Communication,* 1979, *17*(2), 41–49.

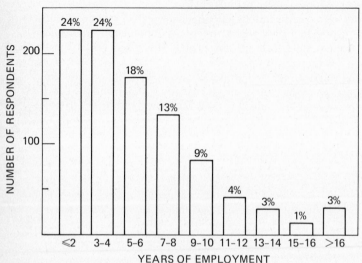

Figure 1 Years of Employment: Results for Entire Sample Considered as a Single Group. (Copyright 1980 Paul V. Anderson)

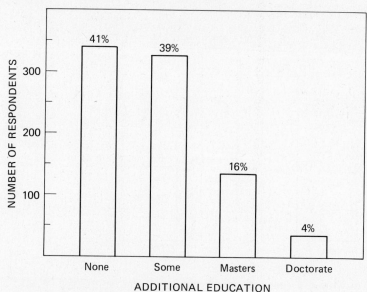

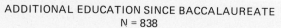

Figure 2 Additional Education Since Receiving Baccalaureate: Results for Entire Sample Considered as a Single Group. (Copyright 1980 Paul V. Anderson)

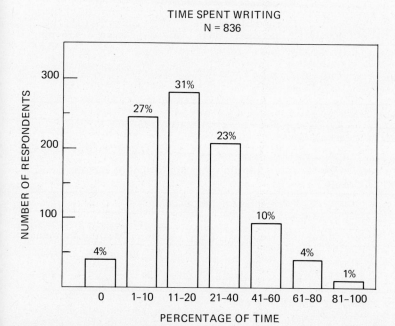

Figure 3 Time Spent Writing: Results for Entire Sample Considered as a Single Group. (Copyright 1980 Paul V. Anderson)

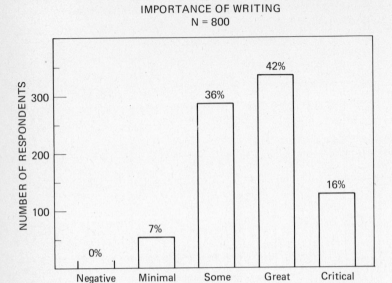

Figure 4 Importance of Writing: Results for Entire Sample Considered as a Single Group. (Copyright 1980 Paul V. Anderson)

TABLE 1 Reader's Knowledge of Writer's Specialty: Results for Entire Sample Considered as a Single Group

Level of Knowledge	N	Never	Rarely	Sometimes	Often	Always	Mean[a]	Grouping[b]
				Percentage of Respondents				
Know Less	794	3	8	31	52	6	2.51	}
Know Same	794	2	9	35	45	9	2.48	}
Know More	794	5	22	35	31	7	2.13	}
Unfamiliar[c]	794	10	29	32	25	3	1.83	}

[a] Responses coded on a five-point scale (0 = Never, 4 = Always).
[b] Brackets enclose means that are not significantly different at the $\alpha = .05$ level of significance.
[c] Row does not add to 100% because of rounding.

TABLE 2 Reader's Level in Writer's Organization: Results for Entire Sample Considered as a Single Group

Level of Organization	N	Never	Rarely	Sometimes	Often	Always	Mean[a]	Grouping[b]
				Percentage of Respondents				
Higher	793	3	6	26	47	18	2.72	}
Same	793	4	9	39	41	7	2.37	}
Lower[c]	797	10	20	35	28	6	2.00	}

[a] Responses coded on a five-point scale (0 = Never, 4 = Always).
[b] Brackets enclose means that are not significantly different at the $\alpha = .05$ level of significance. (In this table, every mean is different than the other two.)
[c] Row does not add to 100% because of rounding.

TABLE 3 Readers Outside Writer's Organization: Results for Entire Sample Considered as a Single Group

| Kind of Reader | N | Percentage of Respondents | | | | | Mean[a] | Grouping[b] |
		Never	Rarely	Sometimes	Often	Always		
Customers	785	37	19	17	20	7	1.41	}
Vendors[c]	782	42	29	19	10	1	0.99	}
Public	794	55	23	12	7	3	0.80	}
Government Officials[c]	793	59	22	13	3	2	0.66	}

[a] Responses coded on a five-point scale (0 = Never, 4 = Always).
[b] Brackets enclose means that are not significantly different at the $\alpha = .05$ level of significance. (In this table, every mean is significantly different from the other three.)
[c] Row does not add to 100% because of rounding.

TABLE 4 How Often Respondents Prepare Eleven Forms of Written Communication: Results for Entire Sample Considered as a Single Group

| Form | N | Percentage of Respondents | | | | | Mean[a] | Grouping[b] |
		Never	Rarely	Sometimes	Often	Always		
Memoranda	793	5	9	23	53	10	2.55	}
Letters	800	5	11	30	46	8	2.42	}
Step-by-Step Instructions	796	7	13	31	44	5	2.28	}
General Instructions	794	11	14	34	38	3	2.09	}
Preprinted Forms (to be filled out by respondent)	793	12	21	33	26	8	1.97	}
Proposals for Funding or Approval of Projects	795	24	18	30	24	4	1.64	}
Formal Reports (with title page or special cover sheet)	793	22	24	33	19	2	1.55	
Minutes of Meetings or Conversations	795	26	27	28	17	2	1.42	}
Scripts for Speeches or Presentations	795	31	27	30	11	1	1.25	}
Advertising	785	70	14	10	5	1	0.54	}
Articles for Professional Journals[c]	791	68	19	9	3	0	0.48	}

[a] Responses coded on a five-point scale (0 = Never, 4 = Always).
[b] Brackets enclose means that are not significantly different at the $\alpha = .05$ level of significance.
[c] Row does not add to 100% because of rounding.

TABLE 5   How Often Respondents Write for Each of Two Reasons: Results for Entire Sample Considered as a Single Group

				Percentage of Respondents				
Reason	N	Never	Rarely	Sometimes	Often	Always	Mean[a]	Grouping[b]
Own Initiative[c]	797	1	8	31	54	5	2.55	⎫
Someone Else's Request[c]	793	1	7	38	47	6	2.49	⎬

[a] Responses coded on a five-point scale (0 = Never, 4 = Always).
[b] Brackets enclose means that are not significantly different at the $\alpha = .05$ level of significance. (In this table, there is not a significant difference between the means.)
[c] Row does not add to 100% because of rounding.

TABLE 6   Importance of Writing: Comparison of Results from Seven Departments

				Percentage of Respondents				
Department	N	Negative	Minimal	Some	Great	Critical	Mean[a]	Grouping[b]
Paper Science and Engineering	92	0	2	27	51	20	2.88	
Chemistry	48	0	6	27	42	25	2.85	
Manufacturing Engineering	139	0	4	30	47	19	2.80	
Office Administration	88	0	6	27	52	15	2.76	
Systems Analysis	266	0	5	38	42	15	2.65	
Home Economics	149	0	10	46	31	13	2.46	
Zoology[c]	58	0	16	43	28	14	2.40	

[a] Responses coded on a five-point scale (0 = Never, 4 = Critical).
[b] Brackets enclose means that are not significantly different at the $\alpha = .05$ level of significance.
[c] Row does not add to 100% because of rounding.

TABLE 7 Reader's Knowledge of Writer's Specialty: Comparison of Results from Seven Departments

				Department			
	Chemistry	Home Economics	Manufacturing Engineering	Office Administration	Paper Science and Engineering	Systems Analysis	Zoology
Level of Knowledge	Same $\overline{X}$ = 2.63	Less $\overline{X}$ = 2.35	Less $\overline{X}$ = 2.61	Same $\overline{X}$ = 2.34	Less $\overline{X}$ = 2.66	Less $\overline{X}$ = 2.64	Less $\overline{X}$ = 2.50
	Less $\overline{X}$ = 2.44	Same $\overline{X}$ = 2.34	Same $\overline{X}$ = 2.56	Less $\overline{X}$ = 2.17	Same $\overline{X}$ = 2.60	Same $\overline{X}$ = 2.52	Same $\overline{X}$ = 2.43
	More $\overline{X}$ = 2.29	More $\overline{X}$ = 1.90	More $\overline{X}$ = 2.22	More $\overline{X}$ = 2.09	More $\overline{X}$ = 2.41	More $\overline{X}$ = 2.06	More $\overline{X}$ = 2.22
	Unfamiliar $\overline{X}$ = 1.74	Unfamiliar $\overline{X}$ = 1.72	Unfamiliar $\overline{X}$ = 1.94	Unfamiliar $\overline{X}$ = 1.71	Unfamiliar $\overline{X}$ = 1.70	Unfamiliar $\overline{X}$ = 1.97	Unfamiliar $\overline{X}$ = 1.67
Results of Analysis of Variance (Randomized Complete Block)	$F(3.140)$ = 9.23	$F(3.440)$ = 19.06	$F(3.416)$ = 23.58	$F(3.246)$ = 5.86	$F(3.273)$ = 26.69	$F(3.672)$ = 38.25	$F(3.171)$ = 10.73
	p = .0001	p = .0001	p = .0001	p = .0008	p = .0001	p = .0001	p = .0001

Note: Brackets enclose means that are not significantly different at the α = .05 level of significance. For this analysis, the responses were coded on a five-point scale (0 = Never, 4 = Always).

TABLE 8 Reader's Level in Writer's Organization: Comparison of Results from Seven Departments

	Department						
	Chemistry	Home Economics	Manufacturing Engineering	Office Administration	Paper Science and Engineering	Systems Analysis	Zoology
Level of Knowledge	Higher $\overline{X}$ = 2.83 Same $\overline{X}$ = 2.46 Lower $\overline{X}$ = 2.13	Higher $\overline{X}$ = 2.44 Same $\overline{X}$ = 2.19 Lower $\overline{X}$ = 2.09	Higher $\overline{X}$ = 2.73 Same $\overline{X}$ = 2.41 Lower $\overline{X}$ = 2.04	Higher $\overline{X}$ = 2.86 Same $\overline{X}$ = 2.10 Lower $\overline{X}$ = 1.49	Higher $\overline{X}$ = 3.02 Same $\overline{X}$ = 2.59 Lower $\overline{X}$ = 2.03	Higher $\overline{X}$ = 2.78 Same $\overline{X}$ = 2.46 Lower $\overline{X}$ = 2.02	Higher $\overline{X}$ = 2.39 Same $\overline{X}$ = 2.34 Lower $\overline{X}$ = 2.12
Results of Analysis of Variance (Randomized Complete Block)	$F_{(2.93)}$ = 5.85 p = .004	$F_{(2.295)}$ = 6.75 p = .001	$F_{(2.276)}$ = 26.24 p = .0001	$F_{(2.167)}$ = 40.93 p = .0001	$F_{(2.181)}$ = 49.60 p = .0001	$F_{(2.445)}$ = 51.33 p = .0001	$F_{(2.113)}$ = 1.63 p = .2

Note: Brackets enclose means that are not significantly different at the α = .05 level of significance. For this analysis, the responses were coded on a five-point scale (0 = Never, 4 = Always).

TABLE 9 Readers Outside Writer's Organization: Comparison of Results from Seven Departments

	Department						
Kind of Reader	Chemistry	Home Economics	Manufacturing Engineering	Office Administration	Paper Science and Engineering	Systems Analysis	Zoology
	Government $\overline{X}$ = 1.11	Public $\overline{X}$ = 1.56	Customers $\overline{X}$ = 1.59	Customers $\overline{X}$ = 1.39	Customers $\overline{X}$ = 1.48	Customers $\overline{X}$ = 1.33	Customers $\overline{X}$ = 1.13
	Customers $\overline{X}$ = 1.04	Customers $\overline{X}$ = 1.51	Vendors $\overline{X}$ = 1.33	Public $\overline{X}$ = 1.38	Vendors $\overline{X}$ = 1.17	Vendors $\overline{X}$ = 0.69	Public $\overline{X}$ = 1.09
	Vendors $\overline{X}$ = 1.02	Vendors $\overline{X}$ = 1.15	Government $\overline{X}$ = 0.56	Government $\overline{X}$ = 0.88	Government $\overline{X}$ = 0.61	Government $\overline{X}$ = 0.47	Government $\overline{X}$ = 0.89
	Public $\overline{X}$ = 0.89	Government $\overline{X}$ = 0.69	Public $\overline{X}$ = 0.53	Vendors $\overline{X}$ = 0.87	Public $\overline{X}$ = 0.40	Public $\overline{X}$ = 0.32	Vendors $\overline{X}$ = 0.73
Results of Analysis of Variance (Randomized Complete Block)	$F_{(3,133)}$ = 0.53 p = .7	$F_{(3,416)}$ = 48.02 p = .0001	$F_{(3,435)}$ = 26.21 p = .0001	$F_{(3,247)}$ = 5.86 p = .0008	$F_{(3,273)}$ = 30.48 p = .0001	$F_{(3,666)}$ = 69.19 p = .0001	$F_{(3,163)}$ = 2.10 p = .1

Note: Brackets enclose means that are not significantly different at the α = .05 level of significance. For this analysis, the responses were coded on a five-point scale (0 = Never, 4 = Always).

TABLE 10 How Often Respondents Prepare Eleven Forms of Written Communication: Analyses of Variance Comparing Responses from Seven Departments

Form	df	F	PR > F
Memoranda	6,786	7.37	.0001
Letters	6,793	9.79	.0001
Step-by-Step Instructions	6,789	3.98	.0006
General Instructions	6,787	1.73	.1120
Preprinted Forms (to be filled out by respondent)	6,786	5.39	.0001
Proposals for Funding or Approval of Projects	6,788	8.81	.0001
Formal Reports (with title page or cover sheet)	6,786	4.90	.0001
Minutes of Meetings or Conversations	6,788	3.63	.0015
Scripts for Speeches or Presentations	6,788	3.24	.0037
Advertising	6,778	23.84	.0001
Articles for Professional Journals	6,784	12.56	.0001

TABLE 11 T-tests Comparing Responses to Questions about the Respondent's Reasons for Writing

Department	df	Mean Someone Else's Request	Mean Own Initiative	t	PR > |T|
Chemistry	47	2.46	2.60	−0.82	.418
Manufacturing Engineering	136	2.45	2.66	−2.69	.008
Home Economics	146	2.35	2.70	−3.51	.001
Office Administration	85	2.57	2.40	0.82	.415
Pulp and Paper Science	90	2.55	2.60	−0.39	.697
Systems Analysis	223	2.60	2.42	2.17	.252
Zoology	56	2.35	2.55	−1.16	.031

Using Personal Computers at the College Level[5]
David W. Bray

In less than a year, the number of microcomputers at Clarkson University has increased from less than 50 to over 1700. This dramatic increase is due to Clarkson's move to incorporate the personal computer into every facet of its academic programs. Early in October 1982, the college announced that in fall 1983 all freshman students, regardless of their major, would be provided with a personal computer. Shortly thereafter, the college began supplying Zenith Data Systems Z-100 computers to faculty who wanted to prepare freshman and sophomore courses using the personal computer. Almost all faculty who had the opportu-

This article by David W. Bray about Clarkson University's plan to provide every entering freshman with a personal computer is noteworthy for the care the author has taken to set forth the standards for selection of a personal computer for students. Note also how clearly the author has introduced his subject and how he closes his introduction with a statement of purpose that forecasts precisely what his article will cover.

[5] *Computer 17* (April, 1984), 36–37.

nity to obtain a computer took advantage of it. Although the idea behind providing the computers was to modify the curriculum for the following years, many began to use the computer immediately.

The full impact of Clarkson's plan will, of course, not be realized for several years. Even the immediate impact was not really felt until the first day of orientation when the freshmen actually picked up their computers and took them to their dormitories. On August 25, 1983, approximately 850 freshman students arrived on campus, and the first day over 700 of them picked up a Z-100 computer, a monitor, and six software packages. The excitement of the students and the sheer number of computers and software packages being transported showed college faculty and administrators that a very new experience in education was occurring. A tour of the freshman dormitories early that evening revealed excited students busily learning their computers. The residence hall staff noted that they had never before seen such quiet dormitories on the first evening of freshman orientation.

The purpose of this article is to describe this plan, discuss some thoughts behind the program's establishment, and speculate as to how it may change the educational process at the college level.

THE CLARKSON STUDENT COMPUTER PLAN

Implementation
By 1986, every undergraduate student and most faculty and graduate students should have a personal computer. These computers will be interconnected by way of a campuswide network that will include the college mainframe computers. Key points of the plan are

Ownership of the student computer resides with the college.
Should the student withdraw from the college without completing degree requirements, the computer will remain with the college and be reassigned to a transfer student.
Upon graduation, the student has the option to purchase the computer. The declared value of the computer at that date will be $200. Upon payment of this sum, ownership of the computer will be transferred to the student.

The student is financially responsible for keeping the computer in good condition after the 90-day warranty has expired. Each student must make a $200 maintenance deposit to be used for any required servicing. The college insures the computers against fire and theft.

A tuition surcharge of $200 per semester is added for those students provided with a computer. This surcharge covers, in part, the additional cost of the enhanced educational program.

This implementation plan has a number of benefits to both the college and the students, some of which are

Because the personal computer is introduced over four years with each new freshman class, the faculty—teaching juniors and seniors—has enough lead time to modify and develop courses for the most effective use of the computer.

Providing the student with the computer, instead of requiring the student to purchase the computer, removes the burden of a rather substantial payment for the computer on top of the initial tuition and room charges. Moreover, a student leaving the college after a short time will not have incurred the expense of purchasing a computer.

The graduating student will have a computer and much self-developed software to help bridge the gap from college to business and industry.

The computer can be a potential means of communication and contact with alumni.

The college is free to select a new model computer as technology advances, since each year new computers are needed for the entering freshman class. Thus, the computing facilities will not become outmoded, often the case when mainframes are involved.

Computer Selection

The introduction of the personal computer into the educational process at Clarkson had been under consideration for several years. During the spring 1982 semester, a group of faculty and students met twice weekly to consider the requirements and specifications of a personal computer and an associated campuswide network. The primary recommendations of this group with regard to the personal computer were

The computer should contain a 16-bit processor. With 16-bit processors coming into the market in force, it would not be wise to begin a new program with a computer operating on an eight-bit processor, even though there is a large amount of software available for eight-bit processor computers.

The computer should be compatible with the IBM personal computer, since a large number of companies are producing software for PC-compatible computers. Universities are also interested in the IBM PC. If this interest was strong enough, the PC-compatible computers might become the foundation for a standard in education. The benefits of such a standard would be enormous; colleges could exchange software and courseware* freely without the current problems of computer incompatibility.

The computer should have a high-quality CRT display capable of at least 80 characters per display line, and resolution high enough for engineering and scientific graphic applications. The 80-character display would allow the computer to provide professional word processing.

The computer should have a professional keyboard to complement the high-quality display for word processing.

The computer should have bit-mapped graphics suitable for engineering and scientific applications. If possible, the graphics would be upgradable to color for special applications.

The computer should have communications ports general enough to support network interconnection, and local printers if a student desires to attach one.

The computer should have software support that includes at least Fortran, Pascal, Basic, Cobol, word processing, and spread-sheet accounting.

After the study, a set of guidelines was prepared and sent to the major personal computer manufacturers believed to have, or be developing, computers that would meet these requirements. These guidelines suggested that the program might be implemented as early as fall 1984. A wide variety

* Instructional packages that contain software and materials such as text booklets, workbooks, and cassettes are often called courseware.

of responses were obtained. The first manufacturer to respond indicated that they had a computer "on the drawing board" that exceeded these specifications and suggested that the timetable could be pushed up to fall 1983. Encouraged by this response, Clarkson decided to attempt implementation of the plan at the suggested time.

In early summer 1982, a formal set of specifications for the computer was prepared and sent to those who received the original guidelines, as well as others who had heard of the project. Bids were received from five companies. From these, the Zenith Data System Z-100 was selected for its price and performance combination, which exceeded the specifications in almost every way. Two features were considered especially important. The first is that the model has a second processor, the 8085, allowing current eight-bit CP/M software to be used if desired. With this additional processor, the concern that a 16-bit computer would limit use of available software was eliminated. The second feature of importance is the inclusion of four available S-100 industry-standard bus slots. With these, the computer could be expanded for use in the laboratory, and software compatibility between the student's computer and the laboratory computers would be possible.

The model Z-100 includes 192K bytes of program memory; high-resolution, bit-mapped graphics with 640×225-pixel resolution; one 5¼-inch, 320K-byte floppy disk drive; three communications ports; and four available S-100 slots. Each student is supplied with the Z-DOS (a derivative of MS-DOS) for use on the 16-bit 8088 processor, the CP/M operating system for use on the eight-bit 8085 processor, Fortran, Pascal, Z Basic (PC compatible), Multiplan, Galahad (a Clarkson-developed word processor), and Cobol for those who need it.

Suggestions for Writing

Selecting a Subject

The following list represents a wide range of possible subjects for practicing interpretative writing, and will probably suggest many additional subjects. In selecting a subject, you will need to make several decisions. Let's consider the first subject in the list as an example of what will need to be thought about. In doing so, let's assume that a decision has already been made as to the approximate length of the completed interpretation.

Your first step, then, would be to decide whether to evaluate only one lawn mower, or to compare several. If you evaluate only one, your second step should probably be to select your imaginary reader. With these two decisions made, you are ready to start obtaining any necessary information you don't have and organizing your interpretation.

Should you decide to compare several lawn mowers, you would presumably compare different brands of the same type, or different types. You might, for example, compare all the brands you could find of self-propelling reel mowers. Or you could compare any reasonable combination of types of mowers. Self-propelling reel mowers and small riding mowers might make a reasonable combination. With this combination, you would not be concerned with particular brands, except perhaps incidentally, but rather with the characteristics of the two types.

As always, however, what is reasonable depends on your decision about your reader. The combination just mentioned would surely not be reasonable for a reader who was an active young person with only 500 square feet of lawn to mow, but it would be entirely reasonable for an elderly person with 5000 square feet to mow.

The choice of reader also, as you will remember, has much to do with the length of the finished interpretation. If you picked a reader who wanted to know about the advantages and disadvantages of all available types and brands of lawn mowers, and who wanted technical details, you would be confronted with an impossibly long project.

With such principles as the foregoing in mind, then, choose one of the following subjects for an interpretation. In your paper, specify the kind of reader you are addressing.

Lawn mowers
Wristwatches
Microwave ovens
Small astronomical telescopes
Shrubs for a hedge or windbreak
Personal computers
Automobile brakes
Trees for lumber
Electrical test sets for the amateur
Small sailboats for the combined purpose of family use and racing
Feed for a particular type of stock animal in a specified locality
Outboard motors
Reels for fishing rods
Preservation and storage of hay
Trail bikes
Gas engines for model airplanes
Garden tractors

Finishes for homemade furniture, excluding paint and enamel
35-mm cameras
TV dinners
Small calculators

Writing Problem Number 1

Write an interpretation of the following information about 19-in. portable color TV sets. To keep your report to a reasonable length, assume that — with the exception of the factors noted below — the sets are all of the same value and quality. All are equipped to work with home computers, video cassette recorders, disc players, games, and the like. All are "cable ready." Take as your objective a recommendation as to which set would be the best choice for a buyer.

Let's consider four different brands: A, B, C, and D. Brand A costs $420, B costs $475, C costs $500, and D costs $400. All have a one-year warranty on the picture tube. On parts, Brands A, B, and D have a one-year warranty, and Brand C a 90-day warranty. On labor, Brands A, B, and C have a 90-day warranty, and Brand D a one-year warranty. The sets differ in fringe reception on the VHF and UHF bands. Brand A is above average on both VHF and UHF frequencies, C and D are good on both, but not outstanding. Brand B is exceptionally good at fringe reception of VHF. All were very good at adjacent-channel rejection except Brand D, which was very poor. All have acceptable automatic color-control circuits except Brand C, which was somewhat deficient.

In planning your report, be sure to consider the advantages of presenting data in tabular form.

Writing Problem Number 2

Assume that you have been asked by a younger friend what field you would advise him or her to choose as a major in a university. Select three different majors about which you are reasonably well informed and write an interpretation of them that will answer your friend's question.

Writing Problem Number 3

Assume that you have been asked by a friend for advice as to which of two personal computer systems, both familiar to you, would be the better for a college student who has a clear need for word processing capability but who is unsure how else the computer system might be useful for general college and personal use. Write an interpretation of relevant facts about the two systems, and choose one of them to recommend.

Introductions, Transitions, and Conclusions

The duties of a writer are somewhat like those of a highway builder. A highway builder must know how to construct a good road, a road that will carry weight. A writer must know how to make a sentence that will carry meaning. Again, a highway builder must know how to lay out a system of roads so that the traveler can go from one place to another easily and quickly. For the writer, this is organization. Finally, the highway builder must know how to devise and locate signs that will keep the traveler properly oriented. The writer's comparable duty is to write introductions, transitions, and conclusions. The purpose of this section is to discuss these last three elements.

In chapters 10, 11, and 12 we offer examples of writing in which there are clear introductions, transitions, and conclusions, and we make suggestions about how to do such writing. We believe you will agree that a route marker in a report, as on the highway, is a good thing.

10

Introductions

The introduction of a technical report has several very definite functions. In fact, it is scarcely an exaggeration to say that the word "introduction" has a special meaning in technical writing, and that you might find it helpful to forget whatever meanings you have associated with the term. The introduction is, of course, the first portion of the text. It may or may not be preceded by a title page, letter of transmittal, preface, table of contents, list of illustrations, and abstract. Whether all or any of these elements are present, however, the introduction should be a complete and self-sufficient unit.

The primary purposes of an introduction to a technical report are to state the subject, the purpose, the scope, and the plan of development of the report. In addition, it is sometimes necessary to explain the value or importance of the subject. Often it is desirable to summarize principal findings or conclusions.

The organization of the introduction and the degree to which any of its parts is developed depend upon circumstances. It should not be supposed that a good introduction is necessarily a long one; sometimes only a sentence or two is sufficient. The organization is affected particularly by the need of stating a key idea in the opening sentence.

We'll discuss the four primary functions of an introduction in the order stated above, concluding with some comments on the problems of initial emphasis and of the statement of the importance of the subject.

As noted above, introductions to technical reports often include a summary of the major conclusions or results that are presented in the body of the report. Since the presence of such an introductory summary does not affect the fundamental character or functions of an introduction, no consideration is given to it in this chapter. A discussion of the introductory summary may be found in Chapter 4.

Statement of the Subject

At the very beginning, the reader should be given a clear understanding of what the exact subject of a report is. How this information is best presented depends, as usual, upon what the reader already knows. To some extent the title of a report is, or can be, a statement of the subject; but almost without exception the subject should be stated again in the introduction, and the title should never be used as an antecedent for a pronoun in the introduction. That is, if the title were "The Arc Welding Process," you should not begin the introduction with the words, "This process. . . ."

The effectiveness of the title of a technical report depends upon making it as informative as possible while still keeping it reasonably short. Titles that are merely ornamental, or even misleading, are a source of constant annoyance, as you have no doubt discovered when using periodical guides. Try to think of the title as it will appear to the reader. We recall a university commencement in which one of the doctoral dissertations listed on the program was entitled, "The Life of an Excited Atom." From remarks we overheard in the audience, it became clear that not everyone understood this to be both a serious and a witty title.

The statement of the subject in the introduction itself may involve one or more of the following three problems: definition of the subject, theory associated with the subject, and history of the subject.

It may prove necessary to define the subject and the terms used in stating the subject. For example, in a discussion intended for an uninformed reader, entitled "Hydroponics — Gardens without Soil," we would want to explain both what the word "hydroponics" means and what gardens without soil are. We might write, somewhere in the introduction:

> The word "hydroponics" is simply a name given to the process of growing plants in a liquid solution instead of in soil. Our chief interest will lie in the commercial application of this principle — that is, in "gardens" in which tanks filled with gravel and a mineral solution replace soil as the source of food for the growing plants.

In this illustration it is primarily the concept of soilless gardening that must be conveyed to the reader; but since that concept is expressed by the unfamiliar term "hydroponics," it is necessary to be certain that the relationship between the term and the subject is clear. In short, the writer should give special attention to making clear both the subject itself and any unfamiliar terms associated with it. If the subject is already familiar in some degree to the reader, the writer should adapt the statement of the subject accordingly.

Sometimes, however, even a well-written formal definition is insufficient, and it may become necessary to give the reader some background information. This background information is usually either theoretical or historical, or both.

For large land surveys, it is occasionally desirable to stop the survey and reestablish the true north direction in order to localize instrument errors. The new determination of true north may be accomplished by sighting on the pole star. For a student of surveying with no knowledge of taking astronomical "sights," a report on the procedures of establishing true north might well begin with a section on the theory of such an operation. Only through a comprehension of the theory could the operation itself (the principal subject) be fully understood. If such a section on theory were short, it could be included in the introduction; if it were long, it might better go into a section by itself, immediately following the introduction. In either case, the writer should remember that the objective is to make clear what the subject of the report is.

The purpose of discussing the history of a subject in a technical report is much the same as that of discussing the theory. It gives the reader an understanding of the total situation of which the particular subject is a part. For instance, in a report on the methods of manufacture of the "buckets" on a jet-engine turbine, a brief history of the development of the jet-engine turbine might help a great deal in showing why the buckets are now made as they are. A warning is in order here, however. Don't allow yourself to start discussing history simply because you can't think of any other way to get started. Ask yourself if the history is clearly contributing toward the basic purpose of the report. If it is, good; if not, out with it.

Like the theory, the history can go either into the introduction or into a section by itself. In fact, theory and history are often combined—which is perfectly all right and natural.

Statement of Purpose

It is imperative that the reader understand the purpose of a report. And remember that we are concerned here with the purpose of the report, not of the subject. The purpose of a drill press is to drill holes; but the purpose of a report about a drill press might be to discuss the most efficient rate of penetration of the drill bit.

There can seldom be any objection to saying simply, "The purpose of this report is. . . ." Frequently the statement of both the subject and the purpose of a report can be accomplished in the same sentence, often the first sentence in the report. If the statement of the scope of the report and of the plan of development of the report can be included in this same sentence without awkwardness or lack of clarity, there is no reason for not putting them there. The fundamental requirement of a good introduction is that it perform the four basic functions; there can be no rules about how they are accomplished, nor can there be rules for a fixed order of these functions.

Statement of Scope

The term "scope" refers to the limits of a subject. The problem in the introduction is to explain what the limits are so that readers will expect neither more nor less than they find in the report.

Limits may be stated in several ways. One way is concerned with the amount of detail; a report may be described as a general survey of a subject or as a detailed study. Another way has to do with how great a range of subject matter is included. For example, a report on standardizing the location of the pilot's controls in aircraft might include all types of aircraft or only one type, like multiengine aircraft. The reader must be told what the range is. A third way is to note the point of view from which the report will be written. There is a great deal of difference, for instance, between announcing that a report is on the subject of the plumbing in a certain hotel, and announcing that the subject is the plumbing in this hotel from the point of view of a sanitation engineer.

These ideas may be of some value in helping you think how to say what the scope of a given report is, but the basic idea to remember is simply that you must keep defining and qualifying your subject until it is certain that the reader will know what to expect.

Statement of Plan of Development

The statement of the plan of development of a report is simply a detailed application of the slogan, "First you tell your readers what you're going to tell them. . . ." It is a simple idea, easy to carry out, and unquestionably one of the most important elements in the introduction. The phrasing may be straightforward and formal: "This report will be divided into five major parts: (1) ——————, (2) ——————, ... " Or it may be more "literary": "The most important aspects of this subject are ——————, ——————, ——————, ... " The manner should suit the situation. Usually the statement of the plan of development comes at or near the end of the introduction.

Other Problems

Two other problems that should be mentioned are the need for a proper initial emphasis and the occasional desirability of an explanation of the importance of a subject.

The first few statements made in an introduction are especially critical because it is on this very limited evidence that readers form an impression of the report as a whole. Their impression as to the content and purpose of the report should be accurate. If they later find that their first impression was wrong, confusion and irritation will be the probable result.

Ask yourself how much your readers already know about the subject. Have they requested this particular report, or will it reach them unannounced? Is it about a subject they are interested in, or a project of which they approve? Or are some of them likely to be, at the outset, indifferent or even hostile? Usually there can be no objection to some variation of the "The purpose of this report . . . " beginning, but it would be a mistake to suppose that this is always true. Consider the following opening sentence:

> When it became apparent, in the fall of 19 ___, that the water supply of
> _____ City would soon be inadequate to support the industry now
> located in the city, the City Council requested the firm of Smith and Rowe to
> prepare a preliminary report on the outlook for the immediate future, together with
> tentative recommendations of measures to be taken.

The initial emphasis here falls upon the urgent need for action. In comparison, an opening consisting of a statement of purpose would be less effective; and an opening consisting of the first sentences of a history of the water supply problem might be quite misleading.

The importance of a failing water supply needs no explanation. But suppose the water supply was adequate and a report was being written to show that steps should be taken to prevent a probable shortage at the end of another ten years. The writer would face the quite different problem of needing to prove that a merely probable event of ten years in the future was of immediate practical interest. The fundamental principle is to analyze your readers and estimate their needs and attitudes. The fourth introduction quoted at the end of this chapter illustrates an extended comment on the importance or value of a subject.

Summary

1. The major functions of an introduction are to state the exact subject of the report, its exact purpose, its scope, and its plan of development.
2. The statement of the subject is primarily a problem in definition, but may require extended discussion of background material, particularly of history or theory, or both. On the other hand, for an informed reader, the subject need only be named.
3. The statement of purpose is often combined with the statement of subject.
4. The statement of the scope of the report may be conveniently considered in three aspects: the "range" of the subject matter, the detail in which the subject is to be discussed, and the point of view from which the subject is to be discussed.
5. The statement of the plan of development presents no difficulties, but is extremely important; it normally appears at or near the end of the introduction.
6. The organization of the whole introduction is affected by the selection of the proper initial emphasis.
7. Sometimes it is desirable to explain the importance of the subject.

Illustrative Material

The following pages contain seven examples of introductions. The third and fourth are examples of students' work while the others, except for the sixth, were written on the job. The sixth is the introduction to an article which appeared in *The Scientific American* (March 1975).

A Method of Calculating Internal Stresses in Drying Wood[1]
(Forest Products Laboratory Report No. 2133)

INTRODUCTION

As wood dries, it is strained by a complex pattern of internal stresses that develop as a result of restraints characteristic of normal shrinkage. Such stresses are found in all lumber during normal drying and are responsible for most of the defects associated with the drying process.

Although such stresses have been known and recognized for many years, no suitable method of calculating their magnitude and distribution has been available. As a result, the development of schedules for drying wood without excessive losses due to drying defects and without unduly prolonging the drying process has been almost entirely by empirical procedures.

In recent years, investigations of the stress behavior and perpendicular-to-grain mechanical properties of drying wood have laid the groundwork for a more fundamental approach to the problem of improved wood drying. However, effective use of such data requires a method of evaluating drying stresses at any point on the cross section of a drying board. Such a method has not been available up to this time.

This report describes a method for calculating the perpendicular-to-grain stresses associated with the drying process and illustrates the application of the method to one condition of wood drying.

This introduction comes first in this group of examples because of the question it immediately raises about the initial emphasis. The question is this: Can you imagine a reader who would need the information in this report but who would be ignorant of the facts in the first paragraph? Before trying to answer this question, let's ask another. If the author wants to prevent *excessive* losses, as indicated in the second paragraph, why emphasize *normal* conditions at the beginning? Shouldn't the emphasis be on the contribution this research has made to the elimination of excessive losses?

One possible revision is to delete the first paragraph and to reword the first sentence in the second paragraph as follows: "Although it has been known for many years that defects appear in drying lumber because internal stresses are developed, no suitable method of calculating the magnitude and distribution of these stresses has been available."

Stability Study of 220-KV. Interconnection between Philadelphia Electric Company, Public Service Electric & Gas Co. of N.J., Pennsylvania Power & Light Co.[2]

The effects of line to ground short circuits on the stability of the interconnection have been investigated by careful mathematical calculation based on the best available data as to line and system characteristics. While, on account of unavoidable differences in actual and assumed conditions, and on account of the methods by which the problem has been simplified for purposes of calculation, extreme accuracy cannot be hoped for, nevertheless most of the essential factors have been

This introduction gives the impression that it was written in haste, with no pleasure. It is certainly no pleasure to read. The second sentence would have a hard time getting by a freshman English teacher. And yet the introduction performs, at least to a limited degree, all of the four major functions of an introduction. The subject of the report is

---

[1] Courtesy Forest Products Laboratory, Madison, Wisconsin

[2] From a General Electric Company report, Engineering General Department (Schenectady, N.Y.), p. 1.

considered and evaluated, and it is therefore felt that the final results obtained are substantially correct.

The report has been divided into three main sections as follows:

  I Results
  II Basis of Study
  III Method of Calculation Employed
  IV Representative Curves and Diagrams

stated in the first sentence, and—for the technical reader for whom the report is obviously intended—the purpose is made fairly clear as well. The second sentence is concerned chiefly with scope. And the plan of the report is perfectly plain. (The little mix-up about how many main sections there are is found in the original.)

Problems of Control of Wheat Rust and White Pine Blister Rust

This report describes problems encountered in efforts to control wheat rust and white pine blister rust by eradication of one of the two hosts required by each disease. As its two hosts, wheat rust requires wheat and barberry. White pine blister rust requires white pine and either currants or gooseberries.

Both diseases have been combatted by efforts to eradicate, in a given area, one of the hosts. These efforts, however, have not always been successful because certain spores of these diseases may survive for a time independently of a host. In addition, it is not always easy to locate and get permission to destroy all the objectionable plants. This latter problem, however, lies beyond the scope of this report.

The life cycle of each disease will be described, and then the circumstances will be explained in which eradication of one of the hosts may fail to eliminate the disease.

This introduction, an example of a student's writing, illustrates several of the elements of a good performance. The initial emphasis is appropriate, the subject and purpose are clear. There is enough background information to indicate the significance of the subject; the scope is limited; and there is a good statement of the plan of development. One criticism that might be made, however, is that a little more does need to be said about scope. Will the description of the life cycle of the diseases be presented in a way suited to the interests and knowledge of the average gardener, or will it be directed to—say—an advanced student of botany? One or two more sentences on this point would be helpful.

Report on the Direct Hydrogenation and Liquefaction of Coal[3]

1. INTRODUCTION

In recent years much time, money, and energy have been spent on the problem of obtaining synthetic liquid fuels. In European countries, where domestic supplies of crude oil are relatively very low, the production of synthetic liquid fuels has become imperative to their self-sufficiency. Today, even in the United States, where reserves of crude petroleum are seemingly very great, scientists are devoting great emphasis to the production of liquid fuels from other sources.

The initial emphasis of this introduction is upon the importance of the subject. The general subject is stated in the first sentence. The first three paragraphs of the report are devoted to the historical background of the subject.

[3] From a report written by Mr. Don R. Moore while a student at The University of Texas, and reprinted here with his permission.

Because of its great abundance and accessibility, one of the principal organic raw materials which has received consideration in recent years as an important source of synthetic liquid fuels has been coal. The known supply of coal in the world today is tremendously great compared to the known reserves of crude petroleum. Although new discoveries of petroleum have boosted supplies, there is little doubt that the supply of petroleum in the United States will run short many, many years — even centuries — before coal supplies are exhausted. Scientific estimates have placed the life of petroleum reserves in the United States at between ten and fifty years while estimates have placed the life of coal reserves well in excess of one thousand years.[1]

This paragraph and part of the next are concerned with theoretical background.

It is because of this possibility of an impending shortage of crude petroleum that the conversion of coal to oil by hydrogenation processes has become so important. As yet, the production of fuel oils from coal is not economically feasible in the United States. Gasoline produced from the direct hydrogenation of coal would cost 22.6 cents per gallon if produced by a plant which had a daily production of 3,000 barrels or between 15 and 16 cents per gallon if produced by a plant which had a daily production of 30,000 barrels; the same fuel produced from crude petroleum by the common thermal cracking refinery process would cost 8.5 cents per gallon.[2] However, in the future it is believed that engineering achievements in the field of coal-hydrogenation coupled with a rise in the price of fuel oils produced from crude petroleum . . . will possibly make the production of gasoline and other motor fuels from coal-hydrogenation economically feasible.

These costs are no longer valid.

Although this conversion of coal to oil appears to be a mysterious and complicated process, it may be discovered from the discussion appearing in the second section of this report that the composition of certain bituminous coals which have been freed from ash resembles the composition of crude petroleum to a great extent.

The actual chemical conversion of coal to oil can be accomplished by either of two hydrogenation processes — the direct, or Bergius,[3] process or the indirect Fischer-Tropsch[4] process. The material presented in this report, however, will concern

Here the scope of the report is limited. The sentence beginning, "The material presented . . ." limits what we earlier called the range of the subject. In the next

only the primary reaction involved in the conversion of coal to oil by the direct hydrogenation process. This report will discuss the conversion from a chemical aspect and will not cover engineering details and difficulties involved in such a conversion by commercial-scale continuous-phase[5] processes.

sentence, point of view and detail are mentioned.

The subject is clarified by definition of terms in the footnotes.

It is the purpose of this report to discuss the mechanism and yields of the primary reaction involved in the synthesis of coal to oil by the direct hydrogenation process, the operating variables involved in the reaction, and the effect of the rank and type of different samples of coal upon the total liquefaction yields from the reaction. These topics will be discussed in the order stated.

The introduction concludes with a formal statement of purpose and plan.

[1] *Synthetic Liquid Fuels,* Hearings before a Subcommittee of the Committee on Public Lands and Surveys, United States Senate, Seventy-Eighth Congress, p. 137.

[2] *Ibid.,* p. 53.

[3] The Bergius Process (named after a German who was a pioneer in the field of coal-hydrogenation) is a process in which hydrogen is forced into the reactive intermediates formed by a thermal decomposition of the complex molecular structure of the coal.

[4] The Fischer-Tropsch Process, devised by the two German scientists, is a process in which the coal is burned to form "water-gas" which is then hydrogenated to form oils.

[5] A continuous production process in which coal is constantly fed to a liquefaction converter and in which the liquefaction yields are constantly removed for further hydrogenation.

Design, Construction, and Field Testing of the BCF Electric Shrimp-Trawl System[4]

Wilbur R. Seidel

INTRODUCTION

In 1961, we—that is, the staff at Gear Research Unit of the Exploratory Fishing and Gear Research Base, Pascagoula, Mississippi—evaluated electric systems that could be used on shrimp trawls to increase the harvest of brown and pink shrimps. Because these shrimp burrow in the ocean floor during daylight, commercial fishing has been restricted to night trawling—that is, to the period when they are not in their burrows. The aim of these preliminary evaluations was to develop a system that would force the shrimp out of their burrows during daylight. If successful, it would allow around-the-clock fishing and more efficient use of harvesting gear.

Initial trials with electric shrimp-trawl systems, though encouraging, were not satisfactory because

Here is an excellent introduction—and yet in our opinion it violates an important principle. That is, we don't think it has a good initial emphasis. We think this deficiency could be taken care of very simply. All that is needed is to start the introduction as a whole with the sentence that now stands as the first sentence of the last paragraph: "The goal of the work. . . ."

This introduction is so well written that there can be little doubt that what we have called its "deficiency" was a matter of deliberate choice and not of carelessness or lack of skill. We

[4] From *Fishery Industrial Research 4,* No. 6, U.S. Department of Interior Fish and Wildlife Service, Bureau of Commercial Fisheries.

the rates of catch during the daytime were much smaller than were those at night. However, they did show that, before a successful electric trawl could be developed, the exact electrical requirements that would cause optimum shrimp response had to be determined. In 1965, Klima studied the response of shrimp to an electric field. His data provided the background information for the needed design of an adequate electric shrimp-trawl system.

The goal of the work reported here was to design a full-scale prototype shrimp trawl that would permit a test of the commercial feasibility of electric trawling during daylight. The aim was not to build a fully engineered production-model trawl, but simply to develop one that would show whether daylight electric trawling is practical. Accordingly, the work was divided into two main parts. The first was concerned with designing and developing a full-scale electric shrimp trawl; the second with testing it under actual fishing conditions.

feel confident that the author of this report has the skill to handle the initial emphasis however he wants to.

Why did he do it this way, then? We don't know. We would say that his way gives a slower, easier pace to the introduction, and that is pleasant. But we still consider it less effective than it would be with the revision we suggest. Of course people's preferences may differ about such things.

Here's a suggestion. If you like the rather slow, indirect beginning in this introduction—fine; but before you try writing introductions in this way, practice getting the initial emphasis exactly on the most important aspect of your subject. Practice until you have proved to your instructor you can do it every time. Then you can experiment, confident that you know what you are doing.

X-Ray Emitting Double Stars[5]
Herbert Gursky and Edward P. J. van den Heuvel

Picture a celestial body no bigger than New York City that spews out energy solely in the form of X rays at a rate equal to roughly 10,000 times the total energy output of the sun. The amount of energy emitted each second by such an object would be enough to propel all 300 million of the world's automobiles an average of 70 miles per day for the next 100 billion years—10 times the estimated present age of the universe. Imagine further that this gigantic outflow of energy shows large fluctuations within a thousandth of a second, and that every few days the radiating object describes a circular path in space with a diameter of some 20 million miles.

These facts should give some idea of the bizarre properties of a newly discovered type of double-star system. Eight of these X-ray binaries have been found so far; each is thought to consist of a fantastically dense, X-ray-emitting source orbiting

This two-paragraph introduction to a fairly long article in a highly respected journal addressed to educated readers is devoted primarily to making clear what the subject is and to gaining the attention and interest of readers. Note the use of familiar comparisons in the opening paragraph and the use of concrete language. The authors do not make a formal statement of purpose (though it is implicit) nor do they state the scope of the article or their plan of development. Should they have?

[5] From the March, 1975 issue of *The Scientific American*, p. 24.

closely around a much larger normal star. Six of the new double-star systems were identified with the aid of a single artificial satellite launched in 1970 from a site in Kenya. The satellite, named *Uhuru* after the Swahili word for "freedom," was built for the National Aeronautics and Space Administration by Riccardo Giacconi and his colleagues at American Science & Engineering, Inc. *Uhuru* has made so many discoveries that it has transformed X-ray observations into one of the most active branches of modern astronomy.

Rapid Method for the Estimation of EDTA (Ethylenediaminetetraacetic Acid) in Fish Flesh and Crab Meat[6]

Herman S. Groninger and Kenneth R. Brandt

ABSTRACT

EDTA, a quality stabilizing additive, is usually applied to seafoods by spraying or dipping, and the amount of EDTA retained by the treated product must be determined by an analytical method. A titration method based on the chelation of EDTA with thorium ion was modified for use in the determination of EDTA in fish flesh and crab meat. The modified method is both simple and rapid and gave about 90-percent recovery of added EDTA from samples of fish flesh and crab meat.

INTRODUCTION

EDTA has been reported to be useful or potentially useful as an additive to seafoods to stabilize color and retard the formation of struvite (National Academy of Sciences, 1965), inhibit enzyme-catalyzed changes in flavor (Groninger and Spinelli, 1968), and inhibit the growth of bacteria (Levin, 1967).

Often, EDTA is applied to seafoods by spraying or dipping the product. The amount of EDTA actually added must then be determined by a suitable quantitative method.

A number of methods have been developed for the determination of EDTA in various materials (Belot, 1964; Brady and Gwilt, 1962; Cherney, Crafts, Hagermoser, Boule, Harbin, and Zak, 1954; Darbey, 1952; Haas and Lewis, 1967; Kratochvil and White, 1967; Lavender, Pullman, and Gold-

This example of an introduction is presented in the context of the entire report of which it is a part. Since it is directed to research chemists in the specialized area of the subject matter of the report, you may find the content not very inviting or informative— unless you happen to share their interests. If, however, you will look at the way the authors have gone about their job of writing, rather than at the technical content, we believe you will discover that this is actually a highly interesting and useful illustration of how to write a report. It is, in fact, a report in miniature, with all the parts of a long report compressed into a small

---

[6] From *Fishery Industrial Research 4*, No. 6, U.S. Department of Interior Fish and Wildlife Service, Bureau of Commercial Fisheries.

man, 1964; Malat, 1962; Vogel and Deshusses, 1962). In all of these methods, the principle of measurement is based on the chelating capacity of EDTA. In general, each method was developed for use on a specific type of product.

Efforts were made to adapt several of these methods (Brady and Gwilt, 1962; Haas and Lewis, 1967; Vogel and Deshusses, 1962) for the analysis of EDTA in fish muscle and cooked crab meat. None were satisfactory. During the testing of the EDTA methods, we found, however, that the method of Pribil and Vesely (1967), which was developed for the determination of EDTA during the commercial synthesis of EDTA, could be modified satisfactorily for use in the determination of EDTA in fish flesh and crab meat.

The purpose of this paper therefore is to report on the modified method. The paper gives the details of the method, the recoveries of added EDTA from fish flesh and crab meat, the precision of the method, and the precautions to be observed when EDTA is used in the presence of interfering substances.

I. Details of the Method

1. Prepare extracts of fish flesh or crab meat by disintegrating 20 grams of material with 40 milliliters of 5-percent trichloroacetic acid for 1 minute in a blender.
2. Filter the mixture through Whatman No. 2^v filter paper.*
3. Collect 2 milliliters of filtrate and adjust the pH to 11.0 with 10-percent sodium hydroxide.
4. To the filtrate, add 5 milliliters of 2-percent calcium acetate and readjust the pH to 11.0.
5. Remove the precipitated calcium phosphate by centrifugation and follow by filtration through Whatman No. 1 filter paper.
6. Wash the precipitate with a dilute solution of alkaline calcium acetate and combine the washings with the phosphate-free filtrate.
7. Adjust this filtrate to pH 3.5 with 0.5 N hydrochloric acid.
8. Add 5 milliliters of 0.2 M acetate buffer, pH 3.5, and from 1 to 2 drops of a 0.16-percent

space—except for a table of contents.

This report, as you may have noticed, is taken from the same source as the shrimp-trawl system example of an introduction. Perhaps not surprisingly, it contains, in our opinion, exactly the same deficiency with respect to the initial emphasis. What do you think about this one?

The many references in this introduction to relevant publications make for slow reading. We ourselves would prefer to use only footnote numbers in the text and to put the names at the bottom of the page or at the end of the article. On the other hand, there are certainly some advantages in putting the names into the text as these writers have done. One of these advantages is that for the chemists reading the report some of the names would probably be familiar and would have the immediate effect of adding information to the background that this part of the introduction is supplying. See p. 136 for further comment.

As we remarked in Chapter 7 (Description of a Process), this report contains a fine illustration of an appropriate use of itemized directions intended for a reader who is thoroughly familiar with the subject matter.

Finally, we suggest that, with the chapter on conclusions and summaries in mind, you take a close look at the similarities and differences among the abstract, introduction, and summary in this report. You may feel that in so short a report the value of a

---

* Use of trade names is merely to facilitate description; no endorsement is implied.

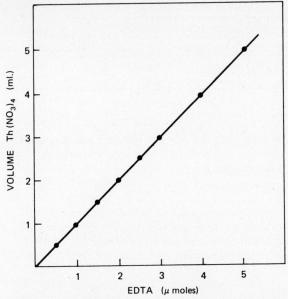

Figure 1 Standard curve for the analysis of EDTA. Samples containing known amounts of EDTA were titrated with 0.001 M Th(NO$_3$)$_4$.

aqueous solution of xylenol orange to an aliquot.

9. Titrate the aliquot with 0.001 M thorium nitrate to a red-violet endpoint.

10. Calculate the content of EDTA from a standard curve prepared by the titration of solutions containing 0 to 5 μM EDTA (Figure 1).

formal summary is reduced almost to the vanishing point. Yet we must admit we enjoy having it there. In any event, the differences in the content of the three elements just named are easy to see in this highly compressed form. Study of this skillfully written miniature report provides many suggestions of much practical value for the writing of longer reports.

TABLE 1   Recovery of EDTA Added to Fish Flesh and Crab Meat

	EDTA Recovered from:	
EDTA Added	**Fish flesh**	**Crab meat**
$\gamma/g.$	*Percent*	*Percent*
37.5	95	—
75	96	95
150	93	90
300	90	91
600	92	92

TABLE 2   Precision of Recovery of EDTA from Fish Flesh

Trials	EDTA Added	EDTA Recovered				Standard Deviation
		Range		Mean		
	$\gamma/g.$	$\gamma/g.$		$\gamma/g.$	Percent	$\gamma/g.$
13	300	246–276		266	89	11

II. RECOVERIES OF ADDED EDTA

Table 1 shows the efficiency of recovery of EDTA added to fish flesh and crab meat. At all the concentrations of added EDTA tested, the recovery appeared to be adequate for the purposes of this determination.

III. PRECISION OF THE METHOD

Table 2 gives a statistical evaluation of the recovery from samples containing 300 gammas of EDTA per gram of fish flesh and crab meat. These recovery results show that the method is suitable for determining the amount of EDTA additive in a fishery product. Also these results compare favorably with those obtained with similar methods.

IV. PRECAUTIONS TO BE OBSERVED

Phosphates interfere with this method. The presence of phosphates gives high results in terms of EDTA, because these compounds combine with the thorium ion. This interference is completely eliminated by removal of the phosphates with calcium.

Citrate also interferes with the method. Usually, however, not enough citrate is present in fish and crab meat to cause a problem. In samples of crab meat in which citrate has been added, this interference can be eliminated by the removal of the citrate from EDTA by chromatography on cation resin. This separation is accomplished as follows:

1. Adjust the filtrate from the phosphate-removal step to pH 8.8 with 0.5 milliliter of 0.2 M tris buffer.
2. Pass the filtrate over a 5- to 6-centimeter AG 50 W-X8 (+ H) column (200–400 mesh).
3. Wash the column with from 10 to 20 milliliters of water containing 0.5 milliliter of 0.2 M tris buffer, pH 9.0.
4. Elute the EDTA with from 10 to 20 milliliters of N hydrocloric acid in 0.5 M potassium chloride.
5. Adjust the eluate to pH 3.5 with 0.5 N hydrochloric acid.
6. Add 5 milliliters 0.5 M acetate buffer (pH 3.5).
7. Titrate the eluate just as you would a regular sample.

The overall recovery obtained when citrate is present is about 80 percent. This loss of 20 percent can be attributed to a portion of the EDTA passing through the Dowex 50 column along with the citrate.

Apparently, metal ions such as Cu^{2+}, Mg^{2+}, and Ca^{2+} do not interfere.

SUMMARY

The Pribal and Vesely (1967) titration method based on the chelation of EDTA with thorium ion was modified for use in the determination of EDTA in fish flesh and crab meat. The modified method, which is simple and rapid, gave about 90-percent recovery of the added EDTA from samples of fish and crab meat.

Phosphate and citrate interfere. Techniques are given for their removal from samples containing EDTA.

LITERATURE CITED

1. Belot, Y. 1964. Determination of EDTA in the presence of complexing metals. Analytical Abstracts 11: Abstract No. 3739.
2. Brady, G. W. F., and J. R. Gwilt. 1962. Colorimetric determination of EDTA. Journal of Applied Chemistry 12: 79–82.

[Eleven additional citations found in the original report are not reprinted here.]

Transitions

11

A transition is an indication of what is going to be said, a reference to what has already been said, or both. It may be a single word, a phrase, a sentence, a paragraph, or an even longer passage. In form, transitions may be quite mechanical and obvious, or unobtrusively woven into sentences with other purposes. We'll discuss what a transition is, how to write a transition, and where to place a transition.

What a Transition Is

As we said above, a transition may be a word, a phrase, a sentence, a paragraph, or an even longer passage. Let's begin with words and phrases. Below are two passages that differ only in the presence or absence of transitional words.[1]

1. Evidently the creation of a plutonium production plant of the required size was to be a major enterprise even without attempting to utilize the thermal energy liberated. By November 1942 most of the problems had been well defined and tentative solutions had been proposed. These problems will be discussed in some detail in the next chapter; we will mention them here.

2. Evidently the creation of a plutonium production plant of the required size was to be a major enterprise even without attempting to utilize the thermal energy

[1] From Henry D. Smyth, *Atomic Energy for Military Purposes,* rev. ed. (Princeton, N.J.: Princeton University Press, 1946), p. 104. Reprinted by permission of the author.

liberated. Nevertheless, by November 1942 most of the problems had been well defined and tentative solutions had been proposed. Although these problems will be discussed in some detail in the next chapter, we will mention them here.

The second version differs from the first only by the addition of two words ("nevertheless," and "although"), and yet it is noticeably smoother than the first. Careful reading of the two passages reveals very clearly the marked effect that two such apparently minor changes can create. Moreover, a definite change in meaning occurs, especially after "nevertheless." This word adds force to the idea that it is remarkable that the accomplishments mentioned were achieved in so short a time.

Numerous words and phrases are frequently used as transitions in this manner. The following is a partial list:

however	in addition
on the other hand	indeed
in spite of	in fact
moreover	as previously noted
furthermore	in comparison
consequently	in the first place
also	secondly
now	finally
so	next
as a result of	then
therefore	in other words
of course	and
for example	but
besides	

Perhaps you feel that such words and phrases as these do not exactly "indicate what is going to be said," as we claimed they do. Yet they do indicate, very often, the logic of the relationship between two units of thought. Such terms as "moreover," and "furthermore" indicate that "more of the same" is coming; "however" suggests that a different point of view is to be introduced or a refutation or qualification offered; "consequently" establishes a cause-and-effect relationship; and so on.

There is another way in which words and phrases serve a transitional purpose, besides the one just described, and that is through the repetition of key terms. Consider the italicized terms in the following. "This experiment can be carried out successfully only under certain *conditions*. These *conditions* are. . . ." The second statement might have begun, "One of these . . .," or "The first of these. . . ." It is a good idea to remember that repetition of the main subject of discussion itself helps keep the reader's eyes on the ball and leads from one thought to another. Suppose that this experiment were Millikan's oil-drop experiment: every now and then in a discussion of it, it would help to substitute "Millikan's experiment," or "the oil-drop experiment" for the term "experiment," or whatever other term might be used as the subject of sentences concerning it.

Sentence transitions and paragraph (or longer) transitions are associated less with stylistic qualities and more with organization than are the shorter ones just discussed. Usually, these longer forms consist of a statement of what has been or will be said. The last sentence in the passage quoted a moment ago is an example of a transitional sentence which is obvious in form. The first sentence of the present paragraph is an example of one less obvious in form. Both, however, serve the same purpose: to provide information about the content of a coming passage. This function may be seen again in the examples below, which represent both the obvious and the less obvious forms.

1. Having considered the economic feasibility of this alloy as a transformer core, we turn now to the problem of hysteresis loss.
2. Even if this alloy is economically feasible as a transformer core, however, there still remains the problem of hysteresis loss.

This sentence, in either form, would serve as a good transition. The first of the two forms calls attention forcibly to the change of topic; the second performs the same function but less obtrusively. The first might be the easier to remember; the second makes smoother reading. A choice between the two forms will follow consideration of this difference.

There is no important difference in principle between a sentence transition and a paragraph transition. In fact, our chief reason for using both terms was to make it perfectly clear that an entire paragraph, as well as a sentence, may be used to make a transition. Naturally, the same differences in form that may be found among sentence transitions hold true for paragraphs, as the following excerpts show:

1. In Chapter 1 and other early chapters we have given brief accounts of the fission process, pile operations, and chemical separation. We shall now review these topics from a somewhat different point of view before describing the plutonium production plants themselves.[2]
2. In previous chapters there have been references to the advantages of heavy water as a moderator. It is more effective than graphite in slowing down neutrons and it has a smaller neutron absorption than graphite. It is therefore possible to build a chain-reacting unit with uranium and heavy water and thereby to attain a considerably higher multiplication factor, k, and a smaller size than is possible with graphite. But one must have the heavy water.[3]

The second example is the less mechanical. In the original text it is followed by discussion of work done on heavy water—and thus it serves as a transition.

So far, we have considered some typical examples of transitions with special attention to their forms, ranging from single words to paragraphs, and from obvious, rather mechanical types to those that are less obtrusive. Finally, we'll note the various purposes or functions that a transition may have. There are six important ones:

1. Smoothing out style, principally by clarifying logical relationships through the use of single transitional words or short phrases. Examples: "however," "on the other hand."

[2] Smyth, *ibid.*, p. 130.
[3] Smyth, *ibid.*, p. 147.

2. Indicating what topics are to be discussed. Example: "This section is devoted to an analysis of the effect of temperature on bearing noise with a given lubricant."
3. Reminding the reader of topics discussed. Example: "It is evident, then, that neither increasing the number of workers nor increasing the speed of the line can, in the present circumstances, increase the output of Final Assembly."
4. Announcing a change of subject. Example: "In addition to the major advantages just described, the proposed changes in design would offer several secondary advantages."
5. Making reference to an earlier or later statement of a similar, related, or pertinent idea. Examples: "As was said in the preceding section. . . ." "As will be shown in Chapter X. . . ."
6. Keeping attention focused by the repetition of key terms. Examples: "The first step in aligning the circuit. . . ." "The next thing to do in aligning the circuit. . . ."

In conclusion we must say that the foregoing discussion has not by any means been an exhaustive statement of what a transition is. To go further, however, would be primarily to enlarge upon the theoretical rather than the practical aspects of the subject.

How to Write Transitions

On the subject of learning how to write transitions, we have two practical suggestions to make, and that is all.

1. Don't hesitate to be quite mechanical about it at first. If you can't think of any better way, just say, "This concludes the discussion of so-and-so. Next, thus-and-so will be discussed." As you continue to practice writing, you will find that you acquire a habit of using transitions, and develop an ability to make them as obvious or unobtrusive as you wish. But don't expect the process to become completely automatic.
2. When you have completed a rough draft of a report, read through it once with the sole purpose of spotting points at which transitions should be added. After locating and marking all such points, go back and write the transitions. Possibly, also, you will want to delete some of the transitions that you originally wrote. It is possible to have too many, and it is possible to get them in the wrong place. This raises the question of how to decide where transitions should appear.

Where to Put Transitions

There is no formula according to which transitions can be located. Every report is unique and presents its own problems. But if there is no formula, there is a principle — and the principle is simply this: don't give your reader a chance to get lost. Again, it's like putting up highway signs. In looking over your report for short transitions, it is wise to keep trying out the effect of adding a word or phrase to a sentence. This experimenting can be done quite deliberately. In checking the location of longer transitions, it sometimes helps to start with the outline. Put an asterisk where the shift in thought is great enough to require a strong transition; then examine the text itself at the points noted.

A special problem that comes up here is the effect of the use of subheads, like the one above, on the handling of transitions. Clearly, the subhead itself is a transitional device, and can be expected to inform the reader fairly accurately of changes in subject. On the other hand, it is easy to overestimate the amount of attention a reader gives to subheads (see the comments on titles in Chapter 10, "Introductions"). Our advise is to write the transition pretty much as if the subhead weren't there. And, almost without exception, our advice is to avoid using the subhead as an antecedent for a pronoun in the sentence that follows it. This point will be immediately clear if you compare the following undesirable sentence with the one that actually appears under the subhead above ("Where to Put Transitions"): "There is no formula for this."

Illustration is probably more valuable than advice with respect to almost every aspect of the art of writing transitions, and so for the rest of our exposition we'll turn to an extended illustration.

Illustrative Material

The material that follows consists of two paragraphs that are fairly representative of good technical writing.[4] They are not loaded down with transitions, but they do make clear, easy reading. (Observe, however, the abuse of the word "such" in the second paragraph.)

We have italicized transitional elements.

Numerous pronouns (not in italics) also serve a transitional purpose. This is one of the problems we had in mind in our previous assertion that our discussion of transitions was by no means exhaustive. Even though pronouns are often clearly transitional in function, it seems best to exclude them from a consideration of transitions because their inclusion would add difficult theoretical problems without adding much of practical value.

The Equivalence of Mass and Energy

1.4. One conclusion that appeared rather early in the development of the theory of relativity was that the inertial mass of a moving body increased as its speed increased. This implied an *equivalence* between an increase in energy of motion of a body, that is, its kinetic energy, and an increase in its mass. To most practical physicists and engineers this appeared a mathematical fiction of no practical importance. Even Einstein could hardly have foreseen the present applications, *but* as early as 1905 he did clearly state that mass and energy were *equivalent* and suggested that proof of this *equivalence* might be found by the study of radioactive

---

[4] Smyth, *ibid.,* pp. 2–3.

substances. He concluded that the amount of energy, E, *equivalent* to a mass, m, was given by the equation

$$E = mc^2$$

where c is the velocity of light. If this is stated in actual numbers, its startling character is apparent. It shows that one kilogram (2.2 pounds) of matter, if *converted* entirely into energy, would give 25 billion kilowatt hours of energy. This is *equal* to the energy that would be generated by the total electric power industry in the United States (as of 1939) running for approximately two months. Compare this fantastic figure with the 8.5 kilowatt hours of heat energy which may be produced by burning an *equal* amount of coal.

1.5. The extreme size of this *conversion* figure was interesting *in several respects. In the first place,* it explained why the *equivalence* of mass and energy was never observed in ordinary chemical combustion. We *now* believe that the heat given off in *such* a combustion has mass associated with it, *but* this mass is so small that it cannot be detected by the most sensitive balances available. (It is of the order of a few billionths of a gram per mole.) *In the second place,* it was made clear that no appreciable quantities of matter were being *converted* into energy in any familiar terrestrial processes, *since* no *such* large sources of energy were known. *Further,* the possibility of initiating or controlling such a *conversion* in any practical way seemed very remote. *Finally,* the very size of the *conversion* factor opened a magnificent field of speculation to philosophers, physicists, engineers, and comic-strip artists. For twenty-five years *such* speculation was unsupported by direct experimental evidence, but beginning about 1930 *such* evidence began to appear in rapidly increasing quantity. *Before discussing* SUCH *evidence and the practical partial conversion of matter into energy that is our main theme, we shall review the foundations of atomic and nuclear physics. General familiarity with the atomic nature of matter and with the existence of electrons is assumed. Our treatment will be little more than an outline which may be elaborated by reference to books such as Pollard and Davidson's* APPLIED NUCLEAR PHYSICS *and Stranathan's* THE "PARTICLES" OF MODERN PHYSICS.

12

Conclusions and Summaries

In this chapter we'll discuss the chief considerations in bringing a report, or a section of a long report, to an end.

One of these considerations is an aesthetic one: how to give a sense of finality and completeness to the discussion. We'll make some comments on this problem, but for the most part we are concerned with the content rather than with the possible aesthetic function of the conclusion or summary. First we discuss conclusions and then summaries.

We've been using these two words, "conclusion" and "summary," together so far, and it may have seemed that one or the other alone would do as well. The reason for retaining both is that we want to use them in different and rather specialized ways. That is, a conclusion, in this chapter, is not going to mean the same thing as a summary.

A conclusion is, of course, an end. In technical writing, however, there are three different kinds of conclusions or ways of bringing a report to an end. As we said above, the idea of a summary is not included in any of these three kinds of conclusions — with the exception that sometimes conclusions are summarized, as we'll explain.

The first of these three kinds of conclusions is what we call the "aesthetic." Its function is merely to bring the discussion smoothly to a stop. This need is felt particularly when there seems to be little point in reviewing what has been said, and yet it seems awkward just to stop. *Often it is wise just to stop;* but not always. For instance, at the end of a description of how to develop film at home, you might want to close with: "Although, as has been shown, developing and printing film is not a difficult process, it is one that affords a great opportunity for experimenting with effects, and thus provides a continuing novelty and challenge. Reasonable caution in carrying out the steps just described will start you on the way to a most pleasant and interesting hobby." Nothing significant in the way of review or of decisions has been said here, but a reasonably graceful conclusion has been made.

The second of the three kinds of conclusions is one in which the results of an investigation or study are stated. For example: "The conclusion reached as a result of this study is that the toe of the dam is being undermined by water flowing through fissures in the limestone bed of the stream." This kind of conclusion is often called "findings," or results."

The third kind is the decision reached at the end of a discussion concerning a choice of action, or concerning a practical problem for which a solution must be presented in the form of a forthright recommendation as to what action should be undertaken. A typical example: "The inescapable conclusion is that the wisest course of action is to fabricate the panels of asbestos rather than to provide heat shielding for the present wooden panels."

Obviously, the difference between the second and third kinds of conclusions is sometimes very slight. And the fact is that the terms "conclusion," "findings," and "results" are used rather indiscriminately for both kinds. The difference between the two kinds does become important, however, when it is necessary that readers understand whether they are merely being given some information (the second kind of conclusion) or told that a decision about a course of action has been made (the third kind).

Conclusions of either the second or third kind should be presented clearly and forcibly. Needless to say, a long report devoted to a complex train of reasoning that terminates in an obscure conclusion is exasperating. Often it is desirable to present the conclusion in a paragraph of its own, with the subhead "Conclusion." In a short report the conclusion may appear only at the end; and in a long report devoted to reaching only one important conclusion, that conclusion may also appear only at the end (except for a possible appearance in the abstract or introductory summary). On the other hand, there may be a series of "conclusions" (findings, decisions) in the body of the report, which are restated at the end of the report and sometimes formally entitled "Summary of Conclusions." The following paragraph is an illustration of this type of conclusion:

Conclusions

Based on the work conducted on the hydraulic unit at this laboratory and reported herein, it is concluded that:

1. The oil in the system under normal operating conditions contained small amounts of air.
2. The amount of air in the oil varied considerably with different operating conditions.
3. There appeared to be no significant difference in the operation of the unit from a force vs. speed standpoint when operated with varying amounts of air in the oil.
4. Excess air introduced into the oil through the pump intake readily dissolved in

the oil when it was subjected to high pressures so that there appeared to be no air mechanically entrained in the oil during normal operation of the unit.

5. The greater efficiency reported for this hydraulic machine is probably due to the fact that it is operated with forces nearer the actual cutting force so that the system is always fairly near equilibrium, thus always placing the maximum load on the cutter, but never overloading it.

6. There appeared to be no significant difference in the operation of this unit when using either Oil A or Oil B.

It is not necessary to number the conclusions, although in this instance it was surely a good idea.

Finally, there are the summaries. A summary is, as we have already implied, a review or concise restatement of the principal points made in the discussion. It is more useful at the end of a report engaged chiefly in presenting a body of information than in an analytical or argumentative report, or in a descriptive report which would not justify anything more than an aesthetic conclusion.

The writing of a good summary requires a very clear grasp of each one of the fundamental ideas of the report. In fact, writing a summary may serve as a test of whether you have actually seen and formulated clearly the fundamental ideas in the report. And a summary should contain no new ideas or information. It is a restatement of information — not an addendum.

The following illustration[1] contains not only a summary but also the paragraph that immediately precedes the summary. The purpose of including this extra paragraph is to show how it is summarized in the last sentence of the summary itself.

Cooperation between the Metallurgical Laboratory and du Pont

Since du Pont was the design and construction organization and the Metallurgical Laboratory was the research organization, it was obvious that close cooperation was essential. Not only did du Pont need answers to specific questions, but they could benefit by criticism and suggestions on the many points where the Metallurgical group was especially well informed. Similarly, the Metallurgical group could profit by the knowledge of du Pont on many technical questions of design, construction, and cooperation. To promote this kind of cooperation du Pont stationed one of their physicists, J. B. Miles, at Chicago, and had many other du Pont men, particularly C. H. Greenewalt, spend much of their time at Chicago. Miles and Greenewalt regularly attended meetings of the Laboratory Council. There was no similar reciprocal arrangement, although many members of the laboratory visited Wilmington informally. In addition, J. A. Wheeler was transferred from Chicago to Wilmington and became a member of the du Pont staff. There was, of course, constant exchange of reports and letters, and conferences were held frequently between Compton and R. Williams of du Pont. Whitaker spent much of his time at Wilmington during the period when the Clinton plant was being designed and constructed.

Summary

By January 1943, the decision had been made to build a plutonium production plant with a large capacity. This meant a pile developing thousands of kilowatts and

---

[1] From Henry D. Smyth, *Atomic Energy for Military Purposes,* rev. ed. (Princeton, N.J.: Princeton University Press, 1946), pp. 128–129.

a chemical separation plant to extract the product. The du Pont Company was to design, construct, and operate the plant; the Metallurgical Laboratory was to do the necessary research. A site was chosen on the Columbia River at Hanford, Washington. A tentative decision to build a helium-cooled plant was reversed in favor of watercooling. The principal problems were those involving lattice design, loading and unloading, choice of materials particularly with reference to corrosion and radiation, water supply, controls and instrumentation, health hazards, chemical separation process, and design of the separation plant. Plans were made for the necessary fundamental and technical research and for the training of operators. Arrangements were made for liaison between du Pont and the Metallurgical Laboratory. [NOTE: this last sentence summarizes the preceding paragraph.]

As you see, the style of this summary is distinctly "choppy." One bald statement follows another. This is probably a good idea. The summary can be regarded almost as a list of the major ideas, and there is little reason to try to escape very far from the form of a list. Indeed, summaries are sometimes broken down into numbered statements, as illustrated below.

Finally, we must say plainly that conclusions and summaries cannot be written by formula. The principles we have discussed are of considerable value, but they are only principles, not prescriptions. As always, the important thing is the successful accomplishment of the function itself, not the particular method adopted.

Summary

1. For convenience, the final section of a report may be classified as either of two types: conclusions or summaries.
2. Conclusions may be subdivided into those that are primarily aesthetic, those that announce the results of an investigation or study, and those that present a decision concerning a course of action.
3. An aesthetic conclusion merely brings the report to a graceful close. The other two kinds of conclusions may appear only at the end of the report (with the exception of a possible appearance in the abstract or introductory summary), or both in the body of the report and at the end.
4. A summary is a restatement of important information. It should contain no new ideas; and, in comparison with the length of the report, it should be short.

SECTION IV

Types of Reports

So far we have been concerned with various fundamental skills and techniques needed in technical writing. Now we turn our attention to the forms of writing in which these skills and techniques are used.

In some organizations there is little formality attached to report writing. Each writer decides what form is best suited to what he or she has to say. Elsewhere, particularly in large organizations, numerous and sometimes elaborate forms are devised and given names. Thereafter, within the organization, these forms are spoken of as types of reports, and young people are given instructions on how to write them. This is exactly as it should be, if the forms devised satisfy the needs of the organization. But it does result in the creation of a tremendous lot of "types." In a casual search that took no more than 30 minutes, we once turned up the following examples of so-called types of reports:

preliminary	feasibility	test
partial	service	examination
interim	operation	examination-trip
final	construction	inspection
completion	design	investigation
status	failure	memorandum
experimental	student-laboratory	short-form
special	industrial-research	periodic
trade	industrial shop	information
formal	evaluative	work

It is possible that the foregoing list could be boiled down to a few funda-mental types. No one, however, has ever succeeded in winning general accept-ance of a working system of classification of reports, and it seems unlikely that any attempt will succeed.

There are, nevertheless, certain kinds of writing so commonly identified as "reports" that acquaintance with them is valuable, both for its own sake and for the purpose of giving the concept of a report some concrete meaning. Outstand-ing examples are recommendation reports, proposals, and progress reports. Others could easily be added to the list: feasibility studies, program evaluations, job descriptions, environmental impact statements. Logically, there are no very clear grounds for designating any of these kinds of writing as reports except when defining "report," in the most fundamental way possible, as a piece of writing intended to serve some practical purpose. And if we start with this fundamental definition of report, then the idea of types of reports requires find-ing a generally acceptable basis of classification of reports—an attempt at which, as noted earlier, no one has ever succeeded.

Should you happen sometime to be asked to write, say, a feasibility study, you might approach your problem in the following two steps. First, consider what special techniques might be required. Since a feasibility study is, as the name indicates, merely a study of whether some project probably could or probably could not be carried out successfully, it is easy to guess that the tech-nique of interpretation would be especially useful. The particular subject matter in the project may suggest other techniques. And second, find out what special characteristics are associated with the term "feasibility study" by the people you are writing the report for. The response may vary from "none" to a detailed plan for the organization of the report. If the response is "none," then of course you can simply use your own common sense. (For an example of a feasibility study in the form of a laboratory report see pp. 364–376, especially pp. 368–369. See also p. 275.) Any assignment to write a given type of report can be approached with the same two steps described above.

For discussion in this Section IV we have selected recommendation reports, proposals, and progress reports. As indicated above, the claim of these kinds of writing to recognition as types of reports rests primarily upon people's habit of thinking of them as types of reports. In practice, this is a strong claim. Subse-quent chapters in this section are devoted to some additional forms that, although not usually regarded as reports in the same sense as those mentioned above, can most conveniently be considered at this point. The additional forms are oral reports, business letters, and writing for professional journals.

13

Recommendation Reports

Introduction

Because the term "recommendation report" is so frequently used in technical writing, both in textbooks and in the field, one would naturally suppose that this is a type of report with an easily identifiable kind of content and organization. In fact, however, any report that contains recommendations is a recommendation report — and almost any report may contain recommendations.

Examination of numerous recommendation reports will show that their basic characteristics differ widely. The bulk of the content of a recommendation report is most often interpretative, but it is not uncommon to find more description than interpretation. There is no standardized organization for recommendation reports, with the exception that the recommendations are usually stated near the beginning, or near the end, or both. The format may be any one of the many varieties in use. The function of a recommendation report, on the other hand, would seem at first to be fixed and stable — that is, to persuade the reader to take a certain course of action; and usually this function is indeed evident in practice. But a consultant might conceivably be indifferent as to whether his or her recommendations were acted upon and reflect this indifference in the tone of the report. We must conclude, in brief, that we are dealing with an ambiguous concept when we discuss recommendation reports.

Nevertheless, the vitality of the idea of a recommendation report, as shown by the wide currency of the term, is a warning that we should not treat the concept too casually. Of course what we are dealing with fundamentally is the situation (in report form) in which the abstract thinking of the laboratory and the office passes over into the realm of practical action. The importance of this action probably accounts for the vitality of the idea of a recommendation report. Furthermore, it would be a mistake to suppose that people don't know what they mean when they use the term "recommendation report." Probably they don't always know exactly; but if your boss or your college instructor tells you to write a recommendation report about something, we strongly urge that you do not stop to itemize the ambiguities we have just been pointing out. Get busy and analyze the problem you have been given, decide upon the proper course of action, and make a forthright recommendation. That kind of procedure is what your boss or your college instructor will have in mind when requesting a recommendation report. In Chapter 14, however, we will consider a specialized kind of recommendation report, the proposal.

In the present chapter we begin with the problems of style and tone, which are of particular importance in many recommendation reports. We then go on to discuss how to phrase recommendations, where to put them, and how to organize a recommendation report.

Reader Analysis and Style

Ideally, the art of persuasion should never enter into the professions of science and engineering. The scientist or engineer would investigate physical laws, or apply them to a specific problem, report the findings, and be through. Readers would need no convincing or persuading; they would be governed solely by logic. Practically, of course, things are seldom so simple.

There are two somewhat different situations to be considered: first, that in which you are given definite instructions to prepare a recommendation report; and second, that in which you volunteer a recommendation. A volunteered recommendation may be inserted into a report written primarily for a different purpose (such as a progress report) or it may be made the chief subject and purpose of a special report.

When you have been instructed to make a recommendation, you may find that it is fairly obvious what action should be recommended, and also that everybody agrees it is the proper action. This is wonderful — and not uncommon at all. But you may on other occasions find after studying the subject that (1) you don't think any action at all should be taken, whereas you either know or suspect that your superior or associates feel that some action is desirable; (2) you think action should be taken, but foresee unwillingness to act; (3) you think a certain course of action should be taken, but expect that a different course of action will be favored; (4) you think action should be taken, but cannot see a clear advantage between two or more possible courses of action.

If you find that the evidence does not indicate a clear-cut decision, the best policy is simply to say so, with an especially thorough analysis of advantages and disadvantages. This does not mean you should not make a concrete recommendation; you should—but not without making clear what the uncertainties are. This situation is found in some measure in the reports printed at the end of this chapter.

When you expect opposition to recommendations you are convinced should be made, you should give a good deal of thought to the tone of your report and to methods of emphasizing the points that clarify the logic of the situation.

Don't let yourself fall into an argumentative tone. We have in our files a report from a research organization that begins with the statement, "This is a very important report." Our own immediate reaction, when first reading this statement, was a suspicion that maybe the report wasn't really very important or the writer wouldn't have thought it necessary to try so unblushingly to persuade the reader it was. It would have been a more effective report had the writer prepared a good, clear introduction, stated the major conclusions in the proper place, and added this sort of statement: "The great importance of this discovery arises from the fact that. . . ."

In general, be forthright in tone or manner, but not blunt. Instead of saying, "the present method is wasteful and inefficient," remind yourself that whoever designed the present method was probably doing the best that could be done at the time, and was most likely proud of the results. You may prefer to write something like "the proposed new method offers a considerable increase in efficiency over the present method"; or, "certain changes in the present method will result in an increase in efficiency."

Putting emphasis on the proper points demands first of all an analysis of the probable attitudes of your reader. If you do not expect opposition, there is little problem here. If you decide that opposition is probable, a good general policy is to discuss first the advantages and disadvantages of the recommendation you think might be preferred to the one you intend to make — being careful to state fairly *all* its advantages; second, to present the advantages and disadvantages of the course of action you prefer; third, to give a summary and recommendation. This approach provides emphasis through relative position—the value of the preferred action being shown after the weaknesses of the alternative have been explained. Emphasis may also be achieved through paragraphing and sentence structure. For example, use of a series of short paragraphs written in short declarative sentences when you sum up the advantages of the preferred course of action will result in an especially forceful impression.

The problem of whether to volunteer a recommendation is most likely to arise when you have a positive suggestion about work in which your own official part is only routine, or about work that is not a part of your official duties. In the long run, there is little doubt that the more ideas you have and the more suggestions you make, the faster you will be promoted and the more fun you will have with your work. But when it comes to volunteering recommendations in writing, two cautions ought to be observed.

First, be very sure that your recommendation is sound and that you have shown clearly that it is sound. (In this connection, a review of the discussion of interpretation in Chapter 9 will be helpful.) Your superiors aren't going to be pleased with mere opinions.

The other caution is to be careful not to give the impression that you are trying to "muscle in" on something. This is likely to be a delicate point, and you'll be wise to think about it very deliberately. To a certain extent this difficulty can be met by avoiding the kind of phrasing used in a formal recommendation (see below) and by presenting your recommendation in the form of a conclusion. Instead of saying, "it is recommended that the temperature of the kiln be lowered 15 degrees and the drying time prolonged to 84 hours," you could say, "better results would evidently be obtained by lowering the temperature of the kiln 15 degrees and prolonging the drying time to 84 hours."

Altogether, then, when you have recommendations to make, your first problem is to determine precisely what course of action or what decision is best justified by the evidence. Your second problem is to estimate your reader's probable attitude toward your recommendations. Your third problem is to prepare a report that will be effectively organized to make clear the logic of your recommendations to the specific reader or readers you expect to have.

How to Phrase Recommendations

To a certain extent it is possible to classify recommendations as formal and informal. An informal recommendation may consist merely of a statement like "it is recommended that a detergent be added to the lubricant," or "therefore a detergent should be added to the lubricant." In a sense, any suggestion or advice constitutes a recommendation. The formality with which it should be presented is determined by its relation to the major problem being discussed and by the tone of the whole report, as stated above. Usually, the more important a problem is and the longer the discussion of it, the more need there is for a formally phrased recommendation.

A highly formal recommendation is illustrated by the following:

> After consideration of all the information available concerning the problems just described, it is recommended:
>
> That the present sewage disposal plant be expanded, rather than that a new one be constructed;
>
> That the present filter be changed to a high-rate filter;
>
> That a skilled operator be employed.

Sometimes each main clause in a recommendation like this is numbered. Sometimes the recommendations are presented as a numbered list of complete sentences preceded by the subhead "Recommendations" and without any other introduction. And, of course, sometimes they are simply written out in sentence form as shown in the "informal" example in the preceding paragraph, without unusual indentation and without numbers.

It is occasionally advisable to accompany each recommendation with some explanation, in contrast to limiting each recommendation to a single statement as illustrated above. An example will be found in the second report at the end of this chapter. This method is often particularly useful when recommendations are stated twice in the same report, at the beginning and again at the end. The explanations usually accompany the second statement.

Where to Put Recommendations

Recommendations almost invariably appear at the end of a recommendation report. If the report is long, and especially if an introductory summary is used, they are likely to appear near the beginning as well, immediately after the statement of the problem. When they appear both at the beginning and the end, however, those at the end are likely to be stated informally, whereas those at the beginning are more formal, usually with the heading "Recommendations."

If the report is long, it is desirable to put the recommendations at the beginning so that a reader may at once find the major results. And it is always wise to state the recommendations at the end so that they will be the last ideas impressed on the reader's mind.

The Overall Organization of a Recommendation Report

As pointed out at the beginning of this chapter, almost any kind of report may contain recommendations, and therefore in some sense may be thought of as a recommendation report. Consequently, there is no single organizational pattern which could be identified as that of a recommendation report. There is, however, a principle that often helps in deciding how to organize a report in which one or more recommendations play an important part. This principle is the distinction between a situation in which standards of judgment must be employed and one involving only a consideration of advantages versus disadvantages.

The concept of standards of judgment was explored in Chapter 9, "Interpretation," and so we will not enter into it again here, except for a brief comment. As the term indicates, standards of judgment are employed when the circumstances governing a decision can be rather clearly specified. For example, if you were engaged as a consultant by a company wishing to buy a fleet of trucks, you might be asked to recommend a truck that would provide a maximum of economy and reliability, and that would have a cargo space of at least x number of cubic feet [cubic meters], and a minimum carrying capacity of at least x number of pounds [kilograms]. No truck failing to meet these specifications for cargo space and carrying capacity could be considered.

In contrast, some recommendations do not involve standards of judgment but rather offer a more subjective choice based on advantages and disadvantages. The line between standards of judgment and advantages and disadvantages cannot be drawn sharply. The concept of advantages and disadvantages is in effect simply a looser, less precise statement of standards of judgment. For example, should you or should you not go to a movie next Friday evening? To some degree, it is true, standards can be found for such a decision. One might be the availability of a movie you really want to see, as distinguished from just going to see whatever happened to be on; another, that the theater in which the movie is being shown is not too far away. But the meaning of these standards isn't very clear. You might feel, for instance, that although there is a movie you have some desire to see, you might really prefer to play tennis instead. And so you would find yourself reasoning that a disadvantage of going to the movie is that you couldn't play tennis. And so on.

Now let's return to the question of organization. When you are making use of standards of judgment, probably the simplest organization is to take up each standard in turn, applying it to as many possible choices as you are considering, and deciding which choice is best according to that standard. For the purchase of a fleet of trucks, the recommendation according to the standard of weight-carrying capacity would be the truck that could haul the heaviest load. (Further discussion appears in Chapter 9.)

On the other hand, if you are making a recommendation on the basis of advantages and disadvantages, the simplest organization is probably to take up each possible choice in turn and discuss its advantages and then its disadvantages (or vice versa; if you're making a positive recommendation, place the advantages at the end and they'll be the last ideas to be impressed on the reader).

Of course, if you have to decide only yes or no for just one possible action, then you have the simplest conceivable situation. We conclude this chapter with a skeletal outline for such a recommendation report.

I. Introduction
   A. Statement of problem
   B. Statement of recommendation
   C. Statement of scope and plan of the report
II. Any necessary comments on source and reliability of data in the report
III. Expanded statement, and explanation, of the recommendation
IV. Discussion of the disadvantages of the recommended action
V. Discussion of the advantages of the recommended action
VI. Summary of conclusions
VII. Restatement of recommendation

If you were choosing between two possible actions, then IV and V in the outline would be devoted to discussion of the disadvantages and the advantages, respectively, of the first possibility; and VI and VII to the advantages and disadvantages of the second. The summary of conclusions would then become VIII, and the restatement of the recommendation would be IX — and so on, if there were still other possible actions.

Illustrative Material

In the following pages two examples of recommendation reports appear. The first is a simple discussion of the value of a terrarium. We'll comment later on the second example, which is from a research laboratory.

Terrarium Gardening[1]
Violet Anderson

A problem often encountered by people who must be away from their homes or apartments for extended periods is how to include living plants in their lives. Many people feel a strong need to have some green, living things around them. But with the exception of varieties suited to prolonged drouth, such as cactus, potted plants cannot survive without attention. For people who want plants, yet who cannot give

[1] Printed by permission of the author.

them frequent care, plant specialists commonly recommend a terrarium, also known as a Wardian case. A few facts about what a terrarium is, about how it functions, and about the advantages it offers make this recommendation understandable.

A terrarium is a garden in a small, enclosed space. The enclosure must be air-tight, or nearly so. A bottle will do, although many people prefer elaborate cases constructed of glass, wood, plastic, and other materials. The reason for making the enclosure nearly air-tight is to prevent the escape of moisture. Once the soil has been prepared in the terrarium, the seeds, slips, or plants have been planted, and the soil has been watered, the terrarium is sealed with a lid or cap. Thereafter, within reasonable time limits, the terrarium is a self-contained world, needing no further attention. Energy is supplied by light; the plants' metabolic processes maintain a carbon dioxide-oxygen balance, and the water evaporates, condenses, and returns to the soil to be recycled. Detailed instructions on planting procedures can be found in almost any library, or in a garden supply store.

A terrarium does present two disadvantages. One is that the choice of plant is generally limited to those which will thrive in an environment of high humidity, and to those which will not quickly outgrow the container. The second disadvantage is that the plants are always inside a glass case; they cannot be smelled or touched, unless the lid is removed. The advantages, on the other hand, include not only the freedom of leaving the terrarium unattended for prolonged periods, but the attractiveness of the case itself, as well as the pleasure of creating a miniature, self-sustaining environment.

A terrarium can be a treasured possession, and a valued experience in the life of almost anyone. But for obvious reasons, a terrarium is particularly recommended for plant-lovers who cannot provide their plants with frequent care.

With the exception that there is no comment on the source of data, the preceding discussion of a terrarium conforms closely to the generalized outline presented on p. 254. The report from the Lindberg Engineering Company, which appears below, is more complex. Like the discussion of the terrarium, the Lindberg report begins with a statement of the problem. It then presents one sentence on the source of the data (under the heading "Procedure"), and continues with a discussion of the data (under the heading "Discussion of Results"). The advantages and disadvantages, or their equivalent in this situation, are moved down into the section entitled "Recommendations." The overall organization of this research report is effective. The introduction is weak, however, and the wording of subheads could be improved.

Determination of the Deposit that Collects on the Element Coils and the Cause of Failure of the Element Coils in Type 2872-EH Furnace, Serial 2162 at Olds Motor Works, Lansing, Michigan

Lindberg Engineering Company, Research Laboratory[2]

INTRODUCTION

The Olds Motor Company of Lansing, Michigan have been experiencing abnormal element failures in their Type 2872-EH furnace. Mr. W. Bechtle of our Detroit

[2] Reprinted by permission of the Lindberg Engineering Company.

Office sent in a peculiar, light brown, fluffy deposit he found on the burnt-out wire coils. He found that this substance was very retentive and difficult to shake off the coil.

His report pointed out that the work being drawn in this furnace was heat treated in an Ajax salt bath and quenched into oil. He stated that although the work was washed and appeared to be clean, small quantities of salt might still be present to cause the deposit on the elements and the subsequent failure. A request was made to analyze the material to determine whether it came from the Ajax salt bath and also to determine if it were responsible for the element failure.

PROCEDURE

Conductivity tests and a complete chemical analysis were made of the coating taken off the heating element that failed.

DISCUSSION OF THE RESULTS

The resistance of the material as received was tested with a sensitive meter, and it was found to be very high and thus not a conductor when cold. Good contact may not have been made when making the above test because of the powdery condition of the material. To verify this and also to check on the solubility, some of the substance was placed in distilled water. The resistance of the distilled water was found to be 80,000 ohms before the substance was placed in it. After the substance was placed in the water and stirred, the resistance dropped to 15,000 ohms, showing very slight solubility and conductivity. Tests at high temperatures were not made because the conditions existing in the furnace with a high voltage of about 450 to 480 could not be reproduced.

The chemical analysis of the fluffy substance removed from the elements is as follows:

Silicon (SiO_2)	28.78%
Iron oxide (Fe_2O_3)	46.70
Alumina (Al_2O_3)	13.06
Nickel oxide (NiO)	1.60
Chromium oxide (Cr_2O_3)	0.30
Barium chloride ($BaCl_2$)	1.88
Barium oxide (BaO)	0.57
Total	92.89%

(Balance may be potassium and sodium compounds that were not determined.)

The presence of barium definitely indicates that the Ajax heat treating salt is getting into the draw furnace. Barium chloride is one of the common constituents of heat treating salts, and there are no traces of this substance in materials used in the construction of the Lindberg Furnace.

The presence of the high percentage of silica, alumina, iron oxide, and other oxides can be easily explained. The silica, alumina, nickel, and chromium oxides come from the dust produced by the wear of the furnace refractory and alloy parts rubbing together under constant vibration. The iron oxide comes chiefly from the scale of the heat treated work or the scale produced in tempering the work. All of this dust is being constantly carried in the recirculating air stream over the work and through the element chamber. Under normal conditions this dust will not stick to the elements because the melting point is very high and the velocity of the air is too great to allow the dust to settle out.

With the introduction of only a slight amount of heat treating salt, however, this condition is entirely changed. The heat treating salt spalls off the work due to the difference in its thermal expansion over steel when heated in the tempering furnace and is carried with the other dust in the air stream. As work is continuously being tempered, the concentration of the heat treating salt particles becomes greater. When the heat treating salt particles are carried over the heating elements by the recirculating air stream, the salt particles strike the element and melt to its surface. (The heating element temperature may go as high as 2000°F. [1093.33 C.], depending on the furnace temperature and the load.) The remaining dust from the furnace refractory and scale from the heat treated work then stick to the element.

The failure of the element then results from the material covering the insulators and making contact to the frame and the element coil. The high voltage of 440 to 480 arcs across, burning out sections of the wire from the element. The material covering the element also acts as an insulator, causing the element to run abnormally high in temperature which also reduces its life.

CONCLUSIONS

1. The material coating the heating element in the furnace is a combination of heat treating salt from the hardening operation and refractory and oxide dust from the tempering furnace. The heat treating salt dust melts when it strikes the elements and the remainder furnace dust sticks to it.

2. The element failure is due to the conductivity of the melted heat treating salt covering the element and insulators and thus shorting to the frame. The high voltage of 440 to 480 arcs across and cuts sections out of the element coil. The substance covering the coils also acts as an insulator, causing an abnormally high element temperature which also greatly reduces its life.

RECOMMENDATIONS

1. The most logical thing to do is to prevent the heat treating salt from entering the draw furnace. This, however, is not quite as easy as one may first think because heat treating salts containing barium are not very soluble in hot water or caustic solutions. The solubility depends upon the amount of barium salts present in the mixture. Barium chloride itself is only slightly soluble in hot or cold water or in caustic solutions. Heat treating salts containing barium are very difficult to dissolve out of blind holes, recesses, and etc.

The only sure way that all traces of barium heat treating salts can be cleaned from the work is by the following procedure: 1st—Clean salt and oil from quenched work in hot Oakite solution; 2nd—Rinse in hot water; 3rd—Clean in an acid solution (one part of hydrochloric to four parts of water, or a standard pickling bath can be used if care is observed); 4th—Rinse in hot water to remove acid.

A better method would be to use a heat treating salt that is readily soluble in water. Such a salt is a compound of a mixture of sodium and potassium chlorides and is the best type of neutral hardening salt that can be used. One such commercial salt on the market is known as Lavite # 130, made by the Bellis Heat Treating Company. This salt has a useful temperature range from 1330°F. [721.11C.] minimum to

1650°F. [898.89C.] maximum. In as much as we are not familiar with the hardening requirements, only the above general suggestion can be given in reference to cleaning all traces of salt from the work.

2. Reduce the voltage that can exist between the coils and the frame by redesigning the heating element and thus have less chance of burning out sections of the element by arcing due to the conductivity of the heat treating salt.

We are now designing an element for this furnace in which the element hook-up will be changed to a series star circuit instead of the delta circuit now employed. Both will operate from the 460 volt line, but the voltage between the coil and the frame over any one insulator will be reduced to about 130 volts instead of the 460 volts on the present arrangement.

This will, of course, reduce the tendency for arcing when the heat treating salt and furnace dust collect on the elements and insulators. This is not, however, a substitute for cleaning the heat treating salt from the work. The subject company must do everything possible to remove the traces of heat treating salt if low maintenance is to be achieved. The redesign of the element will act as a safety factor to prevent excessive burn outs in case some heat treating salt gets into the furnace occasionally.

3. If it is found impractical to follow out recommendation 1, and the new redesigned coil we are supplying for recommendation 2 does not reduce the maintenance, then consideration should be given to rebuilding the subject furnace to gas fired instead of electric.

Suggestions for Writing

The following subjects are suitable for recommendation reports.

A. For a report of approximately 300 words:
 1. A particular desk lamp.
 2. A recipe for a dish adapted to a particular occasion (for example, a one-dish meal, a birthday cake, a barbecue).
B. For a report of approximately 600 words:
 1. A particular elective course, to be taken for personal interest only and not to increase professional competence.
 2. A particular "educational" TV series.
C. For a report of approximately a thousand words:
 1. A particular place in which to spend a vacation; specify the length of the vacation.
 2. A particular hobby.
D. For a report of approximately two thousand words:
 1. The contents of an emergency survival kit.
 2. The wear and care of contact lenses.

14

Proposals

Introduction

In this research- and development-oriented age in which we live, the proposal is unquestionably one of the most important types of technical writing you are likely to be called upon to do. Just exactly what is this important presentational form? In its most basic sense, a proposal is a written offer to solve a technical problem in a particular way, under a specified plan of management, for a specified compensation. Put in a slightly different way, a proposal is a document designed to persuade a "customer" that the individual or organization presenting it can provide a useful service or product and — in the probable event that competition is involved — can provide it more satisfactorily than any competitor can.

 Some proposals are quite brief. Often these are prepared by a single author, who may be the person who is to provide the service or product. Closely related to this type of proposal is the proposal a student might submit to a professor for an academic project to be undertaken for college credit. An example of such a proposal is included in the illustrative material at the end of this chapter.

 Often a proposal is a very long report. Instead of being prepared by a single individual, it may require a considerable number of writers assisted by a number of other people who contribute information. Most of this chapter is not a discussion of the kind of report you may produce alone, but rather an explanation of a kind of large-scale project in which you may be involved.

One highly specialized form of proposal that is often prepared by a single author is an application to a government agency for funding of a research project. We won't discuss this specialized form, but you can obtain some advice about it by writing to the Superintendent of Documents, U.S. Government Printing Office, Washington, D.C., 20402, for "Winning a Research Bid" (Catalog No. HE 5.212:12033-A).

In this chapter we explain what a proposal is, briefly review the major steps in the preparation of such a report, consider the evaluation of a proposal, and comment on the principles of organization involved. Although you may seldom or never be included in the evaluation of a proposal as the subject is dealt with in this chapter, knowledge of how it is often done may help you judge your own.

What a Proposal Is

As we remarked in Chapter 13, a proposal is a specialized form of recommendation report. Keep that concept in mind as we consider our other statement that the proposal is a "written offer to solve a technical problem in a particular way, under a specified plan of management, for a specified compensation." Let's take a closer look at each part of this definition.

The *written offer to solve a technical problem* describes, often in minute detail, the design or plan proposed, sometimes along with some discussion of alternate plans and designs. Strictly speaking, it is this written offer to solve the technical problem that is known as the "technical proposal."

The *specified plan of management* mentioned in the definition above is commonly called the "management proposal." In general, the management proposal explains to the prospective client precisely how the entire project will be managed, tells who (often by name) will manage it, and suggests a time schedule for completion of the phases of the project. One of the important purposes of the management proposal is to assure the customer that the work will be done by competent personnel during every stage, from prototype design study through manufacturing if the proposal is for "hardware," or from initial exploratory study to final solutions if the proposal is for "software." ("Software," in this context, refers to a report that provides answers to a basic problem or problems.) Moreover, the management component of the proposal makes certain that the lines of responsibility for quality and reliability — and for efficient communication between customer and supplier — will be firm and clear.

The phrase *for a specified compensation* refers to what is called the "cost proposal." This part of the proposal gives a detailed breakdown of costs in terms of labor and materials. Usually the specified compensation is a sum of money, but it need not be. A mining firm, for example, might propose to construct a road and some recreational campgrounds on a section of government property in exchange for access and mineral rights in the area.

Often, all three of these basic elements of a proposal are contained between one set of covers. Indeed, the entire proposal may be brief enough to be contained in a letter addressed to the intended "customer." Sometimes, however, each of the three elements is so long that it is bound as a separate document. And sometimes the customer requires that the three elements be in separate volumes because evaluation of the different elements will be conducted simultaneously by different groups of evaluators. More will be said about evaluation later. In form and appearance, except for the fact that sometimes it becomes very long, the proposal usually adheres to the general characteristics of a formal technical report. An example is found at the end of this chapter.

Proposals are commonly referred to as *solicited* (invited) or *unsolicited* (uninvited). The former, by far the more common variety, are submitted in response to an *invitation to bid,* sometimes called a "bid request," a "purchase request," or a "request for proposal." Government agencies, however, give these terms a special meaning. An "invitation to bid" from a government agency means that the agency is giving exact specifications for the product or service it is seeking; a "request for proposal" indicates that it has a problem for which it would like the bidder to propose a possible solution or solutions, and that it is willing to discuss the question of what the exact specifications should be. Government agencies publish invitations to bid in appropriate publications. Invitations for bids or proposals from private companies or organizations are either published in journals or mailed directly to selected companies.

Unsolicited proposals usually are prepared by a company in the hope that the excellence of the idea or plan they are proposing will persuade the potential client of the need for the service or product being proposed. Sometimes the unsolicited proposal is preceded by an informal inquiry of interest outlining the idea or plan. If successful, the unsolicited proposal results in what is often called a "sole-source" procurement. Such noncompetitive proposals are commonly restricted to those rather rare circumstances in which it is believed the proposing organization possesses a unique capability. Ordinarily, contracts for research and development are not let without competitive bidding.

Finally, we wish to emphasize the crucial importance the proposal bears to a company's success in meeting competition. A poorly conceived and ineptly presented proposal has an immediate and brutal effect because it fails to get a contract and thus results in less income for the company. Over a period of time, almost every employee of any organization that bids for contracts through proposals may be personally affected by the degree of skill with which the staff can write them.

With the foregoing characteristics of a proposal in mind, let's take a quick look at what happens, typically, when a proposal is prepared. Suppose a firm that manufactures jet engines for aircraft wants a new high-speed wind tunnel in which to test its engines. The firm sends out a number of "invitations to bid" to companies listed in its source file, companies whose work in this field the firm knows and respects. If it has no such file, it might advertise in some of the trade and professional journals. Accompanying each invitation may be a set of specifications together with a general discussion of the firm's needs and, possibly, a formal statement of work (list of items the firm wants done). A deadline may also be indicated. The specifications would, of course, set down the requirements the wind tunnel must satisfy.

People in the companies receiving the invitations to bid study the specifications and bid papers to determine whether to submit a proposal. Capability, availability of qualified staff and facilities, already committed schedules of work, and the possibility of profit are among their chief considerations in making their decisions. Companies deciding to bid then assign staff members to the various tasks involved in the preparation of the proposal. Two of the most important tasks, for example, are the development of a suitable design and the preparation of a cost estimate. During this period, each of the competing companies may confer with representatives of the engine manufacturer to acquire as thorough an understanding of the problem as possible (assuming the customer is willing to grant this favor to bidders). In fact, the engine manufacturer may convene what is called a pre-proposal briefing conference for representatives of the interested companies so that a thorough discussion of the problem can be held.

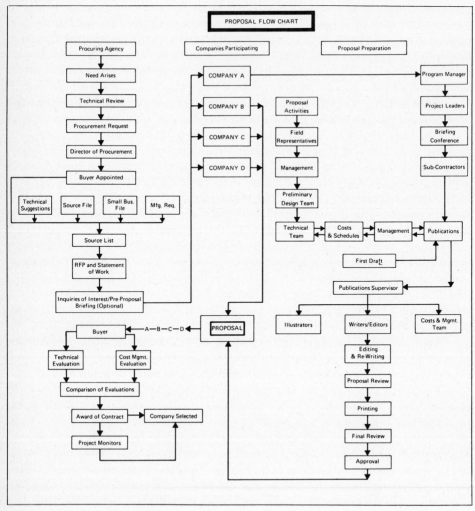

Figure 1 Proposal flow chart.

As each of the competing companies makes its final decisions, it prepares its written proposal. In this proposal, the company presents its design for the wind tunnel and explains how it would do the job. If the specifications have been quite detailed, the proposal may concentrate almost exclusively on how the tunnel would be built, rather than on what kind of design it would have. In any event, by the time the proposal is complete, a great many people will probably have contributed to it in one way or another. Proposals concerned with large projects are almost always team efforts, not the work of a single author, though there may be a proposal coordinator or manager and an editor charged with the responsibility of making the proposal conform to a single style and format throughout. The final effort is to make sure that copies of the proposal are mailed in time to meet the specified deadline for consideration.

When the engine manufacturing firm receives the various proposals, it designates certain of its staff members to evaluate them. Usually the evaluation team will include one group to judge the relative merits of the technical proposals, another to judge the cost proposal, and still another to judge the management proposals. Out of this evaluation process there finally emerges a decision as to which company gets the contract.

Innumerable variations are found from the general situation described in this hypothetical case. The work being proposed may range from the design and production of a small, simple device to projects of enormous complexity and cost. The number of people involved in the preparation of the proposal may range from one to hundreds. And the proposal itself may be a brief letter or an elaborate set of bound volumes. See the Proposal Flow Chart (Figure 1) for an indication of what happens with an elaborate procurement cycle.

You can easily imagine the importance attached to the proposal. Few experiences can be more frustrating to a company's executives than to feel sure they have developed a better technical design or plan than their competitors, only to see a competitor get the contract. The frustration is especially painful when the cost of putting out the proposal has run to tens or even hundreds of thousands of dollars.

Preparing a Proposal

Producing a good proposal consists essentially of the following stages: (1) making a preliminary study; (2) developing a plan or making an outline, including decisions about what to emphasize or focus attention on; (3) writing a rough draft and planning illustrations and layout; and (4) reviewing and revising. The process as a whole is in a sense completed in the evaluation procedure followed by the company to whom the proposal is submitted.

Making a Preliminary Study   In the broadest, most inclusive sense, the preliminaries to the drafting of a formal proposal begin with efforts on the part of those company representatives who have direct contact with customers. Such contacts naturally provide information as to what a customer's needs are, and also make clear to the customer what the proposing company's capabilities are. An effective proposal may have its real origin in these relationships, long before a formal request for bids is received. Once a request is received, a decision to bid or not to bid must be made, if this decision has not already been made during the period when the company was trying to get itself on the bidders' list.

Of course many people may have a hand in making the final decision to bid or not to bid. What is the role of technical staff personnel in this process? They begin by studying the request for proposal, commonly referred to as the RFP or sometimes the "bid request" or "purchase request." They study the specifications, if any have been supplied, and they study any other available material that may help them define the technical problem and determine a plan of attack on it. If this work leads the technical staff to recommend making a bid, and if other departments or divisions of the company are in agreement, then members of the technical staff cooperate with representatives from purchasing and accounting to prepare a cost estimate and begin the technical proposal.

The foregoing summary of the work of the technical staff preliminary to the writing of the proposal can be broken down into the following phases or steps: (1) detailed study of the invitation to bid, of the specifications, and of any related papers or information such as notes of a briefing conference or correspondence with the procuring agency or company; (2) study of background information, such as the reports of field representatives who have called on the procuring agency or company; (3) careful analysis of the probable competition; (4) strategic evaluation of the technical design or program to be presented; and (5) preparation of a tentative schedule for completion of the various phases of proposal preparation, which should allow enough time for careful review and editing of the finished document.

Developing a Plan A second stage in the preparation of a proposal, either a long or a brief one, is the writing of an outline. This skeletal representation of the material offers an opportunity for checking whether all necessary information is included; it gives a visual means of determining whether the parts of the discussion are in balance and in the most effective sequence; and, of course, it serves as a convenient prod to the memory in the writing process itself. When several people are collaborating in writing a proposal, an outline becomes an indispensable help in avoiding confusion. As the writing proceeds, a need for changes in development may be seen, possible even additions of information, and the formal outline is revised accordingly. An important consideration in the writing of the outline is the fact that it will finally appear in the proposal itself as the table of contents and as subheads within the text. Every entry in it should be intelligible to people who did not participate in the preparation of the proposal.

Writing a Rough Draft The third stage is the writing of the rough draft of the proposal itself and the planning of illustrations and layout. In principle, nothing new is involved here — except graphic aids, which will be taken up in Chapter 20. The techniques of definition, classification, and interpretation are likely to be especially helpful in the introductory portion of the proposal, where the problems presented are clarified and possible solutions considered. The techniques of describing mechanisms and processes often play a particularly important part in the main body of the report.

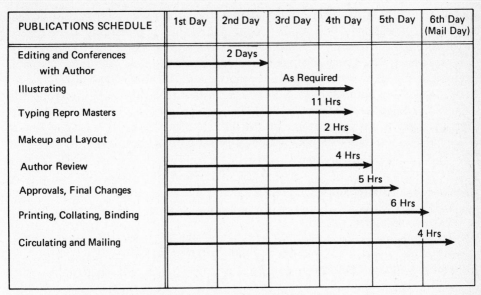

PUBLICATIONS SCHEDULE	1st Day	2nd Day	3rd Day	4th Day	5th Day	6th Day (Mail Day)
Editing and Conferences with Author		2 Days				
Illustrating			As Required			
Typing Repro Masters				11 Hrs		
Makeup and Layout				2 Hrs		
Author Review				4 Hrs		
Approvals, Final Changes					5 Hrs	
Printing, Collating, Binding						6 Hrs
Circulating and Mailing						4 Hrs

Figure 2   Example of an editing and production schedule chart.

In the preparation of a long report to which numerous staff members contribute, one of the chief problems is the coordination of the writing by all the individual staff members. Usually a single person is chosen to serve as coordinator. One other unusual problem that appears in the writing of proposals is "exception" taking; an exception is a deliberate decision not to attempt to comply with or to change a specification made by the customer. Too many exceptions, or exceptions that are not satisfactorily explained, result in the customer's charging the proposal with being "nonresponsive"; this charge is bad news for the company submitting the proposal. No exceptions at all are permitted on invitations to bid (in contrast to requests for proposals) from government agencies.

Reviewing and Revising   The final stage in the preparation of an effective proposal consists of review and revision. For a brief report, this job is either that of the author or jointly that of the author and a technical editor. For a long report to which a number of staff members have contributed, somebody has to be assigned this responsibility, usually the coordinator or a technical editor, or both. The first task is to review the rough draft to see where it could be made more effective by revision; the next is to make sure that the revisions are actually carried out and are well handled. Figure 2 is an example of a schedule for this entire process, including the final typing and production.

Evaluating a Proposal

When the proposal has been completed, it is submitted to the customer for evaluation. Normally, the evaluation procedure employed follows the pattern of the proposal itself. That is, three separate evaluations occur: (1) adequacy of the technical content, (2) feasibility of the management plan, and (3) acceptability of the cost or price information. In some procurement agencies and companies, a weighted point system is assigned to each of these elements, sometimes on a 100-point basis and sometimes not. We know of one agency that set up 2500 points as the perfect score, with the largest number of points, 500, allocated to technical content and the smallest number, 200, assigned to the bidding company's interest in the job. What-ever the system, it is fair to state that the lion's share of "points" goes to technical competence. A second example of an evaluation point system is as follows:

	Points
Problem understanding	25
Soundness of solution	25
Compliance with requirements	15
Design simplicity	10
Ease of maintenance	10
Capabilities and qualifications of company	15

The preceding evaluation point system applies to the technical content, chiefly; it does not include, obviously, any points for a cost evaluation. Another evaluation scheme assigns point values to the following criteria, with "technical aspects" roughly twice the value of any other single criterion.

Technical aspects
Relative costs
Time schedules
Technical competence of bidder
Management competence
Financial responsibility
Interest in proposed work

Still another agency reports the following scheme:

Scientific/engineering approach
    Understanding of problem
    Soundness of technical approach
    Responsiveness to requirements
    Special technical factors (unique aspects)
Bidder qualifications
    Pertinent experience
    Management organization for project
    Adequacy and availability of facilities
    Special or unique qualifications
Customer experience with bidder (if applicable)

Although we do not have precise point values in the last two examples, which were presented in a conference on proposal writing, technical competence and soundness were given the greatest emphasis.

Whatever point system or relative value scheme is made use of, discussions of proposal evaluation make it abundantly clear that the evaluators look for evidence of the following:

1. Understanding of the customer's problem or need
2. A sound and concrete technical solution to the problem
3. Compliance with customer's requirements or specifications, or a suggestion for improvement in the product or plan if exceptions are taken to requirements
4. Recognition of the possibility of solutions other than the one proposed, together with knowledge of their strengths and weaknesses
5. Clarity of presentation
6. Realistic and reasonable pricing
7. Financial responsibility
8. Sound and intelligent management planning
9. Adequacy of qualified personnel and facilities
10. Realistic time schedules for proposed work
11. An effective communication plan
12. Effective quality and reliability procedures

Although the faults of proposals are, largely, implicit in a listing of strengths such as the one above, the following are the most frequently heard criticisms:

1. Failure to understand customer's problem
2. Failure to explain adequately or satisfactorily the deviations from specifications or requirements
3. Overemphasis on company product or products (too much space devoted to successful performance on other contracts and too little to concrete solution of *this* problem)
4. Oversimplified technical treatment
5. Excessively optimistic estimates of performance and schedules
6. Inadequate test information
7. Too many vague generalities and sweeping statements
8. Misinterpretation of specifications
9. Too much "window dressing"—too much attention to attractive packaging of the proposal
10. Imbalance in presentation: too much space devoted to problems of interest to bidder but not to customer
11. Uncertainty of tone: overcautiousness about ability to carry out program
12. Ineffective presentation: wordiness, obscurity, irrelevancies, weak or nonexistent transitions
13. Unrealistic cost proposal
14. Weak performance record on past contracts
15. Insufficient detail in technical approach and management plan; vagueness about assignment of expert personnel

Organizing a Proposal

There is no single "best" way to organize a proposal. Examination of successful proposals reveals a great variety of patterns of organization. These differences seem to reflect both the character of specific problems encountered in the preparation of the proposals, and the individual thought processes of the authors. On the other hand, there are a few kinds of subjects that are found in most proposals, and recognition of what these subjects are helps in understanding what the possible patterns of organization are.

Naturally, most proposals begin with a statement of what the problem is, and they commonly turn next to a formal recommendation of action to be taken or to a summary of the solution proposed. Often, these elements are combined in an introductory summary (see Chapter 10). But this way of beginning is by no means a rule, and the organization of the remainder of the proposal cannot be predicted. The subjects taken up after the introductory elements have been presented, however, are likely to include the following: a detailed description of the product or service being proposed, an explanation of the plan of management for the project (together with identification of personnel and individual résumés of their careers), a cost analysis for the project, and a statement of the company's facilities for the work and experience with the kind of work involved.

Sometimes it is known that several companies will be offering a product that meets all the specifications contained in the invitation to bid. The organization of the proposals submitted by all these companies will then probably emphasize the price of the product, its relative quality, and the service each can provide. On other occasions, a company may believe that it can come closer than any of its competitors to meeting all the ideal standards for the solution of a problem; in this event, it would probably organize its proposal to emphasize these standards of judgment. And, of course, many other situations are encountered in practice.

To provide just one concrete example of a possible organization of a proposal, from the many that could be presented, we will conclude this discussion with an outline of one submitted by a company competing with several other firms. But please remember: *this is only one example; it is not a universally accepted pattern of organization.* There is no universally accepted pattern of organization.[1]

I. Introduction
   A. Statement of the problem
   B. Recommendation, or summary of proposed solution
   C. Scope and plan of the report
II. Discussion of the recommendation or of the proposed solution in the light of specifications or standards of judgment
   A. Discussion according to standard No. 1
   B. Discussion according to standard No. 2

---

[1] The literature on various phases of proposal writing is becoming quite extensive. *Technical Communication,* the quarterly publication of the Society for Technical Communication, has published many articles on the subject, including one with a bibliography of more than one hundred references. The Society has also published, in its Anthology Series, a collection of some 16 articles, entitled *Proposals and Their Preparation* (Washington, D.C., 1973). An especially useful treatment of the subject is by Frank R. Smith in *Handbook of Technical Writing Practices,* Vol. I, edited by Stello Jordan and published by John Wiley & Sons, Inc. (Wiley-Interscience Div., 1971). For additional references see the section on Technical Publications Problems in the Bibliography, Appendix A.

III. Management plan
    A. Organization
    B. Personnel
IV. Cost analysis
 V. Company's capabilities
    A. Facilities
    B. Experience
VI. Summary

Illustrative Material

In the following pages are two examples of proposals, first a student's proposal for a formal report project, and second, a portion of a lengthy proposal for a government contract. The student proposal, shown here in its entirety, is quoted by permission; names and addresses are obviously fictional.

1237 Westwood Drive
Sarnia, Texas 78999
March 10, 1985

Professor John J. Doe
Department of English
Texas Institute of Technology
Sarnia, Texas 78999

Subject: Proposal for formal report project

Dear Professor Doe:

For my formal research report assignment, I propose
to research the problem of computer security and
produce a formal report that would be of value to office
managers who have an interest in maintaining the
confidentiality of electronically stored records.

Since I am completing the requirements for a major
in computer science and am contemplating graduate
work in Business Administration, I have a strong
academic interest in knowing just how secure computerized information is and how secure it can be. I also
have an occupational interest in the subject because
computer security has become a much-discussed
subject in the office where I hold a part-time job as a
Data Processor.

Because of the limitations of space and time, I would
like to focus on just the matter of unauthorized access
to sensitive information and not get involved in such
matters as the loss of information due to mechanical or
electronic malfunctions.

As a result of taking all the required courses for a
degree in Computer Science and presently being
enrolled in six hours of electives, I feel confident of
my ability to understand the bulk of the literature I
will encounter in an in-depth study of the proposed

Professor John J. Doe   —2—   March 10, 1985

topic and will also have access to some information on
the subject that is not likely to be available in the
school library.

Rather than spend a great deal of time trying to
show the seriousness of the problem of insufficient
security, I want to assume a reader who already is of
the opinion that lack of adequate security can be
financially dangerous. Thus I would like to devote
only the introduction of the report to showing how
serious the problem is. The main body of the report
would be divided into just two sections--one for the
controls that are already widely practiced and one for
those that are too new, too expensive, or too complicated to have become widely adopted. In the first
section, I will investigate the reasons why the established methods aren't as successful as they should be.
In the second, I will evaluate the relative merits of the
more exotic controls. This is my present plan of
organization:

> COMPUTER SECURITY: PROTECTION FROM
> UNAUTHORIZED ACCESS

I. INTRODUCTION
   A. Examples and Consequences of Broken Security
   B. Reasons for Increased Concern for Security
      1. Massive shift to electronic storage of records
      2. Corresponding increase in number of people
         trained to use computer systems

II. CONVENTIONAL METHODS OF PROTECTION
   A. Physical Security
      1. Offices and data processing rooms
      2. Keyboards, monitors, and terminals
      3. Telephone lines
      4. Software
   B. Electronic Security
      1. Passwords & identification numbers
      2. Magnetic identification cards

Professor John J. Doe   —3—   March 10, 1985

III. UNCONVENTIONAL METHODS
　　A. User Recognition
　　　　1. Fingerprints and palmprints
　　　　2. Voice recognition
　　　　3. Retina scans
　　B. Encryption
　　C. Type and Frequency of Request Analysis
　　D. Transaction Reporting

IV. SUMMARY AND CONCLUSIONS

Among the sources I expect to make use of are the
following:

Becker, Hal B. Information Integrity. New York:
　　McGraw-Hill, 1983.
Berhard, Robert. "Breaching System Security." IEEE
　　Spectrum, 19 (June 1982), 24-31.
"Beware: Hackers at Play." Newsweek (September 5,
　　1983), pp. 42-48.
Eason, Tom S. and Douglas Webb. Nine Steps to Effec-
　　tive EDP Loss Control. Boston: Digital Press, 1982.
Hoyt, Douglas B. Computer Security Handbook. New
　　York: Macmillan, 1973.
Kolata, Gina. "Flaws Found in Popular Code." Science
　　(January 28, 1983), pp. 369-70.
Michelman, Eric H. "The Design and Operation of
　　Public Key Crypto-systems." AFIPS Conference
　　Proceedings. 1979 National Computer Conference.
　　48, 305-311.

　　I realize that some of these sources are a bit dated.
I'm sure there is a more recent edition of Hoyt's
Computer Security Handbook, for example. I intend to
use the most recent sources of information available
for this project.
　　In addition to the printed sources, I plan to interview Professor J. N. Fry. He is on the Graduate Faculty
of the Computer Science Department and has just

Professor John J. Doe   —4—   March 10, 1985

returned from a three-week trip to Los Angeles where
he was engaged as a consultant by an amalgamation of
software developers. I will ask him to review the
enclosed outline to see if there are any significant
omissions that I am presently unaware of.

I have two friends at the office where I work who
have agreed to read my rough draft to make sure I
have eliminated all the computer jargon that might
keep the report from being sufficiently clear to an
office manager who has not had specialized training
with electronic networks and data processing systems.

If this proposal is approved, I will begin further
research immediately and follow this time schedule:

ACTION	First Week	Second Week	Third Week	Final Week
(XX = Number of hours budgeted)				
Collect bibliography	XXXX			
Take notes	XXXXXXXX			
Organize notes & prepare initial draft		XXXXXXX		
Do supplementary research & revise draft			XXXXXXXX	
Prepare final draft				XXXX
Proofread, touch-up, bind, and submit the final draft				XXX

If the proposed project is not acceptable for any
reason, please indicate what changes would need to be
made to make it acceptable. Because of the preliminary
research I have already done and my strong interest in
the subject, I feel sure I can produce a respectable
report on the topic.

Sincerely yours,

Vivian Blair

Vivian Blair

The following proposal was prepared by the staff of the Atlantic Research Corporation for submittal to the Bureau of Aeronautics of the United State Navy. As you will see from the table of contents for the proposal, we have included the first three main sections only. Section IV gives a detailed breakdown of estimated costs in terms of direct labor, overhead, materials, communication and travel, and special capital equipment. Section V tells, by name, who will supervise and direct the work. Section VI itemizes the special pieces of capital equipment that will be needed to carry out the project. The last section describes briefly some projects the company has carried out in which experience applicable to the proposed work was gained. Detailed contract experience and personnel résumés are given in the appendixes.

Examination of the text of this proposal reveals that Section III accounts for about half of the total number of pages in the report; the reason for this proportion is quite clear. Since this is to be a research project, a principal strategy in the preparation of the proposal was to persuade the reader that the staff of Atlantic Research Corporation had a thorough grasp of the problem and would be able to work effectively toward its solution.

Design and Development of an Aviation Fuel Contamination Detector[2]

TABLE OF CONTENTS

Design and Development of an Aviation Fuel Contamination Detector

I. OBJECTIVE

The ultimate objective of the program will be to furnish the Bureau of Aeronautics with a prototype model of an instrument system which will determine accurately and reliably the degree of contamination in aviation fuels and which will meet the

---

[2] Quoted by permission of Atlantic Research Corporation, a division of The Susquehanna Corporation, Alexandria, Virginia.

service conditions and service requirements set forth in the General Information section of Request for Proposal No. PP-412-125-59.

II. BREAKDOWN OF THE WORK

The breakdown into three phases, including the eight items to be delivered, given in the Request for Proposal is suitable for this development project, and the work proposed here would follow the specified breakdown.

Phase I — Feasibility Study

The objectives of the feasibility study are: (1) to select the one, or perhaps two, physical or chemical phenomena most likely to be useful as the basis for the desired instrumentation system; (2) to examine the selected phenomena in some detail to determine the feasibility of making use of them for the desired measurements; and (3) to outline the experimental program required for the instrument development.

To achieve these objectives the work would include, but not necessarily be limited to, the following: (1) the collection of detailed information about the contaminants as found in the practical situation; (2) an examination of published literature to obtain as much information as possible about the physical and chemical properties of the contaminants and the fuels; (3) analyses of the various physical phenomena by which the presence of particulate contaminants suspended in aviation fuel can be detected and measured, in terms of the known properties of the fuels and contaminating materials; and (4) an examination of any instrumentation system proposed for development with regard to safety, reliability, and accuracy of operation in shipboard use.

It may turn out that not enough is known about the contaminants as found in fuel lines or that the parameters needed for an effective quantitative analysis of the performance of a proposed system are not reported in the literature. In this event, some experimental work may be required before the above-stated objectives can be achieved in full.

Phase II — Developmental Model

The objectives of Phase II are to design, develop, fabricate, test, and deliver a developmental model of the contamination detector.

The work of Phase I will have outlined the experimental program, and the precise nature of the work cannot be specified in advance of the study. In a general way, however, the work may be outlined as follows: (1) basic experimental investigation of the one, or perhaps two, phenomena chosen as the basis of the instrumentation system, to determine the effects of various parameters and to obtain an estimate of the signal output likely to be achieved in the final system; (2) tests of a measurement system, simulating in the laboratory the practical situation by using artificial contaminants and a small-scale flow system; (3) laboratory tests of a measurement system using actual contaminated fuel; (4) design, fabrication, and test of the developmental model.

We anticipate that the investigation of the basic phenomena and the design of the sensing element will require considerable work, but that the design of the associated electronic gear to provide the required display and alarm functions can make use of established techniques.

Phase III—Prototype Model

We assume that the developmental model will be given service tests and that these tests will suggest desirable improvements. The work of Phase III will involve, then, the design and test of a modified system, the design of the final system for production, and the fabrication and delivery of the prototype model.

III. PROPOSED METHODS OF SOLUTION

A. General Discussion

Under ideal conditions, there would be no great difficulty in making continuous measurements of the concentration of particles of a given material suspended in a fluid. The conditions presented by this problem are far from ideal, however. A multi-component system is involved in which the relative and absolute concentrations of the components, the particle sizes, and the particle-size distribution all vary. It seems unlikely, therefore, that the measurement of a single quantity, an absorption coefficient for example, will meet the requirements. It is more likely that several sensing elements will be required whose outputs will be electrically combined to give a single display. Another possibility is that a single sensing element can be used to make a specified set of measurements under different conditions, the set of measurements being repeated several times a second to give effective on-line operation.

A further complication may be presented by the fact that different lots of aviation fuel can vary somewhat in physical properties, even though the applicable specifications are met, and that this variation might be displayed by a given measurement system as a contamination. In this event, a comparison system might be adopted in which the output of a sensing element in the main fuel line is compared with the output of a similar element in an uncontamimated fuel sample and the difference in output displayed as contamination. The uncontaminated sample would be prepared continuously from the main fluid stream by filtration, with appropriate safeguards to insure that the reference sample is actually uncontaminated.

A rapidly moving stream of fluid may contain bubbles of air or vapor which could be falsely reported as contamination. Even turbulence might have a similar effect on certain measurement systems, particularly if an instrument of high sensitivity is used. If these effects are troublesome, it may be necessary to sample the main fluid stream continuously, evaporate the liquid in the sample, and measure the solid content of the resulting aerosol. A separate measurement of liquid water content would be required in a system of this kind.

In summary, we envision a multi-element sensor with the outputs combined to give a single display and the possibility that special continuous sampling techniques may have to be adopted to eliminate perturbing factors. Although we do not wish to minimize possible difficulties which may arise in this development, it does appear that the problems which can be foreseen have solutions which can be adapted to instrumentation requirements. The most important part of the project is the selection of the principle on which the sensing elements will operate so that an accurate, reliable, and safe instrument will result.

B. Principles of Operation

Our initial study of the problem has indicated a number of phenomena which should be investigated further in Phase I of the project. The discussion here is intended only to illustrate the possibilities, without implying that one of these may be successful or that these are the only phenomena to be investigated.

1. Attenuation of Sound: A study of the transmission of ultrasound in aqueous suspensions, recently reported, shows that the attenuation is large in the frequency range from 15 mcps to 30 mcps for aqueous suspensions of lycopodium spores (representative of organic matter) and of quartz sand (representative of inorganic matter). The mean diameter of the lycopodium spores was about 31 microns and of the quartz sand was about 2 microns. The attenuation was found to be strongly dependent on the number of particles per unit volume of suspension. Suspensions in which the particle concentrations were one or two parts per million by volume were tested, and in this concentration range the attenuation coefficient was approximately a linear function of the number of particles per unit volume. Considering a path length of one centimeter, the intensity of the beam was reduced by more than ten percent when the number of particles per unit volume was doubled.

The attenuation of sound depends on the temperature, the frequency, the particle size, and the nature of the particle, as well as on the particle concentration. These are complicating factors which must be considered in detail in designing an instrumentation system based on sound absorption.

The temperature range of the fuel in the practical situation may be small enough so that its effect can be neglected. If the effect cannot be neglected, a comparison system may solve the problem. Alternatively, a measurement of the temperature and a corresponding adjustment of the output meter of the instrumentation system may be operationally acceptable.

The effects of frequency and of particle size are inter-related. If the wave-length of the sound is short compared to the particle size, the particles contribute to the total attenuation principally by scattering. If the wave-length of the sound is long compared to the particle size the particles contribute to the total attenuation principally by frictional effects. In any event, the attenuation will depend on the particle size distribution, and we would expect that measurements would have to be made at more than one frequency, either by using a broad-band noise generator or by sweeping a single frequency over a given frequency band.

Figure 1 illustrates the principle of a possible system using a sweep frequency generator to drive a crystal sound-generator. The signal is received by a second crystal, after transmission through the fuel, and is amplified, detected, and displayed by appropriate electronic circuitry. A control system can be provided so that the amplitude of the detected signal in the pure fuel is independent of frequency or has any other desired frequency dependence. Introduction of particles into the system will markedly increase the attenuation in a certain frequency range, depending on the relation between the wave-length of the sound and the particle diameter. It may be possible to adjust the frequency range of the sweep and the frequency response of the system so that the output will be a measure of particle concentration, independent of particle size distribution. [Note: Figure 1 is not shown.]

The experimental conditions of the referenced work are close enough to those expected in measuring fuel contamination, and the reported effects are so large, that we regard the use of sound attenuation as the most promising principle of operation of all those we have examined.

2. Light Scattering: The light scattered by a suspension of particles is determined by the number, size, shape, and refractive index of the particles, the refractive index of the suspending fluid, and the wave-length of the light. In principle, then, light scattering can be used to measure particle concentration.

Light scattering phenomena are similar in some respects to sound attenuation phenomena, and much of the discussion of sound attenuation given above applies in a general way also to light scattering. In this case, also, one cannot expect a single measurement, of turbidity for example, to be adequate because of the effects of particle size distribution and of different materials. The additional measurements required might be made at different frequencies or at different angles with the incident beam. There is also the possibility of using polarized light.

Our preliminary examination of the problem leads to the following tentative conclusions for planning purposes: (1) a light scattering system can probably be devised; (2) a light scattering system will have all the complications of a sound-absorption system plus the additional instrumentation complication of combining an optical system with an electronic system; (3) the light scattered by some of the contaminating particles will not be great, because the refractive index is too close to that of the fuel, and sensitive electronic equipment will be required. For these reasons we regard a light scattering system as less likely to be successful than a sound attenuation system and therefore would give priority to investigation of the latter in this project.

3. Electrical Properties: Neither sound attenuation nor light scattering effects depend on the volume of the contaminating particles in any simple way, and it would be worth some effort in Phase I of this project to examine phenomena which do depend on volume. If one can be found which would give a sufficient output, the instrumentation would be much simplified.

The dielectric properties of a composite medium consisting of particles of one material dispersed throughout another material depend on the volumes and dielectric properties of the two materials, but do not depend on the particle size distribution. The measurable properties include dielectric constant, resistivity, and loss factor. If the concentration of one component is a few parts per million, the change in dielectric constant and resistivity is usually of this same order of magnitude. In certain cases, however, the loss factor is markedly changed with the addition of contaminants.

We expect the addition of particulate contaminants to a fuel would cause only small changes in dielectric properties and that such measurements would be difficult to make on a routine basis. Nevertheless, the advantages are great enough to warrant investigation.

A Suggestion for Writing

Write a three- or four-page proposal to undertake a project that will involve research and result in the production of a formal technical report (of about 5000 words) that will present the results of your research to a particular audience. Your proposal should be in the form of a business letter or a memorandum.

This assignment has two basic purposes: (1) to provide your instructor with the information needed to help you produce a high-quality report, and (2) to enable you to demonstrate your understanding of what a good proposal does and what it should contain.

Remember that a proposal is a written offer to do something. A good proposal will normally specify what is to be done, who is to do it, why it is to be done, how it is to be done, when and where it is to be done. An important consideration of the "why" is the matter of how you and the person (or organization) receiving the proposal stand to benefit from the proposed action. An important consideration of the "how" is the matter of convincing your reader that you are able to deliver what you propose. Keep in mind that you are offering to do something but that the reader is under no obligation to accept your offer.

To increase the likelihood of acceptance, you will naturally wish to make your offer as attractive as possible without exceeding your capability to deliver. The more clearly the proposal addresses the reader's needs, the more attractive the proposal is likely to be.

For this assignment, your writing instructor will need two kinds of information: (1) information about the research to be conducted, and (2) information about the report that is to result from that research. Below is a list of some of the more important questions that a reader of your proposal is likely to need answers for:

1. What subject do you wish to investigate?
2. What specific aspect of that subject do you wish to focus on?
3. What are your reasons for wanting to research this topic — and for writing a formal report on it?
4. What qualifications do you have for undertaking such a project (both the researching and the writing)?
5. How do you intend to acquire information about the subject? For a library research project, you should give a detailed account of your search for published materials: what you found in the card catalogue, the periodical indexes you consulted — and the time period covered — reference works examined, etc. You should include in your proposal (or attach separately) a tentative, preferably annotated, bibliography of the sources you have so far located, and you should explain what additional bibliographical resources you intend to make use of for additional information.
6. How is the report to be organized? The plan of organization is likely to change significantly as a result of your finding information that you don't presently have, but a tentative outline at this point can help to focus your research and persuade your reader that you know where you are headed.
7. What sort of audience will the report be designed for? Your instructor may, of course, specify whether you are to write for an informed or uninformed reader, whether your reader might be expected to have an interest in the subject, etc.
8. How will you budget the time available for research and writing so as to be sure the report will be completed on time?

If each of these questions has been dealt with satisfactorily in your proposal, the probability of acceptance is very high. Note, however, that the decision to approve or disapprove your proposal may not be an indication of the quality of the proposal itself. It is possible to write an excellent proposal for a poor project and to write a very poor proposal for a very good project.

15

The Progress Report

 Introduction

One easily distinguishable type of report is the progress report — distinguishable because of its purpose and general pattern of organization. This chapter explains how to prepare a progress report.

The progress report's main objective is to present information about work done on a particular project during a particular period of time. It is never a report on a completed project; in some ways it is like an installment of a continued story. Progress reports are written for those who need to keep in touch with what is going on. For instance, executives or administrative officials must keep informed about various projects under their supervision to decide intelligently whether the work should be continued, given new direction or emphasis, or discontinued. The report may serve only to assure those in charge of the work that satisfactory progress is being made — that the workers are earning their keep. Not the least important function of the progress report is its value as a record for future reference.

It is neither possible nor worthwhile to list here the extent of the activities on which progress reports are made. Any continuing, supervised activity may have progress reports made on it — anything from research projects in pure science to routine construction jobs. Nor is it possible to be dogmatic about the frequency with which such reports appear; often progress reports are made on a monthly basis, but sometimes the week may be the time unit, or the quarter-year. Regardless of the time element, the period covered in the report has little to do with the way the report is organized and presented.

Organization and Development

About the best way of getting at the problem of what should go into a progress report, and how, is for the writer to consider what the readers will want to find in the report.

Common sense tells us that the readers will want to know at least three things: (1) what the report is about, (2) what precisely has been done in the period covered, and (3) what the plans are for the immediate future. Quite naturally, they will want this information given in terms they can readily understand, and they will expect it to be accurate, complete, and brief. Great emphasis is often placed on brevity.

The foregoing suggests a pattern of organization as well as some clues regarding development of the report. From the standpoint of organization, there should be three main sections: a "transitional" introduction, a section giving an account of progress made during the current period, and a "prophetic" conclusion.

The Transitional Introduction   In the first of these sections, the transitional introduction, the reporter must identify the nature and scope of the subject matter of the report and relate it to the previous report or reports. The writer may be expected to summarize earlier progress as a background for the present account. Finally, if circumstances warrant — or if it is expected — the writer may present a brief statement of the conclusions reached in the present unit of work and, possibly, some recommendations. This latter function is especially applicable in progress reports on research projects. It is not so pertinent in an account of the progress on a construction or installation project.

In serving as a transition between the current report and the preceding one, this part of the report need not be lengthy, for it is essentially a reminder to the readers — a jog to their memories. Reading it gives them an opportunity to recall the substance of the previous reports so they can read the present one intelligently. The title may partially bridge the gap between reports, for it may name the project and number the report. Something like "Boiler Installation in Plant No. 1, Progress Report No. 5" is characteristic. But even such a descriptive title is not enough, and many reports do not bear such titles (see the example reprinted at the end of this chapter). The discussion — or the briefing — is needed to hook the current report securely onto the preceding one.

The Body of the Report   With the introduction out of the way, the reporter must next tackle the body of the report — the detailed account of current progress. The first point that needs to be stressed here is the importance of making this part of the report complete, accurate, and clear. This is much easier said than done, mainly because it is easy to forget the reader. Remember that the report is not a personal record for the writer, but contains information for some particular reader or readers about the work done. If you keep this in mind, you should have very little trouble.

The second thing that needs to be said concerns organization. Although some progress reports are organized chronologically with subsections covering parts of the overall period (a monthly report might have four subdivisions, each being a running narrative account of the work done during a week's time), most of them are organized topically. For instance, a report of progress made on a dam construction job contained the following subdivisions: (1) General [interpretative comments], (2) Excavation, (3) Drilling and Grouting, (4) Mass Concrete, and (5) Oil Piping. A report of progress made on the production of an aircraft model contained these topical subdivisions: (1) Design Progress, (2) Tooling, (3) Manufacture, (4) Tests, and (5) Airplane Description. The sample progress report included in this chapter provides another example. But these illustrations should not be regarded as prescriptions. The important thing is that the development of the main section of the report should grow logically out of the subject matter itself and the requirements of those who want the report.

Giving a careful, detailed account of work done may require the presentation of quite a mass of data. Usually, such data, particularly numerical data, cannot be presented in the conventional sentence-paragraph pattern; it would be unreadable. Tables, of course, are the answer. But since you will want to make your reports as readable as possible, you will do well not to interrupt your discussion with too many tables. It is better to put them in an appendix at the end of the report and confine yourself to evaluative or interpretative remarks about the data in the body of the report itself. But don't forget to tell the reader that the tables are in the appendix. For instance, the report on a dam construction mentioned in the paragraph above contained in its appendix a table giving an estimate of quantities of material used, one on unit and concrete costs, and another giving the type and number of employees along with the amount of money paid out for each. Here is the first of these tables:

Estimate of Quantities — Week Ending April 25, 19____

Bid Item	Description	Unit	Previous Total	This Period	To Date
1	Mass concrete	cu yd	787,686	18,792	806,478
2	Steel reinforcing	lb	2,369,350	29,883	2,399,233
3	Black steel pipe	lb	213,107	666	213,773
4	Cooling pipe	lin ft	317,417	188	317,605
5	Electric conduit	lb	367,480	309	367,789
6	Copper water stop	lb	35,424	856	36,280

The presentation of data such as this in connected reading matter would be difficult, to say the least; moreover, this is a short table — each of the others contained four or five times as much data. Although tables are a great convenience and sometimes a necessity, remember that they should not be allowed to stand alone without comment.

The Conclusion With one exception, the requirements of the conclusion to a progress report will depend entirely on the nature of the work reported on. If progress on research is being reported, for instance, it may be necessary to present a careful, detailed statement of conclusions reached — even though these conclusions have been briefly stated in the introduction. It may also be desirable to make recommendations about action to be taken as a result of present findings or about future work on the project. On the other hand, it is not likely that a report on the progress made on a simple machine installation would require formal conclusions or recommendations. But you are not likely to have trouble with this problem, for the nature of the subject matter will suggest naturally what should go into the last section.

There is one thing, however, which you will do in almost all progress reports, regardless of subject matter, and it is suggested by the term "prophetic" used earlier. You must tell your readers approximately what they may expect the next report to be about and what its coverage, or scope, will be. Along with this forecast it may be advisable to estimate the time necessary for completion of the entire project. Here is an important caution: Don't promise too much. It is very easy for the inexperienced worker to overestimate the amount of work that can be covered in a forthcoming period. You will naturally want the forecast to look promising, but you will not want it to look so promising that the readers will be disappointed if the progress actually made does not measure up to your prediction.

A final word of advice: Be brief but complete, and use the simplest terminology you can.

Form

We have discussed the content and presentation of the three main parts of the progress report — introduction, body, and conclusion. There remains the problem of form. Two forms are used for progress reports, the choice depending on the length and complexity. They are the letter or memorandum form and the conventional or formal report form. The first is used for short reports submitted to one individual or to a small number of persons. The second is used for longer reports, submitted perhaps to an individual but more often for circulation to a number of company officials and perhaps to stockholders and directors as well.

The memorandum has the conventional "To," "From," and "Subject" heading (see p. 307). The letter report has a conventional heading, inside address, and salutation. Many are in military form (especially those written on government contracts for research and development projects; see the illustrative form on p. 285). The opening paragraph makes reference to the preceding report and identifies the nature and scope of the present one. The remaining parts of the report are usually labeled by means of marginal headings, which correspond to the subject-matter divisions. The conventional ending is the complimentary closing, "Respectfully submitted," followed by the signature. This form is especially suitable in those organizations where the report serves primarily as a means of "accounting for" the reporter's activity. Besides, it has the advantage of the personal touch.

The letter, however, is not suitable for long reports of progress on elaborate projects submitted for wide circulation to sponsors or directors. For one thing, the letter loses its identity as a letter if it extends over a large number of pages, especially since marginal subheadings are usually employed. There may be, of course, a letter of transmittal. But the report proper will follow the pattern decribed in the chapter on report format (Chapter 19).

Illustrative Material

The material on the following pages includes three illustrations: (1) a student's report, in memorandum form, of progress on the preparation of a term report; (2) an illustration of a typical form for a letter-form report; and (3) a "formal" progress report from a research organization.

The second of these examples, the letter-form report, illustrates the form only; we will comment on it here so that most of a page can be given to the example itself. The contents of such letter-form reports from the Applied Research Laboratories are classified information, and therefore we cannot present more than the form itself. One such report, however, contained seven subdivisions in the "Summary of Activity" division, each dealing with progress in a specific area. In addition, two brief reports of trips made for conferences with individuals in cooperating organizations were attached. The report as a whole was five pages long. Two unusual elements in the form shown are that there is a "Subject" heading, as in a memorandum, and that there is no salutation.

Example 1

To: Professor _____

From: _____

Subject: PROGRESS ON TERM REPORT FOR WEEK ENDING _____

In the previous progress report, completion of research on the term report was announced, and also the writing of part of the Introduction.

During the past week three things have been accomplished. Some changes have been made in the outline of the report, the note cards have been reorganized accordingly, and the Introduction and Section II have been completed. The revision of the outline was done after writing had been started, when it became evident that the original plan for the report could be improved upon. The new outline is as follows.

THE FISCHER-TROPSCH PROCESS
  I. Introduction
 II. The Process
        Production of Gas
        Purification
        Contact Chambers
        Product Recovery
III. Products of the Process
        Permanent or Residual Gas
        Gasol
        Gasoline
        Diesel Oil
        Wax

Fatty Acids
Alcohols
IV. Factors Affecting the Process
V. Conclusion

The original outline was presented in the report for the week
ending ——————.
In the coming week, the term report will be completed.

Example 2 Typical format for a letter-form progress report.

THE UNIVERSITY OF TEXAS AT AUSTIN
APPLIED RESEARCH LABORATORIES

POST OFFICE BOX 8029   IOOOO FM ROAD 1325   AUSTIN, TEXAS 78712

AC 512, 836-1351

Serial No.
(Date)

Sponsoring Agency
Attn: (Code or Office Symbol and name of Project Engineer as appropriate)
(Contract Number)
Contractual Line–Item Reference (e.g., as identified on DD Form 1423)
(Address of Sponsoring Agency)

Subject: Research and Development Technical Performance Summary,
        (Applicable Monthly Period)

       This is the – – – research and development summary submitted under
Contract – – – , Project – – – . . . .

Summary of Activity

(Normally this section is limited to two or three very concise paragraphs summarizing
activities for the month. Reference is made to attachments such as Trip Reports,
Visit Reports, Technical Opinions, Technical Memoranda, and Professional Resumes
of newly-assigned personnel.)

       Summaries of Activity for Specific Tasks. (Reasonably detailed, but brief,
nontechnical summaries of individual tasks, identified by title, which received attention
during the reporting period are presented. Depending upon the level of effort, from two
to twelve or fourteen such summaries may be treated.)

Summary of Man-Hours

(Man-hours expended during the reporting period are tabulated by classification of the
investigators or by name. A cumulative total of man-hours through the reporting date
may be required.)

Items Requiring Coordination

(Specific items requiring action and/or coordination on the part of the Sponsor are
identified.)

Work Schedule for Next Reporting Period

(A projection is made of specific action items to be considered during the ensuing period.
Also, planned trips and expected visits are identified.)

                             (Signature of Responsible Official)
                             Name and Title

– – – / – – –
Enclosures (n)

Example 3

Flight Dynamics of Ballistic Missile Interception[1]
Study Coordinator: Dr. James Ash

STATEMENT OF THE PROBLEM

This study concerns the problem of active missile defense against a ballistic missile. The study is confined to the elementary case of interception of a single ballistic missile by a single defensive missile. The primary objectives of the study are to:

a) Determine what information is required for the specifications of an interceptor system for an ICBM.

b) Specify the performance requirements for an AICBM system.

c) Formulate the mathematical description of the system.

d) Compare and evaluate various systems under various flight situations.

The results of this study should furnish information necessary for the planning of more complex regional defense systems against actual multiple ICBM attacks.

CURRENT PROGRESS

Study 14 terminated 31 August 19____. The results are contained in a forthcoming final report entitled, WADC TR 59-516, *Flight Dynamics of Ballistic Missile Interception*. A résumé of that report follows.

The problem of an active unitary interceptor system operating against a ballistic missile is studied to determine the most suitable functional forms of the system. Particular attention is given to the intercontinental ballistic missile, and it is assumed that detection and tracking are accomplished by radiation means from friendly territory. Analytical methods and procedures are presented for the investigation of the ballistic missile vs. countermissile duel with consideration of the ballistic missile approach speed and angle, detection range and tracking range capabilities of the interceptor system, preparation time of the inter-

The first two paragraphs of this progress report are a statement of what the report is about. An unusual feature here is the fact that the clearest statement of the relationship between this report and previous work is contained in the first sentence of the second major section, rather than at the beginning. Moving that sentence ("Study 14 terminated 31 August 19____.") to the beginning of the report would be an improvement —perhaps in the following form: "This report is concerned with work done under Study 14, which was terminated 31 August 19____." Then the first sentence under "Current Progress" might begin, "The results of Study 14 are contained. . . ."

The general organization of this progress report is characteristic of the type. At the end, however, the author is evidently suggesting what work could be done if the contract were renewed, rather than promising what will be done under a contract still in force.

[1] University of Chicago, Laboratories for Applied Sciences. Work sponsored by United States Air Force under Contract No. AF 33(616)-5689.

ceptor missile, and lethal radius of the interceptor warhead.

Methods of computation for reaction time and range relationships are developed for both minimum-energy and nonoptimum ICBM elliptic trajectories. Refinements of the Keplerian elliptic trajectory for the effects of air drag and nonsphericity of the earth are considered for accuracy computation. Expressions have been developed for the effects of observational errors on the predicted orbital elements. The re-entry phase has been considered and equations are provided for the estimation of path deflections and energy emission due to air drag. Frequency distributions of probable United States targets have been compiled to provide estimates of expected ranges and azimuth angles in the event of an ICBM attack. The geometrical limitations of detection and tracking, and visibility zones for observation stations have been graphed in relation to the parameters of the ICBM trajectory and reaction time.

The AICBM radar factors, including the shortrange requirements and dependence on attenuation, the propagation factor, antennae, power sources, radar cross-sections, noise effects, and the present and projected capabilities of AICBM radar systems are discussed and analyzed. Expressions and graphs are formulated for the computation of expected elevation angle error, range error, and Doppler velocity errors for radio propagation through successive layers consisting of the troposphere, the ionosphere, and free space. The possibility of improved detection by means of low-frequency radio waves is examined, and a theoretical analysis of plasma shock waves has been made in an effort to understand the physical mechanism underlying recent low-frequency reflection observations. In addition to radar, the feasibility of detection and tracking by means of passive infrared techniques, including consideration of the radiation from the rocket exhaust, optical limitations, and atmospheric effects have been examined.

The general dynamic equations of the interceptor rocket have been formulated with consideration of the effects of variable mass and changing of inertia as the fuel is expended and of the secondary effects of Coriolis and centrifugal accelerations due to the motion of the earth. Generic interceptor equations of motion suitable for flight simulation studies are

given. The theoretical background for the estimation of high altitude drag by means of free-mole-
cule flow considerations is reviewed. The interceptor performance parameters and their
interrelationships involving weight ratios, specific
impulse, slant range, and reaction times have been
developed for both the ballistic and powered
phases, as well as a method for computing
dispersions by means of perturbations about the
basic trajectory.

FUTURE WORK

The restrictions under which Study 14 was
pursued include (1) free flight phase of a ballistic
missile trajectory, (2) single interceptor-ballistic
missile duel, and (3) land-based interceptors and
land-based ballistic missiles. Future work would
call for a relaxation of these hypotheses to include
more realistic situations involving a more detached
analysis of some specific interceptor systems
operating against specific ballistic missiles in certain selected flight situations.

The publication of WADC TR 59-516 obviates the
necessity for a classified supplement to this
Quarterly Report.

An appendix lists trips and visits
made during the quarter and
another lists technical notes and
reports issued under the research
contract.

Suggestions for Writing

1. Write a progress report giving an account of the progress you have made to date
   on a long report assignment. If you are writing a research report, include in your
   progress report an account of library research (indexes consulted, books available
   on the topic, general reference works, and so on), note taking, making of illustrations, rough draft, and the like — anything pertinent. Include a statement
   about what remains to be done and a prediction of the anticipated date of
   completion. Additional reports may be made later on this same project.
2. If you are engaged in any sort of extended laboratory experiment in one of your
   technical courses, make a progress report on the work accomplished to date.
   Assume that it is being made to someone unfamiliar with the technical nature of
   the subject matter. Use conventional report form — title page, table of contents,
   and so on — but omit the letter of transmittal, since the introduction can perform
   its function.
3. Assuming that your technical writing course is the project, write a report of the
   progress made during the preceding month. Do not forget to include in the
   beginning section a brief statement or synopsis of earlier progress that (you will
   assume) has already been reported. Put this in letter-report form and address it to
   a hypothetical educational adviser.

16

Oral Reports

Introduction

The purpose of this chapter is to make a few practical suggestions about talking with people. For the most part it will be concerned with talking rather formally to an audience, but some attention will be given also to conferences. This chapter is not a substitute for a course in speech, nor for reading a good textbook on speech[1]— both of which we strongly recommend. Rather, our suggestions are merely a brief introduction to a broad and important subject. Emphasis will be given to speech problems especially common in the technical field.

Most of what has been said earlier in this book about the organization and language of technical writing applies to speaking on technical subjects as well. The discussion that follows will be confined to factors that appear only because the form of communication is oral, rather than written.

Making a Speech

To be an effective speaker you must know how to use your voice properly and how to maintain a good relationship with your audience. These subjects aren't as formi-

[1] See Howard H. Manko, *Effective Technical Speeches and Sessions* (New York: McGraw-Hill, 1969) or Michael Spitzer, Michael Gamble, and Teri Kwal Gamble, *Writing and Speaking in Business* (New York: Random House-Knopf, 1984).

dable as they sound. Actually, the chief need of the novice speaker is simply the application of common sense — and practice. In addition to the subjects mentioned, we will comment on transitional material, graphic aids, and the question period at the end of a talk.

The Voice It is impossible to become a polished speaker without making speeches. Practice is unquestionably the most important single element in acquiring skill. Advice on how to speak is often ineffective until practice begins to lend meaning to it.

Fortunately, there is one aspect of speech making that each of us practices every day, at least to some degree. We all talk. We all say words. So we might as well practice saying words in a way that is pleasant to hear and easy to understand. Here are four suggestions that are helpful, whether you are talking to one person or a hundred.

1. *Relax.* Tenseness causes the muscles in the throat to constrict and raises the pitch of the voice. Your lungs are a pair of bellows forcing air through the vocal cords. The force is applied by muscles in the abdomen. When you are relaxed and speaking naturally the vocal cords vibrate easily.
2. *Open your mouth.* Speaking with your mouth insufficiently opened is like putting a mute on a trumpet. This fault is one of the commonest causes of indistinct speech. If you find you're having trouble being heard, you'll probably feel that you look ridiculous when you first start opening your mouth wider. Look in a mirror. Watch other people.
3. *Use your tongue and lips.* We remember a student who announced he was going to explain how to graft ceilings. It sounded ominously political, until it turned out he meant "seedlings." You can't say a "d" or a "t" without using the tip of your tongue. Nor can you say "b" or "p" without using your lips. Repeat the alphabet and notice the muscular movements required for the different sounds. It's a mechanical problem. As with opening the mouth, you may feel foolish if you start using your tongue and lips more than you have been doing. Of course you may sound foolish, too, if you overdo it. Make the sounds clearly, but not affectedly. Listen to and watch a good TV announcer.
4. *Avoid a monotone.* It's hard for a speaker to interest an audience, or a companion, in a subject in which the speaker doesn't sound interested, and there is nothing interesting in a monotonous drone. Enthusiasm is naturally shown by a variation in the pitch of the voice to match the thought being expressed. See how many shades of meaning you can give the following sentence by varying the pitch of your voice: "You think he did that?"

Some other suggestions that are also related to the problem of using your voice to best advantage are these:

1. Pronounce syllables clearly. Don't substitute "Frinstance" for "for instance." Be sparing of the "I'm gonna because I gotta" style of pronunciation.
2. Give a little attention to the speed with which you talk. Moderation is a good principle: neither very fast nor very slow.

3. Try to talk along smoothly, with fairly simple sentence structure, and without repeatedly saying "you know" or "and-uh," and without long pauses between groups of words. It is probably best not to think much about sentence structure in your first few speeches. Concentrate on what you have to say and keep going. But you can practice good sentence structure every day in conversation.
4. Speak loudly enough so that everyone you are addressing can hear easily, but don't blast people out of their seats.

Your Relationship with Your Audience   The audiences you can expect to address as part of your professional work will be made up of people who are seeking technical and economic information, not a show. Typically, you may expect to address fellow members of professional societies, fellow employees conferring on special problems or meeting on special occasions, prospective clients, and so forth. Aside from reports made in college classes and seminars, your first speech is likely to be before a chapter of a professional society.

With such audiences, your relationship should be unaffected and unassuming, but at the same time confident and businesslike. You should by all means avoid anything approaching what is sometimes called florid oratory. Say what you have to say as directly and simply as possible.

As for posture, the best advice is to be natural—unless nature inclines you toward sprawling limply over the table or lectern. Stand up straight, but not stiff, and look at the audience. If you feel like moving around a bit, do so; but don't pace, or walk away from a microphone if a public address system is being used. If you feel like emphasizing a point with a gesture, go right ahead, but don't make startling or peculiar flourishes that will interest the audience more than what you are saying. In general, it is wise to move slowly. Don't do anything (like toying with a key chain) that will draw attention away from what you are saying.

Above all, act like a human being, not a speech-making automaton. Try to convey to the members of your audience a feeling of interest in your subject; show that you enjoy talking with them about it. A particularly useful device is to bring in occasional references to personal incidents involving yourself or your co-workers, incidents that have some relation to your subject and may be used to illustrate a point. People are always interested in other people, and an appropriate personal anecdote may warm up and give life to an otherwise dull body of information.

What can be said about preparing for such a performance? You can choose among three basic possibilities. You may read your speech from a manuscript, you may memorize the speech from a manuscript, or you may deliver it "off the cuff," using a few notes if necessary.

The last method, with or without notes, creates an impression of spontaneity and naturalness that is greatly to be desired. The use of notes is not a significant barrier to this impression and is a considerable support to self-confidence. Very often, however, custom calls for, or sanctions, the reading of a paper. This method is especially desirable when the material to be presented is complex, as it is likely to be in meetings of professional societies. The possibility that you would need to commit a speech to memory, word for word, is very remote.

If the speech is given extemporaneously and notes are used, it is generally wise to put them on small cards, to type them or write them clearly, and to indicate only major headings. Too much detail in the notes might result in confusion. You might lose your place.

The initial preparation of the speech, whether notes are used or not, is like the preparation of any report. That is, an outline should be made first (some differences in content will be noted later). If the speech is to be memorized, or read, the outline is used as a guide in writing the manuscript. If the speech is to be delivered more spontaneously, the writing step is omitted and the outline becomes a guide for practice (to a friend or relative) and a basis for the notes.

Naturally, you will want to learn all you can about your subject. Make it a point to know more about every phase of it than you expect to reveal. This extra information is like armor between you and the fear of running out of something to say when you get up to speak.

It may be helpful to read something aloud, in private, and at your normal speaking rate, to count the words per minute and thus estimate the number of words you'll deliver in the time allotted to your speech. But remember that almost everybody uses more words to cover a given subject when speaking than when writing. Don't underestimate the length of your talk and keep your audience longer than they expected: they won't like it.

Finally, there is the problem of nervousness. When confronted with the necessity of making your first speech, you may feel that your problem is not to speak well but to speak at all. Nervousness is best regarded, however, as a nuisance that will diminish with experience. Most people never do get over feeling a little trembly when they arise to speak. There are two sources of comfort with regard to this matter. One is that you are almost certain to find that after you have been on your feet a few minutes, the going is easier. Speak slowly at first, and pause for a good breath now and then. The other comfort is that your nervousness will be less apparent to the audience than you think. We once sat in the front row of an audience to which a young engineer was making one of his first speeches. We were thinking what a fine job he was doing, and what composure he had, when we just happened to notice that the knees of his trousers were vibrating at what we roughly estimated to be ten cycles a second. He made an excellent impression, and it is doubtful that anybody else in the room knew that his knees were shaky. Speak whenever you have the chance; experience will put you at ease.

Maintaining an effective personal relationship with an audience is exceedingly important in making a speech, but so is maintaining clarity. In this connection, transitional devices deserve a comment.

Introductions, Transitions, and Conclusions   Two problems faced by a speaker, but not faced by a writer, are that an audience cannot be expected to give unwavering attention to what the speaker says, and that the audience cannot turn back to review an earlier part of a speech. Consequently, the speaker is under a heavy obligation to provide clear introductions, transitions, and conclusions. There is nothing new in principle here, and you need expect no special difficulty if you give careful thought to the matter. Sometimes, if you are using notes, it is helpful to indicate points at which transitions are needed. A glance at the headings on the card will supply its content.

A third problem that should be mentioned is the possible need for a more dramatic introduction in a speech than would seem necessary in a written report. We said earlier that, for the kind of audience you are likely to have, you should be supplying information rather than putting on a show. That statement holds true; nevertheless, it is almost inevitable that a speaker will find it desirable to heighten interest by using certain devices (still far short of putting on a show) that would seem out of place in a written technical report. One such device has already been mentioned; that is, simply an attempt to make the whole delivery animated and enthusiastic; a second is the use of personal anecdotes; a third is the use of graphic aids to lend drama and emphasis to your discussion; a fourth is the use of a dramatic introduction or conclusion. It is worth noting here, however, that caution is always necessary in an attempt to be dramatic or humorous. If you feel uncertain of success, don't try.

Graphic Aids   Philip H. Abelson,[2] a distinguished scientist and editor of *Science,* wrote a very blunt statement about the problem of communications in science. What he had to say contains a particular comment on graphic aids:

> . . . when it comes to communicating, few scientists are skillful. The majority cannot even effectively convey scientific information to each other. This is true of verbal presentations, in which decade after decade most scientists use slides that cannot be read beyond the front row of the audience.

What can be said about graphic aids that will be helpful in making them genuinely useful and not the source of annoyance reflected in the comment above? Four principles are especially worth keeping in mind:

1. Use graphic aids if you can — as long as there is no special circumstance that would make them inappropriate. There are almost unlimited possibilities as to types: graphs, tables, flowsheets, objects that can be held up by hand or specially mounted, slides, moving pictures, sketches on a blackboard. If you draw sketches on the blackboard, do it beforehand if possible. Try not to have to let the audience sit in silence while you draw, but don't talk to the blackboard as you draw.
2. Don't use too many graphic aids. If you keep popping up with new gadgets, the total effect may be spoiled.
3. Make sure that all your graphic aids are properly located and are big enough to be seen by everyone in the audience.
4. Keep your graphic aids simpler than would seem necessary in a written report on the same subject. And don't use any aid (like certain types of graphs) that some members of your audience won't understand.

We can't go further into the big subject of graphic aids here, but the book noted at the beginning of this chapter (Manko, *Effective Technical Speeches and Sessions)* con-

[2] Philip H. Abelson, "Communicating with the Public," *Science,* 194 (Nov. 5, 1976), 565.

tains an excellent discussion of the subject, including practical advice about selection of types of projection equipment.

Answering Questions   You may be asked to answer questions after you finish your speech. Naturally, the best preparation for this part of your performance is to acquire such a thorough knowledge of your subject that you can answer any question promptly and precisely. Obviously, such omniscience lies beyond the reach of most of us. What, then, is to be done?

In the first place, prepare yourself as thoroughly as you can, and then try not to worry about the question period. Chances are it won't be half the ordeal you might imagine. If you don't know the answer to a question, say so. A simple statement to the effect that you are sorry but you just don't know the answer is preferable to an attempt to bluff or to give an evasive answer.

In the second place, be considerate of the questioners. They may be a little nervous too, and may ask foolish questions or put a question with unintentional sharpness.

In the third place, don't try to answer a question you don't understand. Ask politely for a restatement.

In the fourth place — and also in connection with the preceding comment — make sure that the audience has heard and understood the question before you answer it. If someone else is chairing the meeting, this person may take care of the problem, repeating questions when necessary to clarify them or to make sure everyone has understood. Otherwise, you should assume the obligation yourself. You can start your reply with some such statement as "If I understand correctly, you are asking whether. . . ."

Finally, it may be necessary for you to bring the period to a close. If a specific length of time has been allotted for questions, the time limit should be respected. You should, however, try to gauge the feeling of the audience. If there is reason to think that most of the audience would like to continue, you can suggest that the time is up but that anyone who cares to may stay and go on with the discussion. In any case the audience as a whole should not be kept against its will merely because two or three persons persist in raising questions. You can usually achieve a graceful halt by declaring that time will permit only one more question.

Conferences

A large portion of almost every professional person's time is taken up by conferences (estimated at 12 percent by one corporation). These conferences may involve only two people, or many people. They may vary from nothing more than informal conversations to highly formal group proceedings. Your preparation for, and conduct in, a conference deserve serious thought.

1. Try to formulate the purpose of a conference ahead of time. Is the purpose to clarify a problem? To single out feasible courses of action? To make a final decision on a course of action? It is easy to permit the words "Let's get together and talk things over" to lull one into a passive state of mind in which problems that should

have been thought out carefully beforehand are not even recognized until the conference is in progress. This is a waste of time and energy.

2. Try to formulate your own objectives before you go to a conference. If you know your own mind before discussion begins, your chances of making a significant contribution are very much greater than if you drift into the meeting like a boat without a keel. On the other hand, you should go equipped with a rudder as well as a keel so that you can change direction if the conversation opens up facts and points of view that had previously escaped you. Don't be stubborn.

3. Estimate the attitudes of the people in the meeting. This is not a new problem: it is simply the principle of "reader analysis" carried into the conference room.

4. Take some time to speculate on how things are likely to go. Try to think of the conference as a structure. An experienced person can lead a group of people through a series of deliberations with an ease and clarity little short of astonishing when viewed in retrospect. A discussion leader can do this because he or she is thinking of the situation as a whole and not letting progress bog down in irrelevancies. Other members of the group cannot direct the discussion quite so effectively, but they can make well-timed suggestions.

5. A last bit of advice is that you give some attention to your oral delivery as you engage in discussion, following the principles suggested earlier. In some respects, more skill and flexibility are required in the conference room than on the lecture platform. The situation is less under the speaker's control, and adjustments must be made quickly. Voice control is particularly important. We are all familiar with complaints about people whose voices are so loud in conversation they can be heard in the next block. It is probably true that certain types of people are actually offended by being addressed in an especially loud voice; but some psychologists assert that there is also a type of personality that is offended by an especially soft voice. People with this type of personality, it is said, tend to feel that anyone who addresses them in a soft voice must dislike them.

At any rate, remember that your voice is an important part of your personality, and in the close quarters of a conference it should be used with care. If you avoid either roaring or whispering, enunciate clearly but not affectedly, and pronounce your words without slurring syllables, you need have no worry.

Summary

1. The best advice we can give you is to take a course in speech, and to speak before groups of people whenever you can. Meanwhile, you can help yourself by making sure that you are enunciating distinctly, with adequate movement of mouth, lips, and tongue; that you are varying the pitch of your voice effectively; and that you are pronouncing words without an irritating or confusing slurring of syllables.

2. You can watch the technique of speakers you hear. Do they use their voices well? Do they interest you in the subjects, and seem interested themselves? Is their posture suitable? Are their speeches well organized? Are introductions, transitions, and conclusions clear, so that you don't get lost? Have they employed

graphic aids to best advantage? Can they handle a series of questions smoothly? Whenever you get a chance to speak, practice these techniques yourself.

3. In conferences and discussions in which you take part, you can practice formulating purposes and deciding upon your own objectives. You can also try to guess the attitudes of the other participants and try to predict the probable course of the discussion. Make mental notes on the general course of the discussion. Finally, use your voice effectively.

Suggestions for Speaking

1. Bring to class an article from a magazine or professional journal and give a brief analysis of its construction. Discuss the introduction, particularly subject, purpose, scope, and plan. Write the main headings in the organization of the article on the blackboard, and comment on the logic of the organization. Discuss the use of transitions. Discuss the conclusion or summary. Don't choose a complex article or one more than 3000 words long. Time: 3 to 5 minutes.
2. Give a short talk based essentially on one of the special techniques discussed in Section II of this book. For instance, describe a simple device like a miniature flashlight. Time: 3 to 5 minutes.
3. Take one aspect of your library research report as your topic, and discuss it in detail. Time: 10 to 15 minutes.
4. Present a persuasive argument in favor of a course of action or a certain way of doing something. Time: 3 to 5 minutes.

Business Letters

You will probably have to do a lot of letter writing. Most professional people do. And the more successful you are, the more correspondence you are likely to have to carry on. This chapter may be taken as a guide to the form and layout of letters, to the handling of style and tone, and to the organizing of a few selected types of letters.

A great many details and refinements in the art of letter writing lie beyond the scope of this chapter. In the future, as correspondence assumes an increasingly important place in your work, you will find it useful to consult books on letter writing such as those listed in Appendix A. For the present, this chapter should be adequate.

The Elements of a Business Letter

The elements, or parts, that normally appear in a letter are the heading, the inside address, the salutation, the body, the closure, and the signature. See Figure 1. Additional elements that appear in some letters are the subject line, the attention line, and notations about enclosures, distribution, and the identity of the stenographer. We will discuss each of these elements before commenting on their overall layout and appearance on the page.

The Heading The heading of a letter includes the sender's address and the date. Business firms ordinarily use stationery with a printed heading containing the name

```
                                            1201 Linwood Avenue
                          HEADING ————→ Peoria, Illinois 61650
                                            February 16, 19--


    Wakey Products, Inc.
    1401A Grand Avenue      ←— INSIDE ADDRESS
    Detroit, Michigan 48239

    Gentlemen: ←———— SALUTATION

    I would appreciate it if you would send me your
    catalogue of home movie equipment, as advertised in
    your Circular 33-C.

    If you handle stereoscopic cameras and equipment,
    I would also like to have any information you can
    conveniently send me.  I am especially interested
    in securing a projector.

    I will certainly appreciate receipt of your cata-
    logue as well as information about the other equip-
    ment I have inquired about.
    BODY OF LETTER ————→

                                  Yours very truly, ←COMPLIMENTARY
                                                          CLOSE
           HANDWRITTEN ——→  Richard Roe
            SIGNATURE

                                  Richard Roe ←— TYPED SIGNATURE
```

Figure 1   Elements of a block-form business letter.

of the company and its address, and frequently other information—the names of
officials, the telephone number, the cable address, the company motto. When
letterhead stationery is used, therefore, the writer need add to the heading only the
date, either directly beneath the printed heading, at the left margin, or to the right of
center so that it ends, roughly, at the right margin.

If you write a business letter on stationery without a letterhead, you will need to provide your street address, the name of the city and state in which you live, the zip code, and the date of the letter, as in the following example. Depending on the overall format you select, which we will discuss later, the return address should be placed so that it either begins at the left margin or ends at the right margin.

4516 Ramsey Avenue
Austin, Texas 78731
October 15, 19—

Note that the zip code number appears after the name of the state.

The Inside Address   The inside address includes the full name and business address of the person written to, just as it appears on the envelope. Particular care should be exercised to spell the addressee's name correctly, and courtesy demands that his or her name be prefaced with "Mr.," "Ms." or an appropriate title. Business titles, by the way, should not precede a name; they may appear after it, separated from the surname by a comma, or on the line below. Compare the following illustrations:

Ms. Jane C. Doe, President   Dr. John C. Doe
American Manufacturing Company   Director of Research
110 First Street   Wakey Products, Inc.
Houston, Texas 77021   1410A Grand Avenue
                                      Detroit, Michigan 48239

In writing the name of the company or organization, take pains to record it just as the company does. For instance, if the company spells out the word "company" in its correspondence, you should spell it out too, rather than abbreviate it. This is simple courtesy.

If you must write a letter to a company but do not know the name of an individual to whom to address it, you may address the company or a certain office or department of the company. Deletion of the complete first line in either of the examples above would leave an adequate address. When a letter is officially addressed to a company but you wish some particular individual or office of the company to see the letter, you may use an "attention line." Placed a double space below the inside address, or below and to the right of the inside address, this line has the word "Attention," or the abbreviation "Att.," followed by a colon and the name of the proper person or department, as shown here:

Wakey Products, Inc.
1410A Grand Avenue
Detroit, Michigan 48239
        Attention: Head, Drafting Department
Gentlemen:

At least a double space should be left between the heading and the inside address. Further comment on this point will be found below in the section "Form and Appearance."

The Salutation   The salutation or greeting is located a double space below the last line of the inside address and flush with the lefthand margin. In formal business correspondence, "Dear Sir" is always acceptable in greeting an individual, but if possible his name should be used. The greeting "Sir" should be reserved for very formal letters, and the even more formal "My dear Sir" can probably be dispensed with altogether; to most persons, it has a stilted, artificial sound. More informal than "Dear Sir" and more suitable when you are not acquainted with the individual you address (but do know the person's name) is "Dear Mr. _____." The latter greeting is used more than all others, with the possible exception of "Dear Sir." In addressing a company, or a group of men, use "Gentlemen." When writing to a woman or a group of women, use the equivalent of the forms just noted (Dear Miss _____, Dear Mrs. _____, Dear Madam, Ladies), or simply Dear Ms.; whenever possible, use her name.

A particularly troublesome problem arises when you don't know the gender of your intended reader or readers. The term "gentlemen" has fallen into disfavor in an age that has become interested in eliminating sexist language. If your letter is to go to an individual and you don't know that individual's name or sex, you can write "Dear Sir or Madam," and avoid sexist assumptions by being cumbersome, but you will sometimes have a better choice open to you. If you know the person's title, or can approximate it, use it rather than a term that would indicate gender. You can do the same thing when you're not sure whether you are addressing an individual or a group. Simply use the name of the group or organization.

Remember that the only acceptable mark of punctuation following the greeting is a colon. The comma is satisfactory in personal letters, but not for business letters. Too often we see an even less satisfactory mark — the semicolon. It is always incorrect.

The Body of the Letter   The body of the letter is, of course, its message, or what you have to say to the addressee. In a general way, we can say that the body of most letters is made up of three parts: (1) the introductory statement identifying the nature of the business the letter is about or the occasion for it, along with references to previous correspondence if appropriate or necessary; (2) the message proper; and (3) the closing paragraph, often a purely conventional statement. The body of the letter begins after a double space below the salutation.

The Complimentary Close   The complimentary close is the formal way of signalizing the end of the letter. See Figure 2. It is ordinarily a conventional expression, which should correspond in formality with the greeting. Standard closings are as follows:

Yours respectfully, or Respectfully yours
Yours truly (not Truly yours), Yours very truly, or Very truly yours
Yours sincerely, or Sincerely yours
Yours very sincerely, or Very sincerely yours
Cordially yours
Sincerely

Cordially yours,   Yours very truly,

John C. Doe   *John C. Doe*

John C. Doe   John C. Doe, President

Yours sincerely,   Yours very truly,

John C. Doe   *John C. Doe*

John C. Doe   John C. Doe
Chief Technical Advisor   Chief Technical Advisor
                                    Research Division

                 Very truly yours,
                 AMERICAN MANUFACTURING CO.

                 John C. Doe
                 John C. Doe, President

Figure 2   Forms of signatures.

The first of the closings listed, or a variant, "Respectfully submitted," is proper for letters of transmittal to superiors, letters of application, or for any letter in which you wish to show special respect to the addressee. "Cordially yours" is suitable only when you are personally acquainted, on a basis of equality, with the person to whom you are writing.

The usual practice calls for a comma after the closing, but the practice seems to be weakening rapidly. Since it is hardly possible for any misreading to occur if that comma is not used, many writers delete it in the interest of efficiency, but the safe, conservative practice is to use the comma. Only the first word of the closing should be capitalized. Although many letter writers like to place the closing so that it ends in alignment with the right-hand margin, accepted practice approves of its being placed anywhere between the middle of the page and the margin, a double space below the last line of the text. Usually it is aligned vertically with the return address if the return address is typed rather than presented by a pre-printed letterhead.

The Signature   Directly below the complimentary close and aligned with it appears the typed signature of the writer of the letter. The typed signature should be placed far enough below the closing so as to allow plenty of space for the handwritten signature. Four to six spaces are about right.

Often the writer will need to include a business or professional title ("Chief Engineer," for instance) and sometimes the name of the company or department of a company for which the letter is written. The business title is placed either above or below the typed signature. The use and location of the name of the company or

WAKEY PRODUCTS, INC.

Education Department

General Office,
Newark, New Jersey 07109

October 17, 19--

American Manufacturing Co.
110 First Street
Houston, Texas 77096

Attention: Mr. Richard Rose

Gentlemen:

In response to your letter of October 12, I am glad to say
that we have several training films now available that would
be suitable for the needs you have described.

I am taking the liberty of enclosing two pamphlets which will
give you an idea of the contents of these films. I am also
requesting that a representative from our Dallas office call
upon you within the next week.

We hope that the enclosed pamphlets answer any questions you
may have about the films; if not, please make inquiries of our
representative when he calls. He should be able to amplify
the information presented in the pamphlets. He can also give
you details about the rental or purchase terms.

We hope to be able to serve your needs in a thoroughly satisfactory way, and we look forward to doing business with you.

Yours very sincerely,

Joe C. Ashford

Joe C. Ashford, Manager
Education Department

JCA: wk
Encl. (2)

cc: Joseph Smith (Dallas Office)

Figure 3   Full-block letter form.

department depend upon circumstances. The name of the company or department should appear below the signature only if it does not appear in a printed heading. But there is one exception. If you use a business title, like "Manager," which indicates your relationship to a department or section but not to the entire company, then the department or section should be stated after or below the business title even if the department is also identified in the heading. You will almost certainly have company letterhead stationery for official correspondence, but you may not have a departmental letterhead. The name of the company may appear *above* the signature, however, if you wish to emphasize the fact that you are speaking only as an instrument of the company and not with personal responsibility. The examples in Figure 2 illustrate various forms.

Miscellaneous Elements   Several other items may be necessary or useful parts of a business letter. They include a notation identifying the stenographer or typist, an indication of enclosures, a distribution list for copies of the letter, and a subject line. The stenographer's identification consists of the sender's and the stenographer's initials, separated by a colon or a slant line. This notation is placed at the left margin, either directly opposite the typed signature or two spaces below. If there are enclosures to the letter, the abbreviation "Enc." or "Encl." is typed just below the identification notation. Many writers indicate the number of enclosures in parentheses after the abbreviation, as "Encl. (4)." If copies of the letter are distributed, the phrase "Copies to" or the abbreviation "cc" is typed at the left margin, below the identification notation, and below the enclosure notation, if there is one, and the names of those receiving a copy are listed below it. If a subject line is used, it appears either just below or below and to the right of the inside address. Most of the items discussed above are illustrated in Figure 3.

Form and Appearance

Although the content of a letter is of first importance, attractive form is also necessary if the letter is to be effective. Good appearance requires that the materials used for the letter be of good quality, that margins and overall layout of the letter on the page be pleasing to the eye, and that the spacing and arrangement of the elements be in accord with accepted conventions of good taste. And the letter must be neat. A typical business letter is shown in Figure 3.

The paper chosen for business correspondence should be a high-quality white bond, 8½ by 11 inches (22 by 28 cm) in size. If the letter is typed, as business letters are, the typewriter ribbon should be new enough so that it will make firm, easily legible letters; if it is handwritten, black, blue, or blue-black ink should be used. Other colors of ink are not generally considered in good taste. Carbon copies should be made with new carbon paper on good quality onionskin paper. If photocopies are made instead of carbon copies, they should be made on a machine that produces attractive work.

Attractive appearance calls for a minimum margin of at least 1 inch (2.5 cm) on all sides of a letter. Margins will have to be increased all around, of course, for letters that do not occupy a full page. Although an experienced stenographer can estimate accurately from shorthand notes about how wide the margins should be set, the inexperienced letter writer will probably have to type a trial effort or two before attractive placement can be achieved. The letterhead, by the way, is ignored in determining overall layout of the letter on the page. It is permissible to allow a somewhat narrower margin at the top of the page than at the bottom, about a 2:3 ratio being acceptable. This means that the center point of the letter may be slightly above the actual center of the page.

Balanced margins on the left and right sides of the page are desirable, but it is impossible to keep the right margin exactly even all the way down the page because of the necessity for dividing words at the ends of lines, or the necessity for not dividing words. Words must be divided between syllables or not at all. In general it is best to avoid divided words as much as possible. A dictionary should be consulted to find the correct syllable division if you feel uncertain.

In letters that are more than one page long, you should write the name of the addressee, the page number, and the date on page 2 and any additional page. This notation appears just below the top of the page (one acceptable form is shown below). The text begins two or three spaces below the notation if it occupies the full page, and about eight spaces below if it occupies only a portion of the page.

American Manufacturing Co.   — 2 —   November 23, 19___

With the exception of very short letters, you should single-space each of the elements of the letter and as a rule double-space between elements and between the paragraphs of the body. This means that the lines of the heading, inside address, and so forth, are single-spaced, but that there is double spacing between the heading and the inside address; between the inside address and the salutation; between the salutation and the opening paragraph of the body; between paragraphs; and between the text and the complimentary close. The rule is not a hard and fast one, however; for pleasing proportions on the page, you may need to triple-space, or more, between the heading and the inside address and between the last line of the body and the complimentary close. Quite short letters may be double-spaced throughout.

Three commonly used styles of arrangement for the elements are the block form, the modified block form, and the full block (or left wing) form. The only difference between the block and modified block forms is that paragraphs are indented in the modified form but are not indented in the block form. Block form is illustrated in Figure 1, modified block form is shown in Figure 11, and full-block form is illustrated in Figure 3. Not so commonly seen is a form adopted by the Administrative Management Society, although it is one with much to recommend it. This form has the following features: (1) All lines begin flush with the left margin; (2) both the salutation and the complimentary close are omitted; (3) in the position where the salutation is usually found is placed a subject line, in capital letters; (4) the writer's name and title are typed on one line, in capital letters, at least three lines below the last line of the text of the letter, to leave space above the typed signature for the handwritten signature. Figure 4 illustrates this interesting form.

Although any one of the three block forms described above is entirely satisfac-

FRITTEN LAPIDARY COMPANY
1241 Crescent Boulevard
LOS ANGELES, CA 90046

September 1, 19--

Mr. Alvin C. Brown
4608 Fomberg Dr.
Austin, Texas 78731

FRITTEN TUMBLER INSTRUCTIONS

You will soon receive a set of memoranda detailing the steps
you should follow in using your new Fritten tumbler to best
advantage. There are five memoranda, each giving instructions
on a stage of the process. The stages are as follows:

1. Stage One: This first stage uses a coarse grit to take off
   the rough edges from the stones you wish to polish. Do not
   expect this stage to produce beautiful stones.

2. Stage Two: This second stage, which makes use of a finer
   grit, will produce very smooth stones with a frosty
   appearance when they are washed and dried. They are still
   a long way from having the shiny, mirror-like finish you
   want.

3. Stage Three: Commonly called the pre-polish stage, this
   stage makes use of a very fine grit (500-600), and it will
   produce stones that are silky smooth and ready for the
   polishing stage.

4. Stage Four: In this next-to-last stage, a polishing agent
   such as tin oxide, will produce the finish that you desire
   so long as the final stage is carefully performed.

5. Stage Five: The final stage consists largely of very carefully washing the stones with enough detergent in the
   tumbler drum to make a thick, creamy solution. If all has
   gone well, you will have some fine, silky sheen stones, ones
   you can see your reflection in and that will serve for
   settings in very attractive jewelry.

We hope that you will find much satisfaction in your Fritten
tumbler and are sure that you will if you follow the instructions set forth in the five memoranda we are sending.

Judy Bateson--CUSTOMER RELATIONS

JB/mjh

Figure 4 Administrative Management Society's simplified letter form.

WAKEY PRODUCTS, INC.

EDUCATION DEPARTMENT MEMORANDUM

17 October 19--

To: Joseph Smith

From: Joe Ashford

Subject: RICHARD ROSE: SALES CALL

Joe, I had a letter from Richard Rose of the American
Manufacturing Company of Houston, dated October 12, in
which he inquired about some of our training films. I sent
him our two pamphlets and told him you'd call within a week.

Would you please arrange to visit him, prepared to answer
any questions about our training film series? Take along
our regular price list--and try to sell him our services.

Drop me a line when you've seen him. Okay?

Joe

Joe Ashford

JA/is

Figure 5 A memorandum.

tory for most formal business correspondence, a simpler and more convenient form exists for interdepartmental and personal communication within an organization as well as for much intercompany and interagency communication. In fact, this form is often used for short informal reports, such as the monthly progress report. (See p. 285 for an example.) Called the memorandum or letter report, it employs the principal features of the military correspondence forms, with "To," "From," and "Subject" printed on the stationery. In some organizations, each department uses stationery of a different identifying color. Many organizations also have the word "Memorandum" printed at the top center of the sheet, about where a company letterhead would appear. See Figures 5 and 6 for examples; Figure 6 illustrates a modified military form.

Usually much less formal than the conventional letter, the memorandum is commonly headed with a date, often expressed in abbreviated form such as 10/17/86 or 10/Oct/86, and simply signed or initialed at the end. Neither a salutation nor formal closing is needed. Two details are worth noting in the modified military form: the use of all capital letters for the subject (or "title") and the use of underscored side headings. Occasionally you will see the side headings numbered for ease of reference in subsequent correspondence.

Style and Tone

In determining the style and tone of a business letter, you should keep three facts in mind: it is a personal communication, it serves as a record, and it is usually brief.

Since a letter is a personal communication, it should be characterized by courtesy and tact. In a sense, a letter is a substitute for a conversation with the person you are writing to, and you should therefore try to be as polite and considerate in your letter as you would be in dealing directly with the addressee. This consideration of the addressee, commonly called the "you attitude," will suggest that your letter should not only be perfectly clear in meaning but also free of any statements that might antagonize or irritate the reader. The "you attitude" thus has two aspects: the general one suggested by the phrase "tact, courtesy, and consideration"; and a mechanical aspect.

To be truly considerate, you need to grasp the reader's point of view. Try to anticipate questions that might be asked and estimate the reader's reaction to your statements. You should examine your sentences to see if they are free of ambiguities, free of words the reader might not understand — in short, to see if the letter will say *to the addressee* what you want it to say. It is important to try to read it from the point of view of the addressee because since you know what you intended to say, you are likely to take it for granted that the words express your intention unless you examine critically everything you have written.

In a more mechanical sense the "you attitude" means substituting the second-person pronoun ("you") for the first person ("I" or "we"). Use of the second person has the effect of keeping attention centered on the reader rather than the writer, and thus helps to avoid any impression of egotism. Suppose a writer has made the following statement in a letter:

MEMORANDUM
                                                        20 June 19____

To: John Ableman, Chief
           Engineering Department

From: Jim Moore

Subject: RENOVATION OF MODEL SHOP AREA

In response to your request, I have studied the Model Shop Area (see earlier memoranda), and I have a number of recommendations to make for its renovation.

Drains and Flooring

Since the Model Shop will probably have a pilot finishing area, the open drains in the
finishing area should be left intact by being covered with boiler plate mounted flush
with the floor. This plating would support the machinery now in the Model Shop and
make it possible to convert the area for pilot finishing or etched board work. The
floor should be covered with tile similar to that in other areas of engineering.

Ceiling Insulation

Acoustical insulation should be installed in the ceiling to absorb noise from the machinery. The type used in the most recent annex to our building is quite effective.

Wall Construction

Since this area is temporary, the walls shown on the layout (see attachment) should
be constructed of plywood or masonite.

Welding Booth

Two special requirements need to be met for efficient welding booth operation: (a) duct
work and blower (a hood is available) for venting fumes from the welding area, and
(b) a 440-volt bus capable of supplying 90 amperes for the Heliarc welder.

Conclusion

So that we may function as efficiently as possible, I respectfully suggest that you approve these recommendations as soon as convenient so that we may get to work on
the remodeling.

                                                        Jim Moore
                                                        Supervisor, Model Shop

Attachment

Figure 6 Modified military-form memorandum.

> I have noticed that your shipments to us have consistently been delayed. We are inconvenienced by these delays and request that you investigate the matter at once.

This rather blunt statement could have been better phrased as follows:

> We are curious about the cause of the rather consistent delay in the receipt of shipments from your company. You will understand, of course, that delay in receiving these shipments is the cause of considerable inconvenience to us, and we are sure you will want to correct the situation as soon as possible.

Perhaps we should add that personal pronouns are entirely suitable to personal communication. Do not hesitate to use "I" or "you" when it is natural to do so. Expressions like "the writer" instead of "I," or "it will be noted" instead of "you will notice" are out of place in correspondence. Be direct and natural.

In addition to being considerate, courteous, and unaffected in style, the letter should be concise. Since it makes no sense to waste a reader's time with nonessential discussion, a letter should be held to a single page if at all possible. But do not make the mistake of believing that brevity alone is a virtue in letters. Too much brevity makes for an unsatisfactory tone. When carried too far, it results in a curtness and bluntness that can be irritating. A further danger of carrying brevity too far is that it may result in a lack of clarity and completeness.

The fact that letters are filed for reference makes it necessary that a letter be clear and complete, not only at the time of writing but also at any later time at which it might prove necessary to look at the letter again. This need for clarity of reference is one reason why most letters of reply begin with a concrete reference to the date and subject of previous correspondence. It also explains why phrases like "the matter we corresponded about last month," are not satisfactory.

Types of Letters

In books devoted exclusively to letter writing, a great many types are discussed, often at chapter length. We'll not have space for lengthy treatment of very many types here; should you find yourself in need of information about types of letters not discussed in this chapter, we suggest that you consult one of the volumes listed in Appendix A, where you will find discussed such types as the following: appreciation, goodwill, remittance, collection, credit, congratulations, announcement, invitation and reply, customer relations, follow-up, acceptance, and resignation — a list by no means exhaustive. Our own discussion will be limited to brief treatment of the following types: inquiry and reply, instruction, order, claim and adjustment, sales, transmittal, and application.

Letters of Inquiry and Reply   Since anyone reading a letter naturally wants to know immediately what it is about, the purpose of the inquiry should be stated in the opening sentence. This does not mean that the salutation must be followed directly by the specific inquiry or inquiries. It means that the purpose of the letter should be identified at once as an inquiry about a specific subject. Thus, you might begin with: "I am writing this letter to inquire whether you have any new performance data for release on the ramjet engine you are developing." This statement could then be followed by concrete, specific questions.

Explaining the reason for making an inquiry is not absolutely necessary unless a response to the inquiry you are making constitutes the granting of a favor. It is always courteous, however, to explain why the inquiry is being made, and when you are asking a favor, it is a courtesy that should not be neglected.

It is exasperating to receive an inquiry phrased in such general terms that no clear notion of what is wanted can be determined. Too often one sees statements like this: "Please send me what information you have on television antennas." The writer of this request probably wants far less information than it seems to call for. Actually, the inquiry would have had more meaning had it been rephrased:

1. What types of television antennas do you manufacture?
2. Can you send me installation instructions for the types you manufacture?

Concrete, specific questions make a reply easier to write. Questions do not necessarily need to be numbered and listed as above, but such a form is perfectly satisfactory and is desirable when several questions are asked. Remember that vague, general requests present an impossible problem to the person who receives the inquiry. We recall a student who wrote a request to a research organization for "any information you have about new aircraft designs" being worked on by the organization. He did not realize that a literal granting of his request might result in his receiving a truckload of reports!

The problems of concreteness and courtesy are both well illustrated by the letter of a graduate student of chemical engineering who wrote for information to be used in a report for one of his courses. The letter he wrote is quoted below, the only change being the deletion of names.

> Dear Sir:
>
> It is requested that literature concerning the history and background, wage and benefit plans, and general research policy of [your] Corporation, and a current statement to stockholders be sent to me at the above address.
>
> This information is to be used in a report assigned in a graduate chemical engineering course. Additional information that you may have available will be appreciated.
>
> This literature will be made available to the University Library after I have finished with it.
>
> Any information you supply will be greatly appreciated.
>
> > Very truly yours,

The first paragraph of the reply received by the author of the above letter went as follows:

> Dear Mr. _____:
>
> We are glad to comply with your rather blunt request that we supply you with a great deal of not altogether inexpensive material to be used in a report in a graduate chemical engineering course. Since you are going to give it to the library after you have finished with it, we are less critical of the tone of your letter than we might

otherwise be. I speak this way to bring to your attention something which may be useful to you later on, since believe me your letter of March 29 could be couched in more gracious terms.

The unfortunate tone of this letter of request is due primarily to the use of the passive voice. The phrase, "It is requested that . . ." is ungracious. Perhaps, "I am writing to ask if you could help me . . ." would have been more cordially received. A comment on the kind of report that had to be written would have helped the reader decide what materials should be sent. The letter should have been rephrased throughout, particularly to indicate the writer's realization that a great deal of help *was* being requested.

Custom suggests that letters of inquiry, especially those in which a favor is asked, close with a statement showing that their writers will appreciate a reply. This is adequately illustrated in the letter quoted above. But remember that good taste suggests that you avoid ending your letter with: "Thanking you in advance, I remain. . . ." If you are in a position to do so, it may be appropriate to offer to return the favor.

In writing a reply to a letter of inquiry, keep these two points in mind: (1) begin your letter with a reference to the inquiry, preferably both by date and subject; and (2) make the reply or replies as explicit and clear as possible. If the inquiry contained itemized questions, it is a good plan to itemize answers. Naturally, the reply should be courteous. See Figures 7 and 8 for illustrations of these two types of letters.

Letters of Instruction   When instructions are to be issued by letter, you will find the following plan of organization suitable:

1. The opening paragraph of the letter should explain the situation or problem that necessitates issuing the instructions.
2. The body of the letter should contain the detailed instructions. Common sense will tell you that these instructions should be clear and definite. Vague and ambiguous instructions often defeat their own purpose by confusing and irritating the reader, and thereby making it less likely that they will be satisfactorily carried out.
3. The conclusion of the letter should suggest any action, other than carrying out the instructions, that should be taken. This may be a request for a report, a conference, or the like. See Figure 9 for an example.

Order Letters   The chief points to bear in mind in writing an order letter is that you should be as explicit as possible, making clear precisely what it is that you want to buy and giving information about catalogue number, model, size, color, weight, finish, and of course quantity and price. Your order should also be specific about mode of payment—whether by check or money order, C.O.D., credit card, or charge account. You may want to give instructions about shipping—parcel post, railway freight, air express, or other. If you are ordering from a catalogue that contains order forms, you should use one of them. With use of an order form, your letter need do little more than transmit the order, with information about mode of payment if that is not shown on the order form.

4608 Fomberg Dr.
Austin, Texas 78731
August 24, 19--

Fritten Lapidary Co.
1241 Crescent Blvd.
Los Angeles, California 90046

Gentlemen:

A newcomer to the hobby of gem making, I was attracted to
your advertisement, in the current issue of the Journal
of Jewelry Making, of a trim saw, especially in view of the
reasonable price given. I wonder if you would be so kind
as to answer a few questions that came to my mind about the
saw? Specifically, I would like to know

    (1) How thick a rock slab will the saw cut?

    (2) How long can one expect a diamond blade to last?

    (3) What kinds of coolants do you recommend?

I would be most appreciative if you could find the time
to answer my questions. I expect that I shall be sending
you an order soon.

            Yours very sincerely,

            Alvin C. Brown

            Alvin C. Brown

Figure 7   Example of letter of inquiry.

FRITTEN LAPIDARY COMPANY
1241 Crescent Boulevard
LOS ANGELES, CA 90046

September 1, 19--

Mr. Alvin C. Brown
4608 Fomberg Dr.
Austin, Texas 78731

Dear Mr. Brown:

Thank you for your interest in our trim saw. We are sure you will find complete satisfaction with it should you decide to purchase one.

We'll try to answer your questions in the order you asked them:

(1) You can easily cut slabs up to ½ inch thick with our saw; with care, slightly thicker.

(2) With reasonable care and a suitable coolant, you should get a year's use, at least, from the blade. Naturally, the hardness of the stone cut affects blade life.

(3) We recommend use of a so-called "white" or "flushing" oil coolant. There are other possibilities, such as anti-freezes of the diethylene variety; you might want to consult your local lapidary dealers.

We certainly thank you for your interest in our trim saw-- and hope you decide to buy one. If your local dealers do not carry them in stock, you may order from us by mail. We are enclosing a new brochure on the current model for you to take a look at. Good luck with your new hobby!

Yours very truly,

Jack Bobson

Jack Bobson
Sales Department

JB/is

Encl: Little Dandy Trim Saw Brochure

Figure 8 Letter of reply to an inquiry.

Instruction Memoranda
  Series A, No. 1

FRITTEN LAPIDARY COMPANY
1241 Crescent Boulevard
LOS ANGELES, CA 90046

To: Purchasers of Fritten Tumblers

From: Judy Bateson, Customer Relations

Subject: INSTRUCTIONS FOR USE OF FRITTEN TUMBLERS: STAGE 1

For you to get the best results from the fine Fritten tumbler you have purchased, it is essential that you follow the instructions contained in the five memoranda sent to you. Failure to follow these instructions carefully will doubtless result in stones without the fine mirror polish and sheen that you desire. This memorandum details the steps you should follow in carrying out stage one of the tumbling process.

1. First, sort out the stones according to size. For a tumbler load, you need stones of approximately the same size. Try a batch about one-half to three-quarter inch in diameter.

2. Wash the stones thoroughly, making sure you have cleaned out all the sand and grit from all crevices. It may help to use a little detergent in this washing process. Weigh the stones.

3. Fill the drum of the tumbler about one-half full, or two-thirds at the very most.

4. Pour in barely enough water to cover the stones.

5. Add one ounce of No. 80 grit silicon carbide and one teaspoon of detergent per pound of stones.

6. Attach the lid firmly and screw down the grooved cover cap tightly.

7. Place the drum on the rollers, making sure that the groove in the cover cap fits into the bracket on the side of the tumbler that faces you.

8. Switch on the electric current and allow the tumbler to operate for a week. Check it from time to time to make sure the drum is not leaking. At the end of a week, the grit should be reduced to a fine slurry.

The stones should now be ready for stage 2, and you should now follow the instructions presented in the memorandum on that stage. Good luck in your new hobby!

                                        Judy Bateson

JB/mjh

Figure 9   Letter containing instructions.

Claim and Adjustment Letters Claim (sometimes called "complaint") and adjustment letters are pretty well defined by their titles. In the first type, the writer lodges some sort of claim, perhaps about defective or incomplete filling of an order. Here, again, the important thing is to be as specific and explicit as you can be about why you are writing, what you expect to be done, and — in extreme cases — what you intend doing if your claim is not satisfied. Remember that you are more likely to receive satisfaction if your letter is courteous and couched in noninflammatory language. In the second, an adjustment letter, a response is made to a claim or complaint letter. It seems only reasonable that it should be prompt, should refer explicitly and precisely to the claim letter, and should state clearly and courteously what action is taken or proposed.

No examples are shown of claim and adjustment letters, since in form they are ordinary business letters.

Sales Letters We feel reasonably sure that no one can tell you with certainty how to write a letter that will work infallibly to make a sale. We would think that such an ability would be worth millions! Nevertheless, it seems reasonable to us to suggest that a good sales letter should have the following characteristics or elements: (1) it should attract the reader's interest; (2) it should set forth clearly the benefits to the buyer of the product or service offered; (3) it should at least try to lead the reader to wish to see the product, or talk to a company sales representative about it; and (4) it should explain what the product or service costs and how it may be paid for. We ourselves are not impressed with "gimmicky" letters, with something-for-nothing come-ons, but we do believe that a lively style, a friendly tone, an honest and sincere — and, if possible, complete — description of the product all help make a good impression. See Figure 10.

Letters of Transmittal The letter of transmittal is a communication from the writer of a report to its recipient. In a general way, it serves about the same purpose for a report that a preface does for a book. Although letters of transmittal are often sent through the mails separately from the report itself, they may be bound in with the report, following the title page. We'll discuss the five primary functions of a letter of transmittal in the sequence in which they usually appear.

1. The letter typically opens with a reference to the occasion of the report or an explanation of why the report is being submitted. There may be a reference to a contract or other authorization of the work being reported on.

2. The letter should state the title of the report being transmitted. This function and the reference function are contained in the first paragraph. They are illustrated in the following sentences:

> In response to your request, dated October 26, 19___, I have investigated the possibilities for a new plant location in the Southwest. The accompanying report, entitled *Advantages and Disadvantages of Five Southwestern Cities as Sites for a New Assembly Plant,* is an account of this investigation and the conclusions it led to.

CHESTERTON SCIENTIFIC COMPANY
1220 McKINNEY AVENUE
DALLAS, TEXAS 75201

September 15, 19--

Professor John Dunn
College of Business Administration
The University of Texas at Austin
Austin, Texas 78712

Dear Professor Dunn:

Ordinarily we avoid direct mail advertising, believing
we can reach most of our potential customers with
conventional magazine and newspaper advertising.
This time we have made an exception, but only for a
very limited number of professional people. We make
this exception because we have just developed a
product of such amazing versatility, convenience, and
usefulness that we felt we should call it to the attention of professional people without delay. This product
is Chesterton's new hand-size calculator--not just like
all the others you have seen, but one with a paper tape
printout!

In your work you doubtless have found a pocket
calculator a great convenience--except for the lack of a
permanent record. We thought you would want to
know that this lack has now been remedied in our new
Mark I model. Not only does it have the new paper
tape and the conventional visual display, but it also
has the following remarkable features:

. Printing speed of 0.8 line per second

. Memory, percent, tax, and discounts in seconds

Figure 10   Sales letter.

Professor John Dunn							— 2 —							September 15, 19___

. Rechargeable batteries and AC adaptor

. Total weight: 0.46 lb

Already in quantity production, you can have this versatile calculator for less than $100. Could we send our local representative around to demonstrate it for you? We feel sure you will be impressed by its convenience and applicability to your needs. Just give us a call or return the enclosed postcard suggesting a convenient time for our salesperson to call.

Yours very sincerely,
Chesterton Scientific Company

Anita Chesterton

Anita Chesterton, President

AC/ir

3. The second paragraph of the letter of transmittal should explain the purpose and scope of the report (unless the opening paragraph has already done so). Beyond this, it should be devoted to any comments about the report that the writer feels should be made to the addressee. It might happen, for example, that the writer had on a previous occasion said that the report would contain information that was finally omitted. This omission should be explained.

Do not hesitate to duplicate in the letter of transmittal those elements that also appear in the abstract or the introduction to the report. A statement of purpose, for instance, could appear in all three places.

4. If the writer has received assistance in carrying out the work with which the report is concerned, and feels that this assistance should be acknowledged, the letter of transmittal is a place in which to name and thank those who helped.

5. Customarily the letter closes with a statement expressing hope that the report is satisfactory.

The five functions mentioned above are illustrated in Figure 11.

Letters of Application Most students feel that no form of the business letter is as important to them as the letter of application for a job. We'll therefore discuss this type of letter in considerable detail.

Since the amount and variety of information an employer will want to know about you is great, it is ordinarily impracticable to include it all within the framework of the conventional business letter. Consequently, applications usually consist of a combination of two elements: (1) a letter, accompanied by (2) a typed data sheet (often called a résumé), or a printed data sheet made available by your college, or a printed application form that has been provided by the company to which you are applying. We'll begin with the letter.

The letter. The letter accompanying a data sheet has four principal functions: (1) making reference to your source of information about the opening, (2) explicitly making application for a job, (3) elaborating on pertinent qualifications, and (4) requesting an interview. These functions are illustrated in Figure 12.

If some person, like a company representative or employee, has told you of an opening, the best way to begin your letter is by making reference to that person by name. Seeing a familiar name, the employer is likely to continue reading and to give your application consideration. If your source of knowledge is an advertisement, refer to that. If you do not know whether an opening exists, you may begin by mentioning your interest in getting into the particular kind of work done by the company, or your desire to be associated with the company. Openings that stress an interest in a particular company must be tactfully written so that they will not sound as if flattery is being employed to gain a sympathetic hearing.

Explicitly stating that you are applying for a job is more than a conventional formality; it permits you to state exactly what work it is that you are after. Applying for a specific job is always better than just asking for employment, and this is particularly true when you are applying to a large corporation in which many technical people are employed in all sorts of jobs, jobs that do not always bear a direct relationship to the academic training of those who fill them. This emphasis on

5127 Clearview Street
Austin, Texas 78730
November 20, 19--

Professor John C. Doe
Department of English
University of Texas at Austin
Austin, Texas 78712

Dear Professor Doe:

The accompanying report entitled <u>Joistile-Concrete Beams</u> is submitted in accordance with your instructions of October 10.

The primary purpose of the report is to present information about joistile-concrete beams for use in floor and roof construction. An effort has been made to cover the subject thoroughly, including the development of joistile, its use and application in beams, investigations and tests of both joistile and joistile-concrete beams, and general specifications for the construction of tile-concrete beams and floor slabs. The section on investigations and tests is limited to the most important and pertinent tests and results.

I wish to acknowledge the information and assistance given me by Mr. James Wood of the Greek Key Ceramics Association.

I sincerely hope that this report will meet with your approval.

Respectfully yours,

Edward Donaldson

Edward Donaldson

Figure 11   Letter of transmittal.

5127 Clearview Street
Austin, Texas 78730
May 14, 19--

Mr. M. A. Lindstrom, Head
Personnel Department
Rhode Island Oil Company
Sarnia, Texas 78999

Dear Mr. Lindstrom:

I talked to your Austin representative, Mr. Clapper,
and he informed me that it is a policy of your Company
to employ college students of petroleum engineering
each summer as assistant gaugers. I should like to
submit my application for that position.

While working with the Marshall Drilling Company, I
became thoroughly familiar with the location of wells
on your Chalker and Sarnia fields. In fact, I happened
to be working with the crew that drilled the first deep
dual-completion well on the Dole lease last summer.
The course which I completed this year in Petroleum
Production covered production gauges and their
functions and operation quite thoroughly.

Please refer to the enclosed data sheet for details of my
education and experience and the names of persons
who have consented to express an opinion about my
ability and character.

I shall be glad to go to your Sarnia office any Saturday
on a few hours' notice. Unfortunately, my schoolwork
makes it virtually impossible for me to go during the
week. I shall be available to begin work June 1.

Yours respectfully,

Malcolm Richards

Malcolm Richards

Figure 12   Letter of application, Example 1.

making application for a specific position may seem to ignore the fact that many graduates are put into a training program upon first being employed by a large organization. If you know that the company you are applying to puts all newly hired people into a training program, make your application for a place in that program and also state what your professional interest is.

The first paragraph of the letter, then, contains at least the second, and perhaps both, of the following two elements: (1) a reference to your source of information about the job that is open, and (2) a statement about what job you want. The next paragraph (or paragraphs) is the hardest to write — and the most important. It is the real body of your letter; it is here that you distinguish your application from others. In it you may single out for detailed discussion something from your training or experience that particularly qualifies you for the job you are trying to get.

Remember that the data sheet or résumé gives the bare details. In an application to a company for a position in a tool and die design department, a mechanical engineer, for instance, may have listed a course in machine design on the data sheet. But merely naming the course does not tell what was done in the course, or whether the engineer's performance was good or bad. Suppose the applicant had undertaken several projects and completed them successfully, and suppose he or she had learned that this experience will help on the job being sought. It would be wise, then, to provide the employer with the details about this course and the projects completed in it. Similarly, merely naming a job on the experience record scarcely does more than suggest the duties, skills, and responsibilities that the job demanded. The letter gives you an opportunity to submit full information about those aspects of your training and experience which best fit you for the job you are after.

You must, of course, carefully analyze the job's requirements and measure them against your own qualifications before selecting something to give details about. It is this elaboration of selected details of your qualifications that makes your letter more than a letter of transmittal for your data sheet. If there is a parallel — even a distant one — between your knowledgeability and the work available, the employer will probably conclude that you are a promising candidate for the job. One final caution: do not state that you feel you are fully qualified for the job unless you supply enough information to support such a claim. And if you present the support, it is hardly necessary to make the claim. After all, very few graduates are fully qualified for any job until they acquire some experience.

Since the immediate objective of a letter of application is an interview, you will want to close your letter with a request for one. You should take pains to phrase this request so that it will convey the fact that you want the interview, not merely that you are willing to have it. Accommodate yourself to the employer's convenience as far as possible. Suggest that you will be glad to appear at a time convenient to the employer. If restrictions on your time make it impossible for you to be interviewed at certain periods, be sure to state what they are and explain them.

If time and distance make it impossible for you to go to an employer for an interview, explain the facts, express your regret, and suggest an alternative, such as being interviewed by a company employee in your vicinity.

If you receive no response from your application within a reasonable time, it may help to write a follow-up letter. In this letter you may inquire whether your application has been received and thus remind the employer of it. If possible, it is a good idea to add some new support to your application. This may be presented as a sort of afterthought. Sometimes the follow-up letter is just what is needed to secure you that extra consideration that results in a job.

The Data Sheet. The data sheet contains about the same information called for on most printed application forms, organized in four sections: Personal data, Education, Experience, References. One form of a data sheet is illustrated in Figure 13; another is shown in Figure 14, a printed form that the applicant fills out. A typical company application form is shown in Figure 15. Note in the typed data sheet (Figure 13) that no reference is made to the kind of work desired, this statement being reserved for the letter itself. You might prefer to put such a statement into the data sheet also.

The personal data section contains enough information to enable an employer to get some idea of what you are like individually. Relevant headings may include age, place of birth, health data (height, weight, eyesight, and hearing), marital status and dependents, recreational interests, organization and society memberships.

Accompanying this data, and placed in the upper right-hand corner of the sheet, may appear a photograph, though it is not required. Indeed — and this fact must be emphasized — in some states the use of photographs on application forms is illegal; so, before you send any prospective employer a photograph, be sure that it is permissible to do so. The law permitting, most employers probably would like to have a photograph. We do know, however, of one interesting instance in which an applicant purposely left his out. As a matter of fact, he made a point of referring to its omission in his letter and explained it by remarking that he was undoubtedly the ugliest man living. Since the prospective employer was impressed by the man's qualifications and curious about this comment on appearance, he invited him to an interview and hired him. He said later, though, that the man's self-appraisal was correct — he was the ugliest man he'd ever seen.

In an application from a person with several years' experience, the section on education will need to contain only the bare facts: schools attended, dates of attendance, and degree (or degrees) awarded. Experience, or what the applicant has done, will be of chief interest to an employer. But for the young applicant, especially the one just completing training, the section on education is more important. Education is what young people have to offer. If you are a student, try to stress those elements in your education that are most relevant to the job you are seeking (see Figure 16). You should in any event state the name of your high school and the date of graduation; the name of your college and the date, or expected date, of graduation; the degree you have, or are seeking; your rank in class; major and minor fields; and a list of the courses pertinent to the requirements of the position sought. It may be desirable to give the number of credit hours earned in each course. Some applicants prefer to classify courses as basic, specialized or technical, and general. If you are using a printed data sheet or application form, these details will be dictated by the form. Such forms are illustrated in Figures 14 and 15.

As a recent graduate, you may have had little or no experience related to the kind of professional work you are seeking. Nevertheless, record the facts about jobs

DATA SHEET

Malcolm Richards
5127 Clearview Street
Austin, Texas 78730
Telephone: (512) 472-8509

Personal Data, May, 19--

Age 25   Health excellent
Height 6 ft, 2 in.   Eyesight excellent
Weight 175 lb   Hearing excellent
Married, no children

Education

Leaton High School (Texas), 19--
Senior Classification in Petroleum Engineering
        in the University of Texas: Grade point
        average, 2.0 of 3.0.
        Degree sought: B.S. in Petroleum Engineering
Important Courses (Semester hours shown):
    Mathematics (15)   Chemistry (16)
    Fluid Flow (3)   Petrophysics (2)
    Engineering   Petroleum Production (3)
        Mechanics (9)
    Engineering Drawing (4)   Petroleum Technology (3)
    Physics (8)   Technical Writing (3)
    Reservoir   Mechanics of
        Engineering (6)   Drilling (4)
Percentage of Expenses Earned: 30%

Experience

Summer of 1976
    "Roughneck" on oil drilling rig under Mr.
    Frank A. Thomas, Marshall Drilling Co.,
    Sarnia, Texas

Figure 13 A typed data sheet.

Data Sheet for Malcolm Richards   page 2

> Summer of 1975
>   Assistant "Shooter" on explosive truck under
>   Mr. James Stone, United Geophysical Co.,
>   San Antonio, Texas
> Summer of 1974
>   Laborer with Gunn Supply Co., under J.A. Gunn,
>   San Antonio, Texas
>
> References (by permission)
>
> Mr. Joseph Wood, President   Mr. James Stone, Manager
> Chamber of Commerce   United Geophysical Co.
> Sarnia, Texas 78999   San Antonio, Texas 78298
>
>   Professor Howard Bitt
>   Department of Petroleum Engineering
>   College of Engineering
>   The University of Texas
>   Austin, Texas 78712

PERSONAL RÉSUMÉ

CAREER ASSISTANCE CENTER
The University of Texas at Austin
Austin, Texas 78712
Cockrell Hall 2.4

Center Photograph
Between Corners

PERSONAL DATA

Name	Last	First	Middle	
	Abelson	Robert	Worden	Use *black* ball point pen — or type —

Age	Birth Date	Height	Weight	Marital Status	Number Children
23	3/14/53	6 ft.	170	Single	

Home Address (street, city, state, zip code)	Home Phone A.C.
4610 Finley Dr., Austin, TX 78731	453-8378 (512)

Watch Your Spelling
BE NEAT

College Address (street, zip code)	College Phone A.C. 512
Same as above	471-4993

U.S. Citizen	If "No", Type of Visa	Physical Limitations	Social Security No.
☒ Yes ☐ No		None	493-48-9883

WORK INTERESTS

Type of Work Desired	Date Available
Technical writing and editing	June, 1977

Work Location Restrictions or Preferences
No restrictions, but prefer southwest or west coast

Brief Statement of Interests, Career Objectives or Long-Range Goals (Graduate students also list thesis title)
Immediate objective is to combine interests in science and technology with desire to write and edit manuscripts in technological fields. Ultimately would like to earn a post as editor of a company house-organ or magazine, especially one addressed to employees and stockholders. Would like to contribute special articles to such a publication.

COLLEGE INFORMATION

Name and Location of Colleges Attended (in reverse order)	Dates From	Dates To	Degree Level and Discipline	Block-Major Interest	Graduation Date	Grade Pt. Avg. Overall	Class Quartile	Grade Basis
The University of Texas Austin, Texas 78712	1973	1977	B.J.	Writing & Editing, esp. tech. writing	June 1977	3.4	1st	A = 4.0

College Honors, Professional Societies, Fraternities, and Activities
Special Honors in Journalism; Kappa Tau Alpha; Feature Editor, Daily Texan

% College Expenses Earned	How Earned
About 20%	Part-time and summer employment

EMPLOYMENT INFORMATION

Work Experience (in reverse order) (Names and Addresses of Employers)	Description of Work	Hours Per Week	Dates Employed From	Dates Employed To
Writing Services, Inc. Jason Brown, Pres. 4611 Brazos, Austin, TX 78712	Assisted in editing brochures, reports	40	June to 1975 &	Sept. 1976
ABC Press, C.B. Acton, Owner 2800 Guadalupe, Austin, TX 78712	Operated duplicating mach. & off-set press	10	Sept. '76 Sept. '75	May '77 May '76
English Dept. R. Abrahams, Chm. Univ. of Tex., Austin 78712	Assistant (grader)	10	Sept. '74	May '75

GENERAL INFORMATION

References (Name, Title and business address—Preferably business reference, past supervisors, or faculty)
Jason Brown (see address above)

Donn Taylor, Instructor of Technical Writing, Dept. of English, U.T. Austin

C.B. Acton (see address above)
J.B. Jackson, Professor of Journalism, Univ. of Tex. Austin, 78712

Other Information (Community Activities, Hobbies, and Interests. Also, organizations outside of College)
Student member, Society for Technical Communication; Reporter, Daily Texan;

Choral singing - member S.P. E.B.S.Q.S.A.; Lapidary work as hobby.

Signature	Date Signed
Robert W. Abelson	March 1, 1977

Figure 14 A printed form of data sheet to be completed by applicant.

FORM 51-0006 (2)

UNITED ENERGY RESOURCES, INC.

AND SUBSIDIARIES

APPLICATION FOR EMPLOYMENT

Please complete either with
pen and ink or typewriter.

PERSONAL

Applicant's Last Name	First Name	Middle Initial or Maiden Name Initial	Social Security No.

Present Address	City	State and Zip Code	Telephone No.

Permanent Address (If Different from Present Address)	City	State and Zip Code	Telephone No.

Birthdate	Height	Weight	Marital Status	Are you a citizen? ☐ Yes ☐ No

Name of Spouse	Where employed?

Person to Notify in Emergency	Address	City	State and Zip Code	Telephone No.

Who referred you to us? Person and/or Organization

What office machines can you operate?   Other
Typewriter ☐   ☐ Take Dictation?
WPM: Yes, WPM: ☐ Adding   ☐ Calculator   ☐ Key Punch   ☐ Data Processing Equip.

EDUCATION

Name and Location of School	Years Attended				Did you graduate?	Courses – Degrees
	From		To			
	Mo.	Yr.	Mo.	Yr.		
High School						
College or University						
Business, Commercial or Vocational School						

Scholastic Honors, Honorary Fraternites, Scholarships

WORK EXPERIENCE (Former employers. Begin with most recent period.)

Company Name and Current Mailing Address	Dates Employed				Occupation	Reason for Leaving
	From		To			
	Mo.	Yr.	Mo.	Yr.		

AN EQUAL OPPORTUNITY EMPLOYER – MALE/FEMALE   (Over)

Figure 15   A printed application form.

REFERENCES (Refer to people familiar with your work performance other than former employers)

Name and Current Mailing Address	Phone No. If Known	Type of Work	Yrs. Known

WORK PREFERENCE

Position or Department Applied For: Other Department and Interested Areas   Salary Range Desired

Were you ever discharged or requested to resign a position? ☐ Yes ☐ No   If Yes, Explain

U.S. MILITARY

Have you completed U.S. Military Service? ☐ Yes ☐ No   Dates of Service   Branch

Do you receive Military Disability? ☐ Yes ☐ No   Are you a member of Active Reserve or National Guard? ☐ Yes ☐ No   Branch

MEDICAL

State all major illnesses or surgery you have had in the last five years.

What physical or chronic ailments do you have?

Are you willing to take a physical examination at our expense? ☐ Yes ☐ No

APPLICANT'S STATEMENT AND AGREEMENT

I agree to the investigation of any and all statements included in this application and declare they are true and complete.

I understand that any misrepresentation, falsification or willful ommission of information contained in this application or in connection with any physical examination shall be sufficient reason for refusal of or dismissal from employment and render void all insurance coverage or other benefits due me from the initial date of my participation.

I agree to take any future examinations, physical or other, requested by the Company as related to employment.

_____   _____
Signature of Applicant   Date

We assure you that your opportunity for employment with this Company will be based only on your merit. Although we have asked your birthdate, we are conforming to those laws prohibiting discrimination because of age. Applications completed by those not employed will be retained in our file for 12 months.

                                              4610 Finley Dr.
                                              Austin, Texas 78731
                                              May 2, 19--

Mr. James B. White
Head, Publications Department
Aerotron Petroleum Corporation
Aerotron Building
Houston, Texas 77098

Dear Mr. White:

Mr. Donn Taylor, instructor in two technical writing courses
I have taken here at the University, tells me that he believes
your company welcomes applications from persons interested in
pursuing a career in technical writing. I am writing to apply
for a job in your department, and I should like to tell you
something about my qualifications.

Although I have taken my professional work primarily in the
Department of Journalism, and will receive my bachelor's degree
in May, I have concentrated my training as much as possible so
as to prepare for a career in technical writing and editing. As
you will see on the enclosed transcript of my university work,
I have taken four courses in technical writing and in technical
editing, one an introductory course, and three at the advanced
level. In addition, I have taken all the courses I could in Journalism that would appear to enhance my ability to perform well
in a technical publications department; these have included copy
editing, intensive writing and editing, feature writing, graphics,
and magazine editing. Along with these courses in writing and
editing, I have tried to round out my training in science by
taking courses in physics, chemistry, mathematics, and computer
science.

My experience, although limited, has at least a tangential bearing on the work I would like to do. As you can see on my
résumé, I have worked for a company devoted to the preparation
of brochures, and I have had the opportunity of becoming familiar
with several types of duplicating processes, as well as with
offset printing. My responsibilities were limited, of course,
but I did have some chance to exercise my skills at editing
and rewriting rough copy.

I would welcome an opportunity of showing you some of the
work I've done and of telling you more about myself. Would it
be possible to arrange an interview? I will be free to come to
Houston any time after the middle of May when I will have
completed my final examinations. I look forward to hearing
from you.

                                    Yours very sincerely,

                                    Robert W. Abelson
                                    Robert W. Abelson

Encl.:Resumé

Figure 16   Letter of application, Example 2.

you have held, including part-time and summer jobs. An employer may be interested in finding out how well you have discharged responsibility and how diligent and cooperative you have been in carrying out duties and working with other people. He or she may believe that an applicant's attitude toward a job and toward fellow workers is not likely to change, regardless of the nature of the work.

List the jobs you have held in reverse chronological order: most recent job first. For each job give (1) the dates of employment, (2) the kind of work done, (3) the name of the company employee qualified to evaluate your services, (4) the name of the company or organization, and (5) its address. Be sure to record the name of your superior and the name and address of the company accurately so that an inquiry will be certain to reach its destination.

One often hears: "It's not *what* you know that gets you ahead; it's *who* you know." We do not subscribe to this cynical remark, but neither do we wish to minimize the importance of having influential people back you up in your application. Often the recommendation of a person whose word is respected is a deciding factor in getting a job. This is perhaps especially true when an employer is considering a number of applications from graduates who do not have professional experience. The academic records alone might not provide a clear basis for a choice. We believe you should devote careful thought to this matter of references and make the most of your opportunities.

At least three, and perhaps as many as five, references should be listed. One of these should be an employer, if possible; one should be a person who has known you personally a long time and who can therefore vouch for your character; and the others should be those of your teachers who can vouch for the quality of your work as a student. Be sure to get permission from each one. It's a good plan to tell each reference, at the time you ask permission, something about the job you are applying for so that a better letter of recommendation can be written for you, one in which emphasis is on qualifications that are pertinent to the job you are after. Be very sure that you show your references the courtesy of spelling their names correctly, and for your own sake be sure to give an address at which they can be reached.

One final word about references. If you are especially eager to get the job you are applying for and feel that you can presume upon the kindness of some of your references, ask them to write unsolicited letters of recommendation, to be sent so that they will be received shortly after your application has been received. This support for an application may be quite effective, and it is comforting to know that a recommendation has been made. A prospective employer may not take the initiative by writing for information about you. We know of one young applicant who went a step further; he asked one of his references to put in a long-distance telephone call in his behalf. He got the job, too.

Conclusion

If you hired someone to act as your personal representative you would want him or her to be pleasing in appearance, businesslike and alert in manner, and intelligent in speech. A letter is your personal representative. Don't be satisfied with inferior specimens. This chapter will give you a start toward good letter writing. A special book on the subject will help you further. Intelligent practice will help you most of all.

Suggestions for Writing

1. If you are writing a library research report for your course in technical writing, you might attempt to supplement the library materials available by writing to several companies for information not obtainable in the library. Be sure, if you write such an inquiry, to explain why you are making it.
2. Write a letter of application, together with a data sheet. To get the most benefit from this exercise, you should make the application as realistic as possible: aim it at a job your present experience and training qualify you for. A summer job is a suggestion.
3. Write a letter of complaint to a manufacturing company about a defect in a mechanical device you have purchased. Then assume you are that employee of the company to whom the letter has been routed for reply and write an appropriate letter in response to the complaint.
4. When you have a report to write, prepare a letter of transmittal to accompany it.

18

Writing for Professional Journals

Publication of articles in professional journals may benefit you in many ways. Such publication is likely to increase your circle of professional acquaintances; it is certain to put an example of your work in the hands of leaders in your field; and it will be a strong stimulus toward mastery of your area of specialization. It may also have direct effect on your advancement, for many firms strongly encourage their employees to publish. (See the accompanying chart from a General Motors discussion of research publication.)

Publication of semitechnical articles in popular magazines is also financially attractive. Technical journalism of this sort interests only a small minority of scientists and engineers, however, and for that reason we'll not discuss it here.[1]

The professional journals — by which we mean loosely any journal published for trained specialists — do not ordinarily pay for contributions. Nevertheless, it is by no means always easy to place an article with them. Usually these journals have many more articles submitted to them than they can possibly publish. You should

---

[1] If you are interested in this field, see *Directory of Publishing Opportunities,* 3rd ed. (Chicago: Marquis— Who's Who Inc., 1975). See also Jay R. Gould, *Opportunities in Technical Writing Today* (Louisville, Kentucky: Universal Publishing & Distributing Corporation, Vocational Guidance Manuals, 1975); John Mitchell, *Writing for Technical and Professional Journals* (New York: John Wiley & Sons, Inc., 1968); Harley Sachs, *How to Write the Technical Article and Get It Published* (Washington, D.C.: Society for Technical Communication, 1976); and Herbert B. Michaelson, *How to Write and Publish Engineering Papers and Reports* (Philadelphia: ISI Press, 1982).

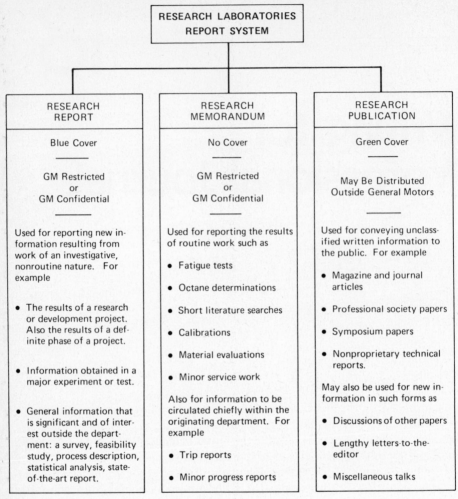

Diagram of General Motors report system.

not be surprised to have your offering rejected or returned with a request for revision. You should not assume that a rejection means you have written a poor article; an article may be turned down for a number of reasons having nothing to do with its quality. Your article may be concerned with a subject the editors feel has already been given all the space in recent issues that they can devote to it. It may be presented in a way that the editors feel would not interest their readers, although it might interest other readers (in this case they may suggest another journal or journals to you); or it may be an article that at another time would have been accepted, but is now rejected simply because the editors have on hand a large number of unprinted articles of high quality. For example, the editors of *Science* had this to say in the issue of July 26, 1974 (p. 346):

In the past few weeks the editors have received an average of 68 Reports per week and have accepted 12 (17 percent). We plan to accept about 12 Reports per week for the next several weeks. In the selection of papers to be published we must deal with several factors: the number of good papers submitted, the number of accepted papers that have not yet been published, the balance of subjects, and the length of individual papers.

Authors of Reports published in *Science* find that their results receive good attention from an interdisciplinary audience. Most contributors send us excellent papers that meet high scientific standards. We seek to publish papers on a wide range of subjects, but financial limitations restrict the number of Reports published to about 15 per week. Certain fields are over-represented. In order to achieve better balance of content, the acceptance rate of items dealing with physical science will be greater than average.

Frankly, an article from an unknown author is less likely to be accepted than one bearing the name of a person with a nationwide reputation. But don't assume that someone without professional distinction should abandon hope. If you have something significant to say, and say it clearly, you have a good chance to have your work published. Don't forget the virtues of patience and common sense.

In this chapter we discuss the problems of choosing a subject, selecting a journal to send your article to, writing the article in a suitable style, and putting the manuscript into the proper form. Our discussion of these topics will, however, be limited to a general survey. Should you become seriously interested in writing for professional journals, we suggest you consult a book concerned specifically with this subject (see footnote 1).

It is often difficult to say whether, in practice, a person selects a subject because of wanting to write an article, or writes an article because of interest in a subject. Ideally, the subject would always come first. Publication would be considered only when a person's thinking and research had developed facts or theories that might be of value to other workers. A famous fictional presentation of this ideal, and a satire of its opposite of seeking fame through shoddy, pretentious, overhasty publication, is Sinclair Lewis's *Arrowsmith*. But, as Lewis said, human motives are seldom unmixed. Granted honesty and sincerity, there is little point in thinking much about whether desire for publication or interest in a subject comes first. In any case, there isn't much we can say about your personal interests.

The only advice we can give here about how to find a subject, in contrast to evaluation of a subject, is to read widely in your field, to acquire a wide acquaintance among your colleagues, to attend meetings of professional societies, and everywhere to use your imagination and to be critical. Be slow to assume that an explanation is correct or that an accepted method is the best method. Out of such caution will come new ideas for research and publication.

When you have an idea that looks interesting, work at it. We once heard a well-known physicist say that one of the chief differences between creative and noncreative people in his field was that the noncreative people simply failed to develop their ideas. A good deal of determination and some stubbornness are called for. Do your own thinking, and allow in advance for opposition to new ideas. But be sure to distinguish between boldness in conceiving new ideas and carelessness in developing them. You should be patience itself in calculating, testing, checking, and criticizing an idea once you have gone to work on it.

It is wise to keep a file of possible subjects for investigation. Here are some questions to ask yourself in evaluating subjects for your file:

1. Is development of the idea within your present ability? Of course you will want to add to your knowledge and skills, but don't take on too much at once.
2. Is equipment available to you for the work that will be required?
3. Do you feel a real interest in the subject? Don't let circumstances coax you into work you don't care for when you could be doing something you'd like.
4. Will the subject open up further possibilities of research and publication, or is it a dead end?
5. Is the subject in a field that has received little recognition? Or is the subject in a field that is overworked? Either possibility may mean difficulty in getting your article published and the merits of your work recognized.
6. Will work on the subject contribute to your ability and success in the kind of career you desire? An article on the fuel cell would probably do little to further the career of a civil engineer.

Such considerations as these are very much worth your attention before you commit yourself to any project that will take more than a few days of your time.

Having decided to write an article, you must begin thinking about where to send it. The first step is to find out what journals publish material of the kind you will have. Of course you should be acquainting yourself with such journals anyhow; familiarity with them is an important part of your professional equipment. (For guides to help you locate the journals in your field, see Chapter 21.) A second step is to analyze the journals you have decided are possible targets. Can you find any articles in them, dating from the past year or two, which are on a subject comparable to yours? You needn't feel that you should find the same subject, and of course you shouldn't expect to find an article that says approximately the same thing that you are going to say. This possibility of duplication, incidentally, brings up another matter. You should be very careful to look at every article that has been published anywhere on your specific subject, no matter how tedious the hunt may be, to make sure that you are not merely repeating somebody else's work, as well as to inform yourself fully on your subject.

When you have a list of the journals that show an interest in the kind of subject you have, it is wise to make your next step a conference with someone who has had considerable experience with professional publications. You may acquire invaluable information about editorial whims, possible places to publish the article, and other matters. It is not uncommon for a beginner's first publication to be achieved through the friendly help of an older, well-known person whose recommendation carries weight. If you work for a company that has a technical editor or a staff of technical writers, it may be that you can get help from them. Many companies encourage their technical staff to seek such help.

Next, you should analyze the style of the journal you have chosen. This analysis should cover two elements — literary style and physical format.

Literary style is perhaps less important in a professional journal than in a popular journal, where appeal to a large, untrained audience demands a vivid presentation. Nevertheless, it is well worth your time to see if there are any special preferences or prejudices in style and general attitude that examination of numerous issues of the journal will reveal. Is the treatment theoretical or practical? Speculative or down to earth? Informal and colloquial, or formal and restrained? You will probably find considerable variety even within a single issue, but usually a distinctive tone will become evident as you read through several issues. Try to get the feel of the journal, and write your article accordingly. Remember that editors are human beings, and the problem of reader analysis is essentially the same in writing for an editor as in writing a routine report in college or on the job. All you know about the editor, however, may be what you can infer from analysis of the articles chosen for publication. As a matter of fact, articles are usually read by several people, usually two or three besides the principal editor who is responsible for the final decision.

Often it is a good idea to send a query to the editors of the journals you have chosen, asking if they think the idea you have for an article is promising. Here is what the editor and publisher[2] of *Automotive Industries* had to say to prospective contributors on this point.

> Before preparing the final text, send AI editors a 150 word outline of your proposed text, stating the subject, the sources of information used, the problem that will be discussed, the illustrations available, the data or charts available, the benefits developed for readers—and your own name, address, occupation, title and business affiliation. Graduate engineers or scientists are also expected to mention their educational attainments and outstanding career achievements.
>
> When the editors have approved the outline, the next step is to accept and *incorporate in your outline,* the suggestions made by the editors.

Automotive Industries is a trade journal. That is, it publishes articles on the subject of what might be called technical news about new techniques and new products. Another kind of journal is that which is devoted more definitely to articles describing original research. An example is *Review of Scientific Instruments.* Generally speaking, editors of trade journals are likely to be more receptive to queries about how promising an idea for an article is than are editors of journals publishing reports of original research.

Analysis of the physical format (form of footnotes, subheads, and the like) preferred by a journal is a simpler task than analysis of the literary style. Very often specific directions are available, and when they are not, the form of articles printed in the journal serves as a model. Some professional societies that publish journals issue pamphlets giving instructions on form. Some journals regularly print short statements about form. Examine closely the journal you are interested in, and if you find no hint of directions to be followed, pick out two or three articles and use them as models. Note particularly such matters as use of subheads, footnote and bibliographical forms, whether or not an abstract is used (and if it is, what type it is), types of illustrations, how numbers are written, and what abbreviations are used.

Your article should be typed, preferably with pica type, and double-spaced.

[2] Hartley W. Barclay, "Letter to Contributors of Technical Articles to *Automotive Industries*" (no date).

You should make at least one copy for yourself. Some journals request that two or three copies of a manuscript be submitted so that several editorial readers can read the article simultaneously. Good clear carbons or photocopies are acceptable, and —in the copies—illustrations can be roughed in; or, if that is not feasible, a brief explanatory note can be substituted.

Here are some general suggestions about manuscript form.[3]

1. Use good paper of standard size, 8½ by 11 inches (22 by 28 cm).
2. Leave a margin of 1¼ inches (3 cm) at the left and 1 inch (2.5 cm) on the other three sides of the page.
3. Type your name and address in the upper right-hand corner of the first page.
4. Type the title of the article about halfway down the first page. Underneath, type "by" and underneath that your name, triple-spaced, like this:

ELECTRONICS

by

John Warren

The empty space in the top half of the page is a convenience to the editor for making notes.
5. In the upper right corner of each page after the first page type the title of the article, followed by a dash and the page number. If the title is long, use an abbreviated form of it.
6. There are various ways of handling illustrations, but if you have no specific directions, the following procedure will be satisfactory. First, be sure you have put a clear title on every illustration (photographs, drawings, charts, graphs— and also tables) and, if there are several illustrations, a figure number. Next, instead of putting the illustrations into the text, collect them in an envelope at the end of the manuscript. To show where they go in the body of the text, write the figure number and the title in a blank space left in the appropriate place in the text. Finally, add to the collected illustrations a typed list, on a sheet 8½ by 11 (22 by 28 cm), of figure numbers and titles (this sheet should not have a page number). Further suggestions on illustrations will be found in Chapter 20, and additional suggestions on manuscript form in Chapter 19. Incidentally, remember that reproduction of illustrations is expensive, and professional journals do not always have an abundant supply of money.
7. Proofread your manuscript with painstaking care, particularly tables and graphs. This job is dull and time-consuming, but it is imperative that it be well done. Ask a friend to read aloud from the rough draft while you check the final copy. A page on which numerous corrections are called for should be retyped. If there are no more than two or three corrections on a page, however, it is permissible to make them neatly between the lines.
8. Mail the manuscript flat. Enclose some kind of stiffener, like heavy cardboard, if there are illustrations that would be seriously damaged by folding. It's a good

---

[3] For more detailed instructions, see Robert A. Day, *How to Write and Publish a Scientific Paper* (Philadelphia: ISI Press, 1979).

idea to mark the envelope "Do Not Fold." Include in the envelope a self-addressed stamped envelope to bring back the manuscript if it is rejected.
9. Resign yourself to a long wait. You may learn the fate of your manuscript in six weeks, but it may take six months. If you've had no word in six months, an inquiry would not be out of order.

If your manuscript is rejected, mail it out again at once to another prospective publisher. But there are two things that should be done first. One is to make sure that the manuscript looks fresh. Editors are never flattered by a suspicion that your manuscript has been sent elsewhere before. The second thing is to consider whether any changes should be made in the article to adapt it to the policies and attitudes of the journal you now have in mind. You should be as careful about this on the second, third, or fourth mailing as on the first.

Of course your manuscript may be accepted the first time out, or it may come back with a request for revision. Whether or not to revise as requested is a matter to be settled between you and your conscience. Chances are that the editor is right. If you disagree, however, and feel strongly about the matter, it may be better to seek publication elsewhere. In any case, don't be fussy about little things. Few editors can resist changing a few commas, at least.

When your manuscript has been accepted, there will probably arrive, in due time, some "proof sheets" or "galley proof." These are long sheets of paper on which the printed version of your article appears. Your job is to proofread these sheets and return them to the editor. You should again get someone to help you.

Corrections on proof sheets should be made with standardized "proofreader's marks." These marks, together with directions for their use, can be found in most good dictionaries. With a few exceptions, corrections should be made only of errors that the printer has committed, because of the expense of resetting type. On the other hand, if you discover that you overlooked errors in grammar or in facts, you should certainly correct them.

You may or may not later on receive the corrected proof sheets to examine. If you do, the checking process should be conducted as meticulously as before, but—with very rare exceptions—only printing errors should be corrected.

Writing an article for publication in a professional journal is fundamentally like any technical writing. The principles of reader analysis, logical organization, and clarity of expression must be observed. There are some special problems: selection of a subject, choice of a journal to submit your article to, handling of the manuscript, and correction of the proof sheets. But these are all problems that can be solved by a methodical approach.

Suggestions for Writing

The most obvious suggestion about writing for professional journals is to write an article and send it to a professional journal. If you have an idea for such an article, fine. Give it a try. Some preliminary work might prove helpful. Select a few journals that seem appropriate and study them. Write out what you find concerning the kind of articles they print, the length of the articles, the readers they aim at, the format of the articles, instructions they provide for contributors, and anything else that is pertinent. It would be a good idea to send a query to the editor of the journal you select as your first target concerning his or her possible interest in your subject; ask for suggestions. With this kind of investigation completed, you can write the article itself with far greater assurance than if you simply sit down and start writing.

Even if you are not interested in publishing an article, the sort of investigation just described can prove highly valuable as a means of getting well acquainted with those professional journals in your field you will want to make it a habit to read regularly. The advice of an experienced member of your profession would be helpful in selecting the journals.

SECTION V

Report Layout

The two chapters in this section discuss report forms and forms of report organization, and graphic aids. Part 1 of Chapter 19 deals with such mechanics of report preparation as the arrangement of a title page and the placement of subheads. Part 2 covers generalized forms of report organization as well as printed forms and laboratory reports. The term "graphic aids" refers to any nonverbal device included in a report: a photograph, a table, or a chart. The subject of the two chapters is similar in that both deal with the appearance of reports.

19

The Format and Forms of Reports

PART 1: THE FORMAT OF REPORTS

Introduction

If you were to make a careful survey of the format of reports prepared by a representative number of companies, you would observe two facts: (1) although all companies do not use the same format, the differences are likely to be minor; and (2) all companies and organizations agree that attractive format is necessary.

While accuracy and clarity are always of paramount importance, remember that a report makes an impression on its reader even before he or she has an opportunity to determine whether its contents are accurate and clear. A well-known engineer once told a story of visiting an industrialist's office and seeing the industrialist pick-up a handsomely bound report just as it was delivered to his desk, leaf through it, and remark that it was a fine job of engineering report writing. He hadn't read the report; he made this judgment solely on the basis of its appearance.

Common sense will tell you that it pays to make your reports look good. The question is not whether attractive format is desirable, but what *is* attractive format. The following pages, therefore, will be devoted to a discussion of typescript standards and the form of the elements of a report, plus some notes on the relationship of form to organization and style. Before presenting the "rules" that follow, we want to say that no existing body of rules for report format can be regarded as authoritative the country over. The ones we present are representative of good practice, however, and will be acceptable whenever you do not have other instructions.

Typescript Standards

When you prepare a typewritten report, you will have to make decisions regarding the choice of paper, width of margins, spacing and indenting, and paging.

Paper   Reports should be typewritten on white paper of high quality, preferably 20-pound (9 kg) bond, 8½ by 11 inches (22 by 28 cm) in size. Additional copies can be made with the use of a photocopier. A good quality paper is essential if a neat, attractive report is to result; inking and erasures, for instance, require good paper. Always use white paper unless you have instructions to the contrary. Some companies use colored sheets to identify certain types of reports or reports from certain departments.

Margins   Margins for the typewritten report should be approximately as follows:

Left Side	Top	Right Side	Bottom
1½ in.	1 in.	1 in.	1 in.
(3.8 cm)	(2.5 cm)	(2.5 cm)	(2.5 cm)

Since the left-hand margin must be wide enough to allow for binding, up to 2 inches (5 cm) may be needed, depending upon the nature of the binding. No reader likes to be forced to strain the binding in order to read the words on the bound side of the sheet. Unless electronic word-processing is used, the right-hand margin cannot, of course, be kept exactly even because of the necessity of dividing words properly, but an effort should be made to keep a minimum margin of ¾ inch (1.9 cm).

Where quotations are introduced into the text of a report, an additional five spaces of margin must be allowed on the left side and approximately the same on the right.

Spacing and Indenting   The text of a report should be double-spaced throughout, except as noted below:

1. Triple- or quadruple-space below center headings.
2. Single-space and center listings (if items are numerous, number them).
3. Single-space long quotations—those that run four or more lines in length.
4. Triple-space above and below quotations and listings.
5. Single-space individual footnotes more than a line long; double-space between notes.
6. Single-space individual entries in the bibliography; double-space between entries; use hanging indention in bibliographical entries of more than one line in length.
7. Single-space the abstract if space demands it; otherwise, double-space it.
8. Usually, single-space material in the appendix.
9. Double-space above and below side headings.

The customary indention at the beginning of a paragraph is five spaces. An additional five spaces (or more if necessary for centering) should be allowed before beginning a listing.

Paging   Use Arabic numbers in the upper right corner, except for prefatory pages and the first page of the body, and pages that begin new divisions. The number should be in alignment with the right-hand margin, at least two spaces above the first line of text on the page, and about ¾ inch (1.9 cm) down from the top edge. The prefatory pages of a report — title page, letter of transmittal, table of contents, list of figures, and abstract — should be numbered with lower-case Roman numerals centered at the bottom of the page, about ¾ inch (1.9 cm) from the bottom edge. It is customary to omit the numbers from the title page and the letter of transmittal, although these pages are counted; thus, the table of contents becomes iii. In the body of the report, it is customary to omit placing the number 1 on the first page, since the title there obviously identifies it as page 1. As for pages that begin main sections of the report, it is probably best to place the number in the bottom middle of the page. Pages of the appendix are numbered as in the body, in the upper right corner. No punctuation should follow page numbers.

Formal Report Form

By formal report we mean a report with all, or nearly all, of the parts that will be described below. Informal reports do not possess all the parts usually included in the formal report, and thus present a somewhat different problem as far as format is concerned. Informal reports will be considered later in this chapter.

The Cover   Ordinarily you will not have to worry about making up a cover for your report, for most companies have prepared covers. These are made, usually, of a heavy but flexible paper, with a printed heading naming the company and the division, and with a space for information about the report itself. This information consists of (1) the title, usually prominently displayed in underlined capital letters; (2) the report number; and (3) the date. Sometimes this information is typed on the cover itself; sometimes it is typed on gummed slips and pasted on, and frequently a window is cut in the cover so that the title block on the title page will show through. In any case, the title should be clearly legible. Triple spacing between the lines of two- and three-line titles is advisable.

Occasionally the name of the client to whom a report is submitted and the name or names of its authors may be found on the cover, but as a general rule only the three items of information mentioned above are recorded. These serve to identify the report for filing and reference; additional information may lessen the prominence of these important facts and detract from the attractiveness of the layout.

If prepared covers are not supplied, you can use a plain Manila folder or one of the readily available pressboard binders. Many companies have covers made of special stock, with an identifying picture or symbolic device.

The Title Page   Besides duplicating the information found on the cover, the title page gives a good deal more. The most significant additional information presented here is the name of the person or persons who prepared the report and their positions in the company or organization. In addition to authorship, the title page of reports from many industrial concerns provides space for the signatures of those who approve, check, and (sometimes) revise the report. Some companies provide space for "Remarks" of those who check the report. Finally, it is not unusual to find a notation of the number of pages of the report. The accompanying illustration (Figure 1) is fairly typical. Note that the title should be all written in capital letters and underscored, and centered about one-half of the way down from the top of the page.

Should you be required to write a report for a company that does not provide a prescribed form for the title page, the model given as Figure 2 is satisfactory. Note that it contains four elements attractively grouped and spaced. The title appears in the upper third of the sheet, underscored and centered, with triple spacing between the lines. Centered on the page appears information about the recipient of the report. On the bottom third of the page appears the reporter's name and professional identification; the last entry is the date of submission. In centering material on the page, do not forget to allow about half an inch for binding.

The Letter of Transmittal   Since we discussed the letter of transmittal at length in another place (see Chapter 17), we simply want to point out here that this part of the report should be meticulously accurate in form and layout. Although it usually appears immediately after the title page, some companies require that it appear as the first item in the report, just inside the cover (or even stapled onto the outside of the cover). Sometimes the letter of transmittal does not form a part of the report at all, but is sent separately through the mails. And sometimes the functions of the letter of transmittal are performed by a foreword.

The Table of Contents   The table of contents of a report is an analytical outline, modified in form for the sake of appearance. It serves as an accurate and complete guide to the contents of the report. The entries in this outline also appear in the text of the report as headings; thus, a reader may easily refer to a particular section or subsection of the report. Every heading in the outline must appear in the text as a heading or subheading. It is not necessary, however, that every subheading in the text appear in the outline.

Except for the Roman numerals for main division headings, the conventional outline symbols (A,B,C, . . . for subdivisions) are omitted in the table of contents, indention alone being used to show subordination. Omitting the capital letters and Arabic numerals results in a neater page. But although this is majority practice, it is by no means unanimous. Some companies retain all conventional outline symbols; some omit all. And some companies use other symbol systems, such as decimal numbering. In any event, it is a good idea not to clutter up a table of contents with minute subdivisions: three levels are enough (this is not intended to suggest, of course, that your *plans* should not be detailed). Examine the accompanying examples in Figures 3, 4, and 5.

REPORT NO.
NO. OF PAGES

JONES CONSTRUCTION COMPANY

WASHINGTON, D.C.

ENGINEERING DEPARTMENT

EVALUATION OF POLYMER IMPREGNATED
CONCRETE FOR SPECIAL APPLICATIONS

SUBMITTED UNDER CONTRACT
PREPARED BY   APPROVED BY

CHECKED BY

                                        DATE
        REVISIONS
            PAGES AFFECTED   REMARKS

Figure 1   Title page, Example 1.

A Report

on

<u>COMBATING THE STALL PROBLEM</u>

Prepared for

The Director of Research
Wakey Products, Incorporated
Detroit, Michigan

by

Mary Morrison
Aeronautical Research Assistant
September 10, _ _

Figure 2   Title page, Example 2.

UNITED STATES DEPARTMENT OF AGRICULTURE
FOREST SERVICE

MANUSCRIPT PREPARATION HANDBOOK

Contents

Figure 3   Table of Contents, Example 1.

Table of Contents

iii

Figure 4 Table of Contents, Example 2.

ABC LABORATORIES

PRODUCTS APPLICATION DEPARTMENT
GREASE AND INDUSTRIAL LUBRICANTS
TESTING GROUP

DETERMINATION OF SUITABLE ROTOR BEARING
GREASE FOR USE IN THE BROWN MAGNETO

INDEX

Figure 5 Table of Contents, Example 3.

You will note that all three specimens have this in common: They provide plenty of white space so that the prominently displayed headings may be easily read and so that the page as a whole presents a pleasing appearance. For the best layout of the page, follow these suggestions:

1. Center and underscore "Table of Contents" at the top of the page. Use either capitals or lower-case letters.
2. Triple- or quadruple-space below the centered "Table of Contents." Double-space between the major items in the contents. If there are numerous subtopics, they may be single-spaced.
3. Begin items preceding Roman numeral I flush with the left margin. These items include the List of Illustrations, Symbols, List of Figures, and Abstract.
4. Indent second-order headings five spaces and third-order headings ten spaces.
5. Use a row of periods to lead from the topic to the page number at the right margin.
6. After the last Roman numeral entry, list items in the appendix. Place the word "Appendix" flush with the left-hand margin, as shown in Figure 4, and indent the individual entries. The bibliography comes first. If nothing besides a bibliography is to be appended, do not use the word "Appendix"; place the word "Bibliography" where the word "Appendix" would otherwise appear. If several appendixes appear, it is common practice to label them "A," "B," "C," and so on, and to give each a title written in all capital letters, or underscored—as with other section titles.

The List of Figures   If a report contains a half-dozen or more illustrations, drawings, or other graphic aids, an index to them should follow the table of contents. Usually called "List of Figures," this page gives the number, title, and page reference of each figure in the report. See Figure 6.

The actual layout of the page is simple. Center the title at the top of the page (allowing for top margin) and underscore it. Triple- or quadruple-space before beginning the list. Figure numbers should be aligned under the word "Figure" and followed by periods. The initial letter of each important word in the titles of figures should be capitalized. Page numbers should be aligned at the right margin, with a row of periods connecting title and number. Double-space between entries, but single-space an individual title requiring more than one line (and remember that a line should not be carried all the way over to the right margin). This spacing will allow for plenty of white space—a requirement for a neat, attractive page. See Figure 6.

There are, of course, some variants of this form. Some companies like to classify nontextual material so that there is, besides a list of figures, an index to tables and perhaps even a list of photographs. Separate pages are not needed for these individual listings unless the length of the listings requires them. When the table of contents is quite short and there are few illustrations to list, both the table of contents and the list of figures may be placed on one page (Figure 5). Informal, short reports containing fewer than five or six illustrations usually omit a formal list.

List of Figures

iv

Figure 6 Layout of a list of figures.

Custom in an organization will dictate whether omission of the list of figures is permissible.

The Abstract The word "Abstract" should be centered and underlined at the top of the page. Allow triple or quadruple spacing after this title and then double-space the text of the abstract itself, maintaining the same margins used for the body of the report. In some cases, where space is at a premium, the abstract may be single-spaced, but double spacing is better.

You may have observed that the term "Abstract" is not universally used; some companies call this part of the report a "Digest," some a "Summary," and some an "Epitome." Whatever it is called, format requirements are the same. For what goes into the abstract, see Chapter 4 on outlines and abstracts.

Headings The topical entries of the outline table of contents also appear in the text of a report as headings identifying the individual portions of the subject matter. All entries in the table of contents should appear as headings in the text, and they should appear exactly as they are worded in the table of contents. The headings serve as transitional devices and enable a reader to find a specific part of a report's discussion with ease. We are concerned with the form and location on the page of the three types: main or center headings, and two types of subheadings.

Main or Center Headings. Main headings name the major divisions (Roman numeral divisions) of a report. Written in either lower-case letters or capitals, a main division heading is underlined and placed in the center of the page, with the Roman numeral preceding it, as "II. *Circuit Elements and Transmission Lines.*" See Figure 7. In formal reports, it is customary to begin new divisions of a report on a new page, just as a new chapter in a book begins on a new page. The centered title should stand a minimum of three lines above the first line of text of the division or the first subheading.

Subheadings. Usually only two levels of subheading are needed beyond the main headings; that is, headings corresponding to capital letter and Arabic numeral divisions, respectively, in the outline.

The capital letter or second-order headings should be placed flush with the left margin. Underline each word separately. Use lower-case letters, but capitalize the initial letter of each important word. Double-space above and below the heading, and do not put any text on the same line as the heading. Don't put any punctuation after the heading.

The Arabic numeral, or third-order, headings should be handled exactly like the second-order headings with three exceptions: (1) indent the heading five spaces; (2) put a period after the heading; (3) start the text on the same line as the heading. See Figure 7.

If it is necessary to use fourth-order headings, treat them like third-order headings, but number them with Arabic numerals.

A variant custom of identifying or numbering sections and subsections is the use of the Arabic decimal system. With this system, the first main division topic is headed with "1," the second with "2," and so on. Subdivisions are headed "1.1, 1.2," and so on. Sub-subdivisions may be headed "1.1.1, 1.1.2," and so on. This

II. <u>Circuit</u> <u>Elements</u> <u>and</u> <u>Transmission</u> <u>Lines</u>

In order to understand the problems and principles of the operation of carrier circuits, it is first necessary to understand the characteristics of circuit elements and transmission lines at both power and carrier frequencies. The most common elements are resistors, capacitors, inductors, transformers and sections of transmission lines.

<u>Circuit</u> <u>Elements</u>

Different circuit elements have separate and distinct properties. Because of this difference, resistors, inductors, capacitors, and transformers will be considered separately.

<u>Resistors</u>. A resistance is an energy-absorbing element. Although the value of the resistance of a material does not vary with frequency, the effective resistance of a resistor or section of wire varies because of the skin effect, or the movement of the current to the outer edges of the conducting resistance. Another important characteristic of a resistance is the phase relation between the current through and the voltage across a resistor. The current is directly proportional to the applied voltage and there is no time lag between a change in voltage and a change in current.

<u>Inductor</u>. An inductor is a circuit element that has an impedance to the flow of current but absorbs no

Figure 7 Layout of headings.

system can become cumbersome (indeed, we once saw a long report with one obscure heading prefaced by eleven digits); but it is used by a good many organizations and it does lend itself to convenient reference. A correspondent, for example, may refer the reader to "section 4.2.1" of a document. Figure 8 shows a suggested breakdown for a section on equipment installation, together with the decimal numbering we have been talking about.

Quotations and Listings   Formal quotations are single-spaced, indented five spaces from the left margin and approximately the same number from the right. Quotation marks are unnecessary; single spacing and extra margin adequately identify the material as a quotation. If the quotation does not begin with the first word of a sentence in the original, the omission of words should be shown by a series of three periods, and any deletion within the quotation should be similarly indicated. Triple-space above and below the quoted matter.

Informal, short quotations that are a sentence or less in length should be run in with the text. As Houp and Pearsall[1] say, "Use quotation marks to enclose quotations that are short enough to work into your own text (normally, less than three lines)."

Formal listings, such as a numbered list of the parts of a device, are mentioned here because they are handled very much like formal quotations: indented an extra five spaces and single-spaced. The list of rules below, under the next heading, illustrates the form.

Equations and Formulas   If you find it necessary to present equations in the text of a report, the following "rules" should be observed:

1. Center each one on a separate line.
2. If more than one line in length, the equation should be broken at the end of a unit, as before a plus or minus sign.
3. Place the entire equation on a single page if possible.
4. Allow three to four spaces above and below, or even more if it is necessary to use symbols of more than letter height: $\int$, for example.
5. Use no punctuation after the equation.
6. Number equations consecutively in parentheses at the right margin.
7. If necessary, define symbols used.

Study the following illustration, adapted from a Civil Aeronautics Authority report[2]:

It is shown that the panel penetration velocity, where failure occurs in the butyral plastic interlayer, varies approximately as the logarithm of the plastic thickness. This can be expressed by the equation

$$T = Ke^{v/c} \tag{1}$$

[1] Kenneth W. Houp and Thomas E. Pearsall, *Reporting Technical Information*, 3d ed. (Beverly Hills, Calif.: Glencoe Press, 1977), p. 402. There is now a fifth edition of this book, published in New York by Macmillan, 1984.
[2] Pell Kangas and George L. Pigman, *Development of Aircraft Windshields to Resist Impact with Birds in Flight*, Part II, Technical Development Report No. 74 (Indianapolis, Ind.: Civil Aeronautics Administration Technical Development), p. 13.

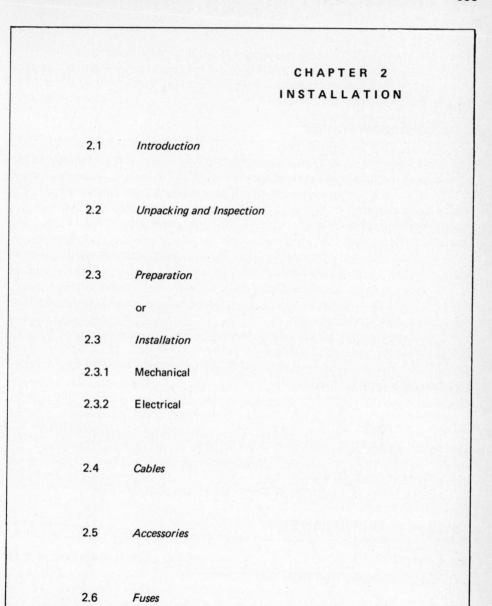

Figure 8   Illustration of decimal system breakdown of topics.

where

$$T = \text{thickness of vinyl plastic in inches,}$$
$$v = \text{penetration velocity of windshield panel in mph,}$$
$$K \text{ and } c = \text{constants.}$$

Informal Report Format

The terms "informal report" and "formal report" are vague, and are descriptive of a tendency rather than of an exact format. In general, form reports, letter reports, and reports designed for circulation only within an organization are called informal. The term usually denotes a short report, say fewer than ten pages, but company practice dictates whether the format of a formal or informal report should be used. You will probably have no trouble finding out which you should use in a specific situation.

A typical informal report has no cover, no letter of transmittal, no title page, no table of contents, and no list of illustrations. If there is an abstract, it appears on page 1, preceded by the title and the author's name and followed immediately by the text. The text is usually single-spaced. The second report at the end of Chapter 13 is an example of an informal report.

With the exceptions just noted, the suggestions for the format of a formal report also apply to an informal one.

The format of a letter report is simply the format of a business letter, except that headings may be used in the text after the first paragraph. The system of headings previously described is satisfactory. Besides the conventional block-form letter, however, a modification of the military letter form is frequently used for informal reports (see Figure 6 in Chapter 17). This form calls for "From," "To," and "Subject" caption lines, and numbered sections or paragraphs. One reason the military form is especially favored for interoffice or interdepartmental memoranda is that the forms may be conveniently printed.

Relation of Format and Style

There is a problem as to whether a well-planned format can perform certain functions that are usually performed in the text. For example, does a table of contents in a report make it unnecessary to say anything about the plan of development in the introduction? Does an abstract make it unnecessary to mention scope in the introduction? Does the use of a system of subheadings make transitions unnecessary?

The popularity of form reports clearly indicates that format can take over certain textual functions if we stretch the term "format" to include the detailed headings printed on a form report blank. In fact, the form report, which is an extreme case of the development of format, indicates both the potentialities and the limitations of the principle of assigning textual functions to format. An intelligently designed form report blank, when filled out by an intelligent person, is highly efficient. That is its strength. Its weaknesses are two. First, it can deal with only a limited number of situations. When something unusual happens, the report writer starts adding explanatory notes. The more initiative the report writer has, and the more unusual the situations encountered, the less useful the form report becomes. Second, the form report is impersonal. It gives the writer almost no opportunity for self-expression. Perhaps it will help you to understand what we mean if you will try to imagine yourself attempting to present, in a form report, a persuasive statement of the advantages of a device you've just invented!

The point of these remarks is this: Yes, sometimes a table of contents makes a statement of plan in the introduction unnecessary, and sometimes a subhead is a sufficient transition; but the further you go toward letting the format take sole responsibility for such functions, the closer you are getting to the form report, which is efficient within a limited range but is neither particularly pleasing nor persuasive. In short, we urge that you recognize the many advantages of a clear and attractive format, but we also urge that you avoid letting it lull you into writing a careless text. Eight times out of ten, you should write a transition even where there is a subhead; you should state the plan of development even when there is a table of contents; and you should clarify the scope even when there is an abstract. If you really want to be understood, try to communicate with every means at your command. To be even more specific about one question that often arises, we might add: Do not be concerned about duplication in the content of the letter of transmittal, the abstract, the table of contents, and the introduction. Such duplication is entirely acceptable and is the common practice.

PART 2: FORMS OF REPORTS

Introduction

Many firms have found that their written reports are most satisfactory when they are organized according to a prescribed form. The form may range from a general one consisting of three or four divisions to the kind of detailed sheets often used in colleges for short laboratory reports. The method of organizing the components of a form varies according to the materials concerned and the preferences of the firm or agency. In Part 2 of this chapter we introduce a few of these special forms, beginning with the highly generalized type, turning next to a highly detailed type, and concluding with an example of a moderately long laboratory report.

Generalized Forms

The principle underlying the policy of prescribing certain major divisions for the organization of a report is usually to make the report convenient to use. Convenience is often achieved by organizing a report in a different way from what would naturally be suggested by the subject matter. In a report on a series of tests, for example, it would seem natural to present the results of the tests toward the end of the discussion, whereas convenience is often served by presenting them near the beginning because some readers will want to know only the significant results without having to examine the rest of the report. Furthermore, a report may go to several readers, some of whom are not technically trained. This situation may be met by providing a preliminary nontechnical statement followed by what amounts to a restatement of the same material but in more detail and with the addition of technical material.

The General Motors Institute, for example, recommends the following organization:

1. The statement of the purpose of the report
2. A summary of the findings, conclusions, or recommendations
3. Supporting expansion of the steps that led to the findings, and so forth
4. Evidence in support of the findings, and so forth

In this organization, parts 1 and 2 are stated in a form intelligible to the nontechnical reader, part 3 is a technical discussion of method and interpretation of results, and part 4 is a presentation of data in support of 2 and 3.

The National Advisory Committee for Aeronautics prefers a more conventional organization:

1. Summary
2. Introduction
3. Symbols [all symbols should be defined]
4. Description of apparatus
5. Test procedure
6. Precision [statement of probable accuracy of measurements, where pertinent]
7. Analysis and discussion
8. Conclusions
9. Appendix
10. References
11. Illustrative material

In a form used by a large construction company, which prefers to remain anonymous, the major components of the form are these:

1. Purpose and scope
2. Summary
3. Conclusions, recommendations
4. Text
5. Appendix

Two general comments should be made about the types of organization that have been shown.

In the first place, almost all firms state that such forms should not be regarded as absolutely binding, but should be modified by the writer if circumstances require. The implication is, of course, that circumstances won't usually require much modification. The flexibility of the forms may be seen in the use of the appendix, a division that appears almost universally. The appendix is primarily intended to contain data that supports the discussion and the conclusions in the body of the report. The decision about how much of the data should go into the appendix and how much should be introduced directly into the discussion must be governed by the particular problems in each individual report. In effect, the requirement of an appendix is a recognition of the principle that the discussion should not be cluttered up with unnecessary details, but that the details should be available to prove the soundness of the discussion. The forms may be regarded as inflexible, on the other hand, in view of the fact that major divisions are prescribed and that a novice engineer or scientist will naturally be reluctant to make any modification of these major divisions.

In the second place, what about the relation of these standardized forms to the whole problem of types of reports? If all the reports issued by a given firm can go into one or two standardized forms, does it follow that one form is actually suitable for several types of reports? Frankly, we are raising this question simply because we thought you might be puzzled about it if we didn't, and not because we believe that it matters. One firm may use a single form for several types of reports; another may require two or three forms. And that is about all we need say here — except to suggest that you notice how easily the forms presented above would accommodate both progress and recommendation reports.

In addition to the forms of reports presented thus far, there is one other commonly used form to be mentioned before we turn to the highly detailed variety — that is, the memorandum report. The form used looks like this:

To: [addressees' names]
From: [author's name, and title after]
Subject: [subject in all capitals]

Usually the sender puts his or her handwritten initials or full name at the bottom of the sheet or after the typed name and title on the "From" line at the top. Sometimes other items are printed on the page, like Subject _____, File _____, Project No. _____, or any other information that will prove useful. Not uncommonly the paragraphs are numbered, often with underscored headings to identify the subject matter. See Figure 6, Chapter 17, for an illustration of headings. The memorandum report is essentially a rather informal communication between acquaintances, often employees of the same firm, about a project with which each is familiar. As with all reports, the content and style are determined by the relationship of sender and recipient.

Detailed Printed Forms

A detailed, printed form is often a great convenience in making routine reports. Thousands of such forms are in daily use. Figure 9 is representative of the type. Also shown are illustrations of forms for keeping a record of report production (Figures 10 and 11).

It might seem that in a form report there would be no problem of reader analysis; but not so. It is wise to think about such matters as symbols, abbreviations, systems of units in stating values, probable accuracy of measurements, and sampling techniques. Don't use a symbol that your reader won't recognize. Remember

ADVICE OF TROUBLE
FIELD REPORT
SERVICE ORDER

ELECTRIC SERVICE DIVISION
FIELD RELIABILITY ANALYSIS FEED BACK
WESTINGHOUSE PROPRIETARY

TO (LOCATION)   (RELIABILITY MGR.)   CONTROL ACCOUNT   JOB NO.

(SERV. OFF.)   (MGR.)   RESPONSIBILITY   SERVICE ESTIMATE   R.R. NO.

(SALES OFF.)   (SALESMAN)

(WORKS)   (ACCTG. MGR.)   CUSTOMER NAME

PRELIMINARY APPROVAL   TROUBLE AT (NAME OF PLANT)   LOCATION

APPARATUS (PRIME)   REPORT TO   TITLE   TELEPHONE

APPLICATION   APPARATUS SOLD TO   LOCATION

STYLE NO.   DWG.   SUB. NO.   ITEM   SERIAL NO.   G.O. (INCLUDE SUFFIX)   S.O.

1. CUSTOMER COMPLAINT - ACTION DESIRED (USE EXTRA SHEET WHEN NECESSARY)

DATE OF TROUBLE OR FAILURE   TIME IN SERVICE EST. ACT.   DATE WARRANTY STARTED   SHIPPING DAMAGE - TERMS   CLAIM ENTERED

FOB DEST. ☐ FACTORY ☐   YES ☐ NO ☐ BY _____

WILL THERE BE A CHARGE CUSTOMER ORDER FOR THIS WORK?   YES ☐   NO ☐   SEND RECOMMENDED BILLING TO SALESMAN FOR DISPOSITION ☐   CUSTOMER ORDER NO.

CREDIT APPROVAL (BY & DATE)   BKG. DIST.   SALES OFFICE   WESTINGHOUSE SALESMAN   DATE WRITTEN

ACCURACY OF APPARATUS IDENTIFICATION WILL BE VERIFIED BY ENGINEER   TO BE FILLED IN BY INVESTIGATING ENGINEER RETURN PROMPTLY (WITHIN 2 DAYS AFTER ARRIVAL)   →   DATE REPORTED ON JOB

PHYSICAL LOCATION OF TROUBLE   PANEL NO.   CABINET NO.   CAGE NO.   LOCATION NO.   SCHEMATIC

ENVIRONMENTAL CONDITIONS WET DRY DIRTY DUSTY CLEAN   CHEMICAL ATMOSPHERE ☐ TYPE   AMBIENT TEMP. MAX. MIN.

2. PROBLEM AS FOUND (GIVE COMPONENT N. P. DATA WHEN REQUIRED)

PROBLEM DUE TO CHECK APPROPRIATE BLOCK(S)   A. DESIGN   B. MFG.   C. COORDINATION   D. PKG.   E. BRACING   F. TRANSP.   G. IDENT.   H. INSTR. MATL.   I. STORAGE   J. INSTALL. PROD.   K. OPER. PROB.   L. MAINT.   M. APPLI.

3. ACTION NEEDED TO CORRECT TROUBLE - IS JOB COMPLETE   YES ☐   NO ☐   EST. MAN HOURS TO FINISH

4. MATERIAL REQUIRED: YES ☐   NO ☐   REQ. NO.   DATE OF FINAL GREEN SHEET

CITY OFFICE   INVESTIGATING SERVICE ENGINEER   DATE   REPORT NO.   OF   PRELIM.   FINAL

CORRECTIVE ACTION (SEND XEROX COPY TO SERVICE OFFICE)   RECORD OF SPLIT DECISION   CONTROL ACCT.   COST

TOTAL COST CLEARED

RELIABILITY MANAGER   DATE CLOSED   FINAL APPROVAL   DATE

(Left margin, vertical: TO BE FILLED IN BY SALES ENGINEER)
(Left margin, vertical: ANALYSIS FEED BACK)
(Left margin, vertical: TO BE FILLED IN BY DIVISION RELIABILITY MGR.)

FORM 35661 B   USE FORM 35662 WHEN ADDITIONAL SPACE IS REQUIRED FOR COMMENTS

Figure 9 Printed service report. (Westinghouse proprietary material. Reprinted by permission of Westinghouse Electric Corporation.)

```
+-----------------------------------------------------------------------+
|                     REPORT PRODUCTION RECORD                          |
|                                                                       |
|  Report Title _____ |
|  _____  |
|  _____  |
|  Author(s) _____  |
|  Classification _____  Report No. _____   |
|  Contract No. _____  Mailing Deadline _____    |
|  Rough Draft Rec'd _____  Completed _____    |
|  Rough Typed Draft Proofed _____   |
|  Rough Draft Checked by Author _____   |
|         Title Page  ____                                              |
|         Foreword  ____                                                |
|         Abstract  ____                                                |
|         Table of Contents _____                                     |
|         Equations Checked _____                                       |
|         Rough Illustrations Prepared _____                            |
|         Tables Checked _____                                          |
|  Illustrations Received _____  Completed _____    |
|  Cover Completed _____   |
|  Edited _____  Date _____     |
|  Editor-Author Conference _____  Date _____     |
|  Final Copy Typed _____  Date _____     |
|  Final Copy Proofed _____  Date _____     |
|  Proof Copy Checked by Author _____  Date _____     |
|  Approval Signatures _____  No. Copies Needed _____     |
|  Assembly and Binding _____  Date _____     |
|  Quality Checked _____   |
|  Mailed _____ Logged _____ Filed _____    |
|  Remarks _____   |
|  _____  |
+-----------------------------------------------------------------------+
```

Figure 10 Report production record.

PRODUCTION RECORD

Publication No. _____   _____ Due: _____ Date _____ Hour _____
W.O. _____ E.O. _____   (Classification)   Editor _____ Engineer _____

Title_____

Copies Required: Consists of: _____ Sections, _____ Appendixes
                                                        Dividers: ☐ All Sections ☐ Appendixes Only
                                                        Appendix: Title: Classification:
_____ Mailing (Complete)   A _____ _____
_____ Without Appendixes   B _____ _____
_____ In T/I Binder (Complete)   C _____ _____
_____ Tissues in T/I Binder   D _____ _____
_____ Total Copies   E _____ _____

Special Instructions: _____ Inserts: _____
_____

Reproduction Class: Brief Class Explanation:

☐ A   Justowriter—Film—Plates—Two Sides—Special Cover
☐ A-1   Justowriter—Film—Plates—One Side—Regular Cover
☐ A-2   Justowriter—Film—Plates—Two Sides—Regular Cover
☐ B   IBM Executive—Masters—Two Sides—Special Cover
☐ B-1   IBM Executive—Masters—One Side—Regular Cover
☐ B-2   IBM Executive—Masters—Two Sides—Regular Cover
☐ C   Justowriter—Xerox onto Masters—One Side—Regular Cover
☐ D   IBM Executive—Albanene—Blueline—T/I Cover
☐ E (Special): _____

Text Rough Draft: Illustrations: Total _____ Date Needed _____

Operation	Received		Completed		Initials
	Date	Hour	Date	Hour	
Editing					
Typing					
Proofreading					
Rechecking					
Project Reviewing					

☐ Approved for Final Reproduction with
    Changes as Marked   ◀

Final Reproduction: ▼

Operation	Received		Completed		Initials
	Date	Hour	Date	Hour	
Typing					
Proofreading					
Correcting					
Pasteup					
Prod. Check					
Edit. Check					
Signatures					
Edit. Recheck					
Prod. Recheck					
Filming					
Xeroxing					
Printing					
Collating					
Quality Check					
Binding					
Delivery					

Fig.	Size	HT	LN	OK	Film	Xerox	Pg. No.
1							
2							
3							
4							
5							
6							
7							
8							
9							
10							
11							
12							
13							
14							
15							
16							
17							
18							
19							
20							
21							
22							
23							
24							
25							

Record Reviewed and Filed:

_____
Production Editor

➡ Return to Editor:

☐ Copies Requested
☐ Excess Printing (Including Tissues)
☐ Original Art (Collated)
☐ Negatives (Collated)
☐ Plates or Masters (Collated)

Figure 11 Production record form.

that an abbreviation that looks clear as crystal to you may be puzzling to somebody who is not intimately acquainted with what you have been doing. Or, if you have taken measurements in the British gravitational system only to discover that everybody else in your organization is using the metric absolute, you'd better convert your values. In brief, give some attention to the needs and knowledge of your reader.

The same attitude of consideration for the reader should lead you to think carefully about what help you can give in the "Remarks" or "Comments" section, if such a section is provided. If certain measurements, for instance, were taken under conditions of unusual difficulty, a short explanation might relieve your reader of undue concern over slightly erratic results.

It is interesting to think about a form report from the point of view of the person who has to prepare one. Here is a comment made by Edward T. Shaw, the engineer at Westinghouse who was given the assignment of preparing the form on p. 360.

> When I was given this assignment, I quickly learned that I had to know specifically what information about a malfunction or a failure is needed by the engineers in the plant where the equipment is designed and made before I could hope to organize a suitable and usable form. In addition, I had to know who would be reading the report. Therefore, I had first of all to concern myself with the purpose of the report; after that, the content; and, then, the readers. And, with nearly 60 plants involved, this was not a simple matter. But I was the principal beneficiary. I learned a lot more about reporting than I knew before I was given the assignment. For this reason, it seems to me that if a student has to dig deeply — as I did — into purpose, content, and readers of a report before he can plan a form, he too will learn a lot more about reporting than he knew before he began the assignment.

Thought of in the terms of what this engineer has to say about the preparation of a form report, the importance of reader analysis in using it becomes clearer.

Laboratory Reports

Laboratory reports can be found in a virtually endless variety of forms and lengths. The example shown on pp. 364–376, a moderately long report, will be used to terminate this chapter. We are reproducing only a portion of the 22 pages in the original. Examination of the Table of Contents will indicate what has been omitted.

This report has been used as a model at the New Jersey Institute of Technology, and the notes at the bottom of some of the pages appear in the original.

Conclusion

The few special forms of organization shown in Part 2 of this chapter provide only a glimpse of the multitude of varieties in existence. It quickly becomes apparent, however, that there is nothing really new in these forms for the person who has a knowledge of the fundamental skills of technical writing.

Experiment No. X-1   Report Submitted By John J. Doe

Date Performed Aug. 25, 1976   Section EB03   Date Submitted Sept. 1, 1976

Instructor Dr. R. A. Comparin   Course Mech Lab II   Course No. ME 405

LABORATORY

NEW JERSEY INSTITUTE OF TECHNOLOGY

The Effect of Intake Passage Length on the Maximum Torque of a Small Wankel Engine

Experiment Title

Performed By Group ___ A ___   , Performed With James A. Jones, Fred B. Smith

Student is not to write below this line

Corrections

Date Received   Date Returned For Corrections

_____   _____

_____   _____

_____   _____

☐ Form   ☐ Calculations   ☐ Discussion

☐ Procedure   ☐ Curves   ☐ Conclusion

☐ Data   ☐ Sketches   ☐ See Pages _____

Comments _____

_____

                                    For further comments see inside cover

☑ Report Accepted   _____ Days Late

☐ Correct and Return   _____ Days Late

Corrected By _____   Grade __4.0__ Date Accepted Sep 7, 76

A Laboratory Report

on

The Effect of Intake Passage Length on the

Maximum Torque of a Small Wankel Engine

Prepared for

Professor Robert A. Comparin
Department of Mechanical Engineering
New Jersey Institute of Technology
Newark, New Jersey

By

John J. Doe
Student
New Jersey Institute of Technology
September 1, 1976

Note: The report must be enclosed in an NJIT folder and stapled along the left-hand edge. It must be typed or written in ink. The appearance of a report is of great importance.

Note: Be formal. Write "John," not "Jack." Write "September 1, 1976," not "9/1/76."

TABLE OF CONTENTS

Note: The sections included in this Table of Contents are representative, not hard and fast. Sections may vary in different reports.

ii

ABSTRACT

A Sachs-Wankel engine, Model KM-48, was tested to determine the
feasibility of increasing the output torque by changing the length of
the intake passage.

Increases in the intake length of up to 17 in. were tested. The
maximum increase in output torque measured in the tests was 5.4 percent
at 3500 rpm.

No wave effects were detected, and the increase in torque was
attributed to inertial effects.

The maximum possible increase in torque was not determined. However,
the tests indicated that the maximum output would occur with an intake
extension greater than 11 in., which may restrict the usefulness of the
technique to special applications.

Note: An abstract is the gist of the report, the essence of the report, the report in a nutshell. It should summarize the purpose, the results, and the significance of the results. It should include everything of importance. It should be complete and self-contained, and should be intelligible without reference to any part of the report itself. The abstract should be as quantitative as possible. It should state the conclusions and any recommendations. Do not omit "the" or "a" in the abstract. Write full sentences. Do not use abbreviations that would not be acceptable in the body of the report. The abstract is double-spaced and is placed on a single page if possible. Remember that some people read only the abstract. The abstract should be written last.

iii

1

I. INTRODUCTION

The intake process in a reciprocating internal combustion engine is not difficult to visualize in a general way. However, on close examination it becomes a complex problem in unsteady flow that includes inertial effects and wave effects. Obert [1]* gives a detailed description of the phenomena involved in the intake process.

Although the Wankel engine is not a reciprocating engine, its operation is similar to the conventional four-stroke cycle, and the same analysis of the intake process applies to both reciprocating and rotary engines.

The wave effects which are present in the intake system are due to the opening of the intake valve, or port. When the valve opens, a rarefaction pulse travels through the intake system to the atmosphere and is reflected as a compression. The compression wave is then reflected as a rarefaction with the result that a pressure oscillation occurs at the intake valve. Obviously, the optimum time to close the valve is when a compression wave is present, and closure when a rarefaction is present should be avoided. The wave phenomena associated with the intake process suggests that tuning may be possible by varying the length of the intake system to take advantage of the pressure increase at the intake port.

---
*Numbers in brackets refer to items in the REFERENCE section.

Note: The introduction should explain what the report is about and why it was written. It should put the experiment into perspective and lead the reader gracefully into the subject. When referring to the Reference section, you may want to use expressions such as "These relationships are presented as Figures 4.8 and 4.16 in Obert [1]," or "This equation is from page 63 of Obert [1]."

2

During the intake process the air in the intake system is accelerated and enters the cylinder with a significant velocity. The kinetic energy of the air can be utilized as a ramming effect to increase the pressure in the cylinder with a corresponding increase in density which will improve the charging of the cylinder. The size and length of the intake passage are important parameters in the optimization of the ramming effect.

A thorough discussion of the wave and inertial effects in the intake process has been given by Broome [2].

The object of this experiment was to determine the feasibility of improving the output torque of a small Wankel engine by changing the length of the intake passage.

II. DESCRIPTION OF EXPERIMENT

The engine tested was a SACHS-Wankel engine, Model KM-48, rated at
8 hp (DIN) at 4800 rpm. The manufacturer's specifications are given in
Appendix A.

Prior to the test, the engine was disassembled to examine the intake
system. The intake area was measured as a function of the output shaft
position, and the results are shown in Figure 1. The length of the intake
passage is 17 inches. (The rotor is cooled by passing the fuel-air mixture
through the rotor before it enters the chamber, hence the long inlet
passage.)

The engine was mounted on a GO-POWER model DY-7D dynamometer for the
test. The dynamometer specifications are given in Appendix B.

The arrangement of the equipment used is shown in Figure 2.

To facilitate changing the length of the inlet passage the carburetor
was remounted using flanges on the engine and the carburetor. These flanges were designed to accommodate standard 2 in. ID automotive cooling water
hose. The total passage length could be easily changed by inserting different pieces of hose.

Note: Each section should begin on a new page.
Note: In the description of equipment, there must be a verbal part with references
to figures or tables. Every figure or table must be labeled with a number and a title,
and must *follow* the reference to it and be close to where it is referred to.
Note: Use the past tense of the verb in describing the experiment. Avoid *I, we, you,*
and contractions such as *can't* and *don't.* Do not use any imperatives (commands).

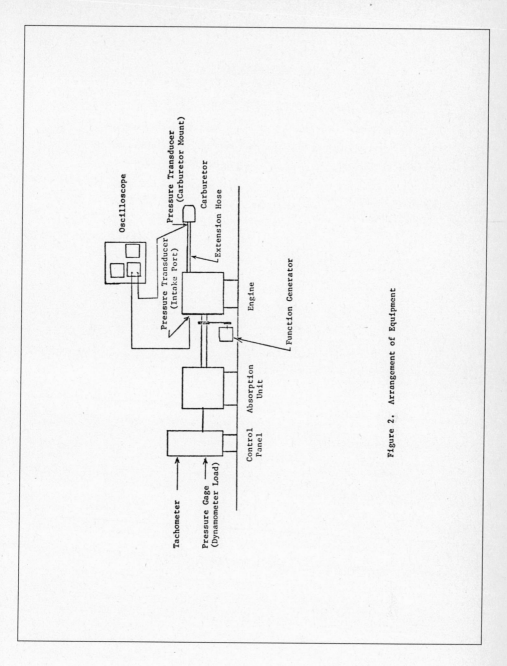

Figure 2. Arrangement of Equipment

7

III. RESULTS

The results of the torque measurements are listed in Table 1.

The dynamometer is equipped with a load cell which has a piston area of one square inch so that the meter reading can be interpreted as pressure in the load cell (psi) or as the force (lb_f) required to restrain the absorption unit. The lever arm of the dynamometer is 0.525 ft. The torque can be obtained from the simple equation

$$T = LP$$

where T = Torque (lb_f- ft)

      L = Lever Arm (ft)

      P = Dynamometer Load (lb_f).

At a speed of 3500 rpm with the 17 in. intake

$$T = 0.525 \times 14.8 = 7.77 \ lb_f\text{- ft.}$$

The power absorbed by the dynamometer is

$$hp = \frac{2\pi TN}{33000} = \frac{2\pi \times 0.525 \times P \times N}{33000} = \frac{N \ P}{10000}$$

where   hp = horsepower, and

            N = Speed (rpm).

At a speed of 3500 rpm with the 17 in. intake

$$hp = \frac{N \ P}{10000} = \frac{3500 \times 14.8}{10000} = 5.18 \ hp.$$

Note: Do not be skimpy with explanations. Write for others as you would have them write for you. Ask yourself what you would want to know if you were reading the report.

Do not say ".525 ft." Say "0.525 ft."

Be careful of significant figures when giving the results of calculations.

8

Table 1. Torque Measurements

Speed (rpm)	Load (psi)	Torque (lb_f- ft)
Intake length 17 in.		
2000	12.0	6.30
2500	13.5	7.10
3000	14.4	7.56
3500	14.8	7.77
4000	14.8	7.77
4500	14.7	7.72
5000	14.2	7.46
Intake Length 28 in.		
2000	12.1	6.35
2500	13.8	7.25
3000	14.8	7.77
3500	15.3	8.04
4000	15.3	8.04
4500	14.9	7.82
5000	14.3	7.51
Intake Length 34 in.		
2000	12.0	6.30
2500	14.2	7.46
3000	15.2	7.98
3500	15.6	8.19
4000	15.5	8.14
4500	15.0	7.88
5000	14.3	7.51

Note: The title goes above the table.

IV. ANALYSIS OF RESULTS

A detailed analysis of the wave phenomena in the inlet passage can be very complicated. See for example, the paper by Taylor, Livengood, and Tsai [3]. However, a simple analysis can be used to determine what wave effects could be expected.

Since the intake port is open for 330 degrees of shaft rotation (see Figure 1), at 3000 rpm the time that the intake is open is

$$\text{time} = \frac{1}{3000} \times \frac{330}{360} \times 60 = 0.018 \text{ sec.}$$

The time for a pressure pulse to traverse the intake passage and be reflected back to the inlet port is approximately

$$t = \frac{2\ell}{a}$$

where ℓ = intake passage length, and

a = acoustic velocity.

Assuming that a = 1000 ft/sec, for the 17 in. intake

$$t = \frac{2 \times 17/12}{1000} = 0.0028 \text{ sec.}$$

During the time that the intake is open, a wave could travel the passage at least six times.

The pressure measurements shown in Figure 4 do not indicate the presence of pressure waves in the intake passage. However, wave phenomena as predicted by this simple analysis have been observed and are reported in references 1 and 2. The complex flow passage through the rotor is apparently

14

V.   CONCLUSIONS

Wave effects were damped by the passage through the rotor, and no change in torque was attributed to wave effects in the inlet.

Inertial effects were sufficient to cause a measureable increase in torque.

It is feasible to increase the maximum torque of the engine by increasing the inlet passage length. However, the long intake lengths required for a small increase in torque would limit the usefulness of the technique.

Note: Every conclusion must be based on information in the report. The line of reasoning that took you from one point to another should be made very clear to the reader. Floating conclusions, that is, conclusions not supported by background information, are out of place. Also, check to make sure that your conclusions are consistent with the kind of result promised in the introduction.

15

VI.   REFERENCES

1   Obert, Edward F., <u>Internal Combustion Engines</u>, 3rd Ed., International
     Textbook Co., 1968.

2   Broome, D., "Induction Ram," <u>Automobile Engineer</u>, Part One: The
     inertia and wave effects, April 1969, pp. 130-133; Part Two: Iner-
     tial aspects of induction ram, May 1969, pp. 180-184; Part Three:
     Wave phenomena and the design of ram intake systems, June 1969,
     pp. 262-267.

3   Taylor, C.F., Livengood, J.C., and Tsai, D.H., "Dynamics in the Inlet
     System of a Four-Stroke Single Cylinder Engine," <u>Transactions of the
     ASME</u>, October 1955, pp. 1133-1145.

Note: Do not include Gray's *So You Have to Write a Technical Report.*
Note: This is a combination of footnotes and bibliography.

Suggestions for Writing

We have offered a number of suggestions in this chapter for handling various "formal" aspects of report presentation. We have also stated or implied that there is nothing absolute about our suggestions because practice is not standardized; different organizations and companies have developed or evolved different formats to suit their own needs. With this in mind, we make the following suggestions for writing.

1. With the help of some of the publications discussed in the chapter on finding published information, locate the library's file of reports. Check out some of these and examine the format (or formats) used. Write a report to your instructor in which you tell how the formats examined differ from that specified for reports in your technical writing course. The material in this chapter can serve you as a guide to what to look for.
2. Write a brief description of the prescribed format (and organizational plan) for laboratory reports in one of your science or engineering courses. Write the description as if for a student who will be taking the course some time in the future.
3. If there is a research laboratory associated with your school in which government contract research is carried out, see if you can find out what report format researchers are obliged to follow in reporting on their research. Write a brief account of what you find for your technical writing instructor. (You can consult the faculty and staff directory to find the telephone number and address of such research facilities.)
4. Devise a prepared format for keeping a record of the progress you make in the preparation of your library research report, making provision for a record from the time you submit a topic for your instructor's approval until you submit your completed report.
5. Obtain some form-report forms from local business firms, industries, or government agencies. Write an analysis of the effectiveness of one of them.

Graphic Aids

20

Introduction

In this chapter our general purpose is to introduce the extensive and important subject of graphic aids. More specifically, our purpose is to discuss some of the common varieties and functions of graphic aids and to consider elementary problems in their construction, exclusive of problems associated with their reproduction. Because the subject of graphic aids is so extensive, we strongly urge you to consult the pertinent volumes listed in Appendix A. We also urge you to become aware of the remarkable capacity that computers have for producing effective graphic aids. The capability of computer hardware and software continues to expand dramatically, and both of them continue to be more "user friendly." The widespread availability of low-cost computers is making it convenient for many writers to produce graphic illustrations that formerly would have required specialized training or particular talent.

The graphic aids discussed in this chapter are charts, drawings and photographs, and tables. The term "chart" covers a broad field, however, and will actually occupy most of our attention.

Before entering into a discussion of the particular types mentioned, we must note two problems in the selection and use of any graphic aid: (1) differentiating between dramatic emphasis and communication, and (2) establishing the proper relationship between the graphic aid and the text.

All graphic aids communicate facts to the reader, but some communicate with much more precision than others. This difference can easily be seen by comparing a curve carefully plotted on coordinate paper with the pictograph often found in newspapers. You might imagine, for example, that a newspaper has indicated the number of workers in a certain industry by a series of drawings of identical uniformed persons, each one representing 5000 workers, except the last one, who is worth only 3000 and is consequently shown as a partial figure. Such a pictograph may be dramatic, but it is not precise. A curve plotted on coordinate paper, on the other hand, can be fairly precise in communicating information. For a technically trained reader, it may also be dramatic, but the dramatic element is a secondary rather than a primary consideration. This difference between precise information and dramatization, qualified by reference to the intended reader, should always be noted in selecting a graphic aid.

Our second general problem is how to establish the proper relationship between the graphic aid and the text. Practically, this usually means deciding how much to say about the graphic aid, and deciding where to put it.

Our experience has been that writers often go to extremes in deciding how much to say. One writer will repeat in words practically everything that is shown in a graphic aid, and another will not even note that a graphic aid is used to complement the text. The second writer will tell you that it's all there in the graph, so why talk about it? You will have to make up your own mind as to which of these offenders is the worse. We suggest that you note your reactions on this point as you read various technical materials. You will probably find yourself most nearly satisfied when the following three practices are observed:

1. If a graphic aid has some bearing on a conclusion to be drawn, no matter how simple, a reference is made to it in the text. An aid used solely for aesthetic or "dramatic" purposes need not be mentioned, however, unless such window dressing is completely out of context, in which case it should be explained in an appropriate legend accompanying it.
2. The significant points shown by an "informational" graphic aid are commented on in the text, but minor details are not mentioned.
3. Some directions are given on the reading and interpretation of a complex graphic aid. What "complex" means depends on the reader.

Finding the most effective location for a graphic aid is usually a simple matter. Informational aids that have a direct, immediate bearing upon conclusions or arguments presented in the text are usually located as close as possible to the pertinent portions of the text. Informational aids of a more general, supporting character are put in an appendix, unless they are so few in number as to offer no serious interruption to the reading of the text. Aids used to dramatize are placed at appropriate points in the text. In general, graphic aids that belong in the text are likely to represent derived data; in the appendix, original data.

If the aid is small enough, it may be placed on a page on which text also appears. Usually it has a border. Larger aids should be put on a separate page. In a typed manuscript they may be bound on either the right or the left edge. If comments on the aid are pretty well concentrated on one page, the aid should be bound on the right edge so that it may face the comment. If there are several pages to which the aid is pertinent, it may be wise to bind it on the left edge and locate it near the beginning of the comments (or place it in the appendix). A page occupied solely by a graphic aid is given a page number if it is bound on the left edge but is not given a page number if it is bound on the right edge.

In connection with the preceding discussion, you may find it helpful to study the use of the tables in the second report quoted at the end of Chapter 9, and the figures in the reports quoted at the end of Chapters 5 and 6. Some eminently practical suggestions about use of graphic aids may be found in Appendix C.

Charts

Introduction   Charts, or graphs, are means of presenting numerical quantities visually so that trends of, and relationships among, numerical quantities can be easily grasped. Although a chart does not, in most respects, permit as accurate or detailed a presentation of data as a table, it has the advantage of making a significant point more readily and in a manner that is more easily remembered. The basic kinds of charts are the line or curve chart, the bar or column chart, and the surface chart. Additional varieties are the circle, or "pie," chart, the organization or line-of-flow chart, and the map chart. Each of these varieties will be discussed. First, however, we must review briefly some elements of chart construction. The elements to be discussed are the scales, the grid, the title, the scale captions, the source reference, and labels or a key.

Figure 1 illustrates the fundamental parts of a chart. Although it is a line chart, it could be easily converted into a bar or column chart by filling in a column from the base line up to the value for each division of the horizontal (abscissa) scale, and to a surface chart by shading the area beneath the line connecting the plotted points.

Most charts have only two scales, a horizontal (often called the abscissa) and a vertical (or ordinate). Typically, an independent variable is plotted on the horizontal scale, and a dependent variable on the vertical. Thus, if we were graphing the temperature rise of an electric motor, we would plot time on the horizontal scale and temperature on the vertical. Ideally, both scales should begin at zero, at their point of intersection, and progress in easily read amounts, like 5, 10, 15, 20. Failure to observe either of these last two principles increases the possibility that the reader will misinterpret the chart. Often, however, the scales cannot be started at zero. Suppose that values on the vertical scale, for instance, begin at a high numerical range, as in plotting temperature changes above 2000 degrees Fahrenheit. It would be impractical to begin the vertical scale at zero if intervals in the scale beyond 2000 are to be small. In such a case, it is occasionally desirable to give the base line a zero designation and place a broken line between it and the 2000-degree line to indicate the gap in the numerical sequence of the scale. This principle is illustrated in Figure 2b, lower left.

Much of the visual effectiveness of a chart depends upon the proper slope or height of the line or bar or area plotted. The idea of movement and trend is emphasized by steepness and minimized by flatness. The American National Standards

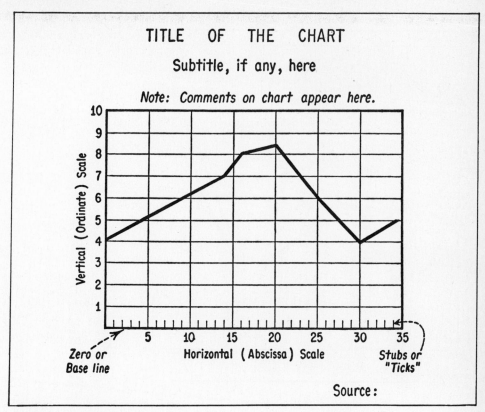

Figure 1 Typical chart layout.

Institute (ANSI) suggests that an angle of slope over 30 or 40 degrees in a curve is likely to be interpreted as being of great significance, whereas an angle of 5 degrees would be regarded as of little significance. It is often difficult to satisfy all the ideal requirements: that is, the proper slope or height, an easily read scale, ample room for scale captions, and a little space between the highest point of the curve or bar and the top of the grid. (These last two points, not previously mentioned, are illustrated in Figure 1.) Sometimes it is desirable to use the long dimension of the coordinate paper for the horizontal scale, to meet the above requirements. If this method still does not solve the problem, larger paper should be used, and a fold or folds made so that the folded chart, when bound into the text, will come somewhat short of the edges of the pages of the text. If you construct a grid yourself, you should (if possible) use "root-two" dimensions (ratio of about 1 : 1.5) for the rectangle formed by the grid. Such dimensions are aesthetically pleasing. (In a precise root-two rectangle, the long side is equal to the diagonal of a square made on the short side.) However, this advice must be qualified by observance of pleasing proportions between the shape of the grid and the shape of the page.

In general, you should use coordinate paper with as few grid lines per inch as the necessary accuracy in reading will permit. The purpose of the chart — the degree

Money Market Rates

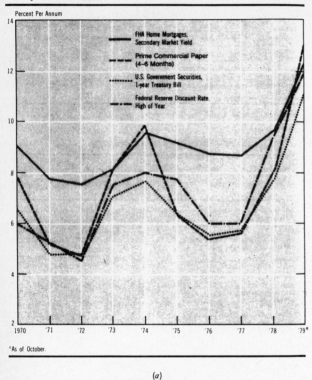

(a)

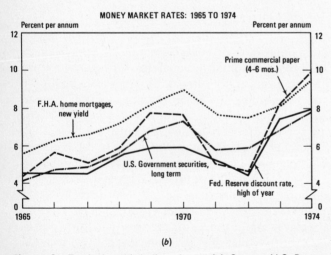

(b)

Figure 2   Typical multiple-line charts. (a) *Source:* U.S. Bureau of the Census, *Pocket Data Book, USA 1979,* Washington, D.C., 1979, Fig. 58. (b) *Source:* U.S. Bureau of the Census, *Statistical Abstract of the United States 1975* (96th ed.) Washington, D.C. 1975, Fig 16.2.

to which it is informational — and the probable error in your data determine the accuracy with which it should be readily possible to read the chart. Sometimes the use of stubs or "ticks," as shown in Figure 1, provides a good compromise between the precision afforded by numerous grid rulings and the clarity and force of fewer rulings. In a bar chart, the grid normally has only horizontal rulings if the bars are vertical, and vertical rulings if the bars are horizontal.

The title of a chart may be placed either at the top or the bottom. Usually, but not invariably, it is placed outside the rectangle enclosing the grid. If there is a figure number, it should appear either above or to the left of the title. In using 8½ by 11 (22 by 28 cm) coordinate paper you will often find it necessary, because of the narrow margins, to draw the axes an inch or so inside the margin of the grid to provide space on the grid itself for the title, the scale numerals, the scale captions, and the source reference if there is one.

The scale captions need no particular comment except that they should be easy to understand. Sometimes the whole effect of a graphic aid is spoiled by one ambiguous scale caption. See Figure 1 for illustration of the placement of captions. Don't forget to note units, like amperes or milliamperes, where they are necessary. The scale numerals or values are written horizontally if space permits.

Source references for graphic aids are written generally in the same manner that text source references are (see Chapter 22). More abbreviation is permissible in the reference to a graphic aid than in a footnote reference, however, because of the need to conserve space. Any abbreviation that will not confuse the reader is acceptable. The placement of the source reference is shown in Figure 1.

It is often necessary to use labels (Figure 2) or a key to identify certain parts of a chart, such as bars or curves representing various factors or conditions. Labels often appear in a blank area with a "box" or border around them, but this is not always possible or necessary. If you are using commercially prepared coordinate paper, it may be helpful to put a box around the label even though there is no white space left for it. Labels for bars may be written at the end of the bar or, if there is no possibility of confusion, along the side. In circle, or "pie," charts, the labels should be put within the individual segment or alongside the "slice" in easily readable form. A "key" or "legend" is simply an identification of symbols used in a chart (see Figure 3). Another element occasionally found is a note, usually in a box on the grid, about some aspect of the chart.

We turn now to consideration of the types of charts.

Line Charts   Of all charts, the line chart (Figure 1) is the most commonly used. Simple to make and read, it is especially useful for plotting a considerable number of values for close reading or for plotting continuous data to show trend and movement. It is usually not so good as the bar chart for dramatic comparisons of amount. For making comparisons of continuous processes, however, the use of several curves on the same chart (Figure 2) makes the line chart superior to the bar chart.

In a multiple-line chart of the kind just mentioned, labels are often written along the sides of the lines, without boxes. When lines intersect, the lines may be broken in various ways (dotted and dashed) to help in differentiating them; or colors or symbols with an accompanying key may be used (Figure 2). Particularly when

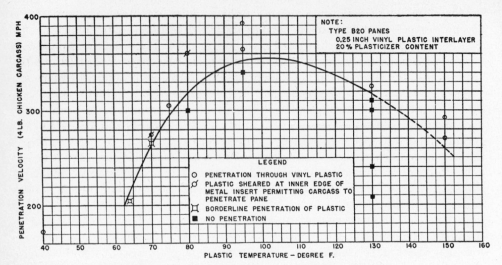

Figure 3 Single-line chart with faired curve: the effect of plastic temperature on the velocity at which the carcass of a bird penetrates an aircraft window. *Source:* CAA Technical Development Report No. 105, Fig. 11.

lines intersect, you should be careful not to put too many lines on a chart, or too many within a small area of the chart. The latter problem can, of course, be alleviated somewhat by the use of an appropriate scale on a large sheet of paper. If comparisons are to be made between different charts, the scales used on the charts should be identical.

Another problem in either single- or multiple-line charts is whether the line connecting points plotted should be drawn straight from point to point or smoothed out (faired) as in Figure 3. If you are showing the trend of a continuous process, like the temperature rise of a motor, it is usually desirable to make a faired curve, but if the process or change is not continuous, fairing the curve may be misleading. For example, if you were plotting an increase in student enrollment in a certain university for successive years, and your data showed enrollments of 10,000, 10,200, 14,000, and 14,300, a fairing of the curve would obscure a significant fact, the sharp increase in the third year, and would also falsely imply that the enrollment was rising steadily throughout each year. Incidentally, where precision is necessary, a point should be plotted by making a very small dot and then circling it lightly with a pencil so that it can later be found easily.

The foregoing discussion has been concerned only with the simplest and commonest of line charts. There are a great many possible variations of elements, including the use of special grids like logarithmic and semilogarithmic, that it is important to know about. Again we urge you to consult the books listed in Appendix A.

Bar Charts   Bar or column charts represent values or amounts by bars of scaled lengths. They are useful for showing sizes or amounts at different times, the relative size or amount of several things at the same time, and the relative size or amount of the parts of a whole. In general, the bar chart is preferable to the line chart for making dramatic comparisons if the items compared are limited in number. Arranged vertically (these are often called "column" charts), the bars are effective for representing the amount of a dependent variable at different periods of time; arranged horizontally, the bars are effective for representing different amounts of several items at one time. See Figures 4 and 5.

Although the bars of a bar chart may be joined, it is more common practice to separate them to improve appearance and increase readability. The bars should be of the same width, and the spacing between them should be equal. The proper spacing depends upon keeping the bars close enough together to make comparisons easy, yet far enough apart to prevent confusion. Another convention of bar chart construction is that the bars are arranged in order of increasing or decreasing length. This convention applies to charts in which each bar represents a component; it does not apply, of course, to those representing a time series.

Portions or subdivisions of an individual bar may be used to represent components or percentages. A single bar so subdivided is usually called a "100 percent" bar chart. Shading and hatching differentiate the portions, along with labels or a key. Darker shadings (often solid black) are used to the left of horizontally placed bars, with lighter shadings or hatchings being used for successive divisions to the right; on vertical bars, the darker shadings are used at the bottom. Colors may be used instead of shadings and hatchings.

One of the more interesting developments in bar-chart making is the pictograph, mentioned earlier in this chapter (see Figure 6). The pictograph substitutes symbolic units, like the figure of a person or the silhouette of a ship, for the solid bar.

School Enrollment, by Level

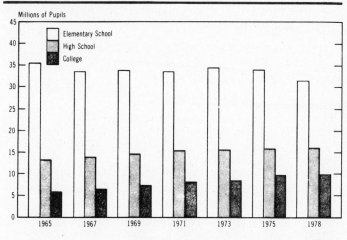

Figure 4 Vertical bar chart. *Source:* U.S. Bureau of the Census, *Pocket Data Book, USA 1979,* Washington, D.C., 1979, Fig. 41.

State Population

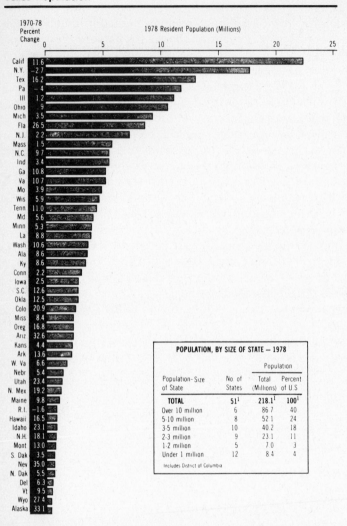

Figure 5 Horizontal bar chart. *Source:* U.S. Bureau of the Census, *Pocket Data Book, USA 1979,* Washington, D.C., 1979, Fig. 30.

The purpose of the pictograph is to increase interest and dramatic impact. The difficulties of preparing this kind of chart make it impractical for most technical reports, but when a report is to be distributed to a large audience of laymen, and when professional help is available for preparing the illustrations, the pictograph may prove highly desirable.

Surface and Strata Charts   A single-surface chart is constructed just like a line chart except that the area between the curve line and the base or zero line is shaded. Multiple-surface, or strata, charts (sometimes called bands or belt charts) are like

The Ten Most Populous Countries: 1979

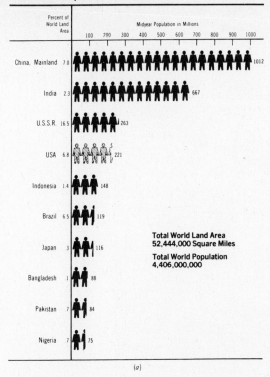

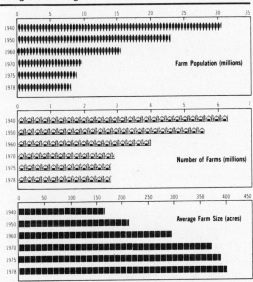

Figure 6 Simple pictographs. *Source:* U.S. Bureau of the Census, *Pocket Data Book, USA 1979*, Washington, D.C., (a) Fig. 1. (b) Fig. 46.

multiple-line charts with the underneath areas shaded in differentiating patterns or colors; that is, the vertical widths of shaded, or hatched surfaces, strata, or bands communicate an impression of amount. They can be satisfactorily used to achieve

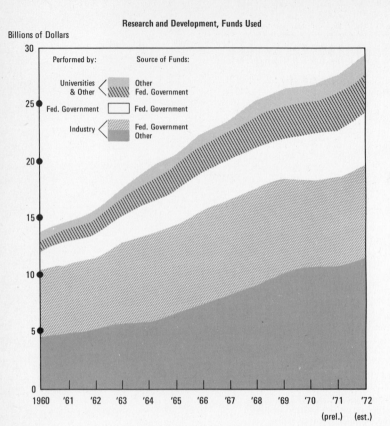

Figure 7 A surface chart. *Source:* U.S. Bureau of the Census, *Pocket Data Book, USA 1973,* U.S. Government Printing Office, Washington, D.C., 1973, Fig. 51.

greater emphasis than is possible with a line chart of the same data when amount is more important than ratio or change. They are not intended for exact reading, and should never be used when the layers or strata are highly irregular or where the plotted lines intersect. Gradual, regular movement or change can best be charted by this means (see Figure 7).

Circle, or "Pie," Charts A circle, or pie, chart is simply a circle of convenient size whose circumference represents 100 percent. The segments or slices show percentage distribution of the whole; see Figure 8(a). Since it is difficult to estimate the relative size of segments, labels and percentages must be placed on or alongside each segment. Although it is not a particularly effective graphic aid, the circle chart may be used for dramatic emphasis and interest as long as the subdivisions are not numerous. Figure 8(b) illustrates an interesting application. One point to remember when you use a circle chart is that the segments are measured clockwise from a zero point at the top of the circle.

Occasionally you may see circles, squares, cubes, or spheres of different sizes

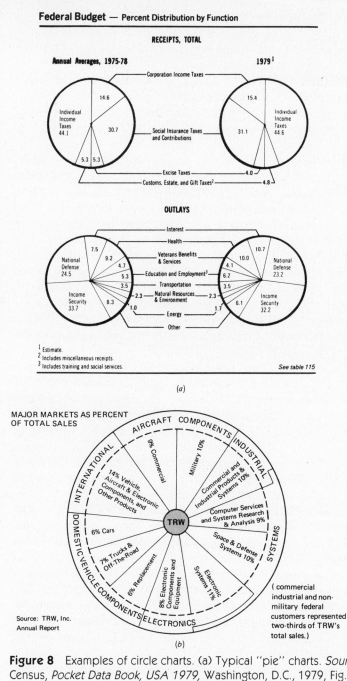

Federal Budget — Percent Distribution by Function

RECEIPTS, TOTAL

Annual Averages, 1975-78   1979[1]

Corporation Income Taxes

14.6   15.4

Individual Income Taxes 44.1   Individual Income Taxes 44.6

30.7   31.1

Social Insurance Taxes and Contributions

5.3 5.3

Excise Taxes ————— 4.0
Customs, Estate, and Gift Taxes[2] ——— 4.8

OUTLAYS

Interest
Health
Veterans Benefits & Services
Education and Employment[3]
Transportation
Natural Resources & Environment
Energy
Other

7.5   10.7
National Defense 24.5   National Defense 23.2
9.2   10.0
4.7   4.1
5.3   6.2
3.5   3.5
2.3   2.3
8.3   6.1
1.0   1.7
Income Security 33.7   Income Security 32.2

[1] Estimate.
[2] Includes miscellaneous receipts.
[3] Includes training and social services.   *See table 115*

(a)

MAJOR MARKETS AS PERCENT
OF TOTAL SALES

AIRCRAFT COMPONENTS
INDUSTRIAL
INTERNATIONAL
DOMESTIC VEHICLE COMPONENTS
ELECTRONICS
SYSTEMS

9% Commercial
Military 10%
Commercial and Industrial Products & Systems 10%
Computer Services and Systems Research & Analysis 9%
Space & Defense Systems 10%
Electronic Systems 11%
8% Electronic Components and Equipment
6% Replacement
7% Trucks & Off-The-Road
6% Cars
14% Vehicle, Aircraft & Electronic Components, and Other Products

TRW

(commercial industrial and non-military federal customers represented two-thirds of TRW's total sales.)

Source: TRW, Inc.
Annual Report

(b)

Figure 8   Examples of circle charts. (a) Typical "pie" charts. *Source:* U.S. Bureau of the Census, *Pocket Data Book, USA 1979,* Washington, D.C., 1979, Fig. 34. (b) A double-ring circle chart showing major divisions and subdivisions. *Source:* TRW, Inc., *Annual Report 1967.*

used to compare amounts. The difficulty of comparing relative sizes, especially of cubes or spheres, makes these devices of no real use. We recommend that they be avoided. The line, bar, or surface chart will do better.

Flowsheets and Organization Charts   A flowsheet is a chart that makes use of symbolic or geometric figures and connecting lines to represent the steps and chronology of a process. An organization chart is like a flowsheet except that instead of representing a physical process, it represents administrative relationships in an organization.

The flowsheet (Figure 9) is an excellent device for exhibiting the steps or stages of a process, but its purpose is defeated if the reader finds it difficult to follow the connecting lines. Flowsheets should generally be planned to read from left to right, and the connecting lines should be arrow-tipped to indicate the direction of flow. The units themselves, representing the steps or stages, may be in the form of geometric figures or symbols. The latter are simple schematic representations of a device, such as a compressor, a cooling tower, or a solenoid valve. Standards for such symbols have been adopted in a number of engineering fields today, and you should make it a point to familiarize yourself with the symbols acceptable in your field. Publications concerning symbols may be obtained from ANSI. These symbols are used, by the way, in drawings as well as in flowsheets. Labels should always be put on geometric figures. Whether labels should be used with symbols depends on the intended reader.

Since a generous amount of white space is essential to easy reading of a flowsheet, and since the flowsheet reads from left to right, the display will often need to be placed lengthwise on the sheet. This makes it necessary for the reader to turn the report sideways to read the chart, but this is better than crowding the figures into too narrow a space. If space requires it, a sheet of larger than page size may be used and folded in.

Flowsheets are usually enclosed in a ruled border, and the title and figure number centered at the bottom inside the border. See Chapter 14, Figure 1, for an example.

Organization charts are very similar to flowsheets (see Figure 10). Rectangular figures represent the units of an organization; connecting lines, as well as relative position on the sheet, indicate the relationship of units. Good layout requires that the figures be large enough so that a lettered or typed label can be plainly and legibly set down inside them, and they must be far enough apart so that the page will not be crowded. Organization charts usually read from the top down.

Colored flowsheets and organization charts are effective for popular presentation. Like the pictograph, such color charts require the services of trained artists and draftsmen.

Map Charts   The map chart is useful in depicting a geographic or spatial distribution. It is made by recording suitable unit symbols on a conventional or simplified map or differentiated area of any sort (like electron distribution in a space charge). It is particularly important in a map chart that the symbols and lettering be clear and easy to read. Geographic maps suitable for use in making map charts are readily available from commercial suppliers. See Figure 11 for an example of a map chart.

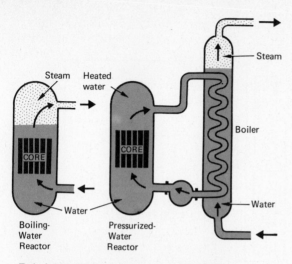

Steam   Heated water

Steam

Boiler

CORE   CORE

Water

Water

Boiling-Water Reactor

Pressurized-Water Reactor

Today's big reactors work this way. Boiling-water type makes steam right in reactor; pressurized-water type, in boiler to which water carries heat.

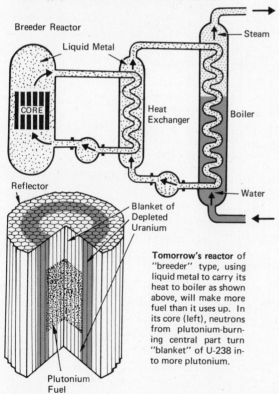

Breeder Reactor

Liquid Metal

Steam

CORE   Heat Exchanger   Boiler

Water

Reflector

Blanket of Depleted Uranium

Tomorrow's reactor of "breeder" type, using liquid metal to carry its heat to boiler as shown above, will make more fuel than it uses up. In its core (left), neutrons from plutonium-burning central part turn "blanket" of U-238 into more plutonium.

Plutonium Fuel

Figure 9   A flowsheet: types of atom-power reactors, *Source:* "World's Biggest Atom-Power Plant" by Alden P. Armagnac, *Popular Science,* September 1969, pp. 95 ff.

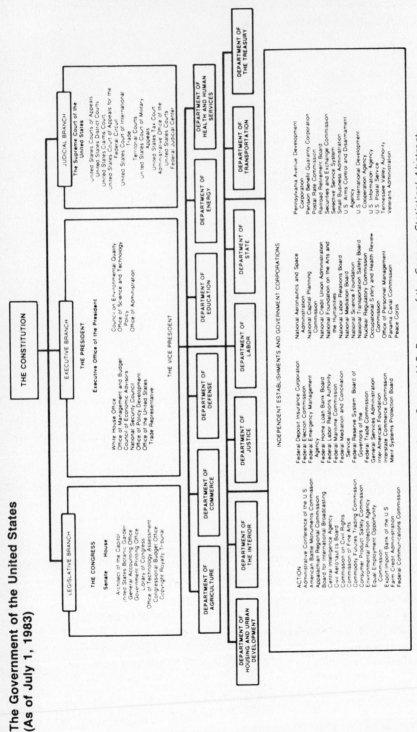

Figure 10 An organization chart. *Source: U.S. Bureau of the Census, Statistical Abstract of the United States 1984 (104th ed.), Washington, D.C., 1983, Fig. 11.1*

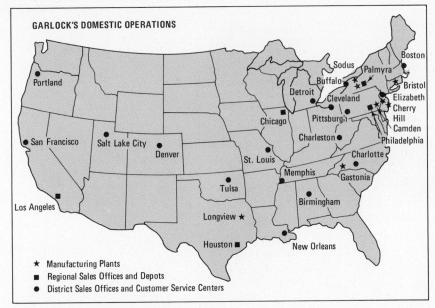

GARLOCK'S DOMESTIC OPERATIONS

★ Manufacturing Plants
■ Regional Sales Offices and Depots
● District Sales Offices and Customer Service Centers

Figure 11   A map chart. *Source:* Garlock, Inc.

Drawings, Diagrams, and Photographs

Drawings and diagrams are especially valuable for showing principles and relationships that might be obscured in a photograph, but of course they are sometimes used instead of photographs simply because they are usually easier and less expensive to reproduce. A photograph, on the other hand, can supply far more concreteness and realism than drawings or diagrams (see Figure 12). We are using the terms "drawing" and "diagram" loosely to refer to anything from a simple electronic circuit diagram to an elaborate structural blueprint or a pictorial representation of a complex mechanical device.

Parts of drawings should be plainly labeled so as to make textual reference clear and meaningful. If the drawing is of a simple device with but few parts, the names of the parts may be spelled out on the drawing itself, with designating arrows. If the drawing is of a complex device, with a large number of parts, letter symbols or numbers with an accompanying key should be used. Figure number and title should be centered at the bottom, inside the border if one is used. If a source reference is necessary, it should appear in the lower righthand corner.

Photographs should be taken or chosen with special attention to how prominently the elements important to your discussion stand out. Very often this principle necessitates the use of an artificial background. A cluttered background distracts attention, not infrequently producing the impression that the photograph was originally intended as a puzzle, with a prize for anybody who could find gear B.

Figure 12 Photograph of a fitting (top) and a drawing of a valve (bottom). *Source:* Hi-Shear Corporation.

Glossy prints are better than flat prints because of their greater effectiveness in reproducing highlights and shadings. Each reproduced print should have an attractive margin of white space. If smaller than page size, prints may be satisfactorily mounted by use of rubber cement. Rubber cement has less tendency than paste or glue to wrinkle the page.

The figure number and title, as well as explanatory data, should be put directly on full-page photographs in black or white ink.

Tables

The table is a convenient method of presenting a large body of precise quantitative data in an easily understood form. Tables are read from the top down in the first column and to the right. The first, or left, column normally lists the independent variable (time, item number, and so on) and the columns to the right list dependent variables (see Figure 13). The table should be designed so as to be self-explanatory, but textual comments on it should be made according to the same principles that apply to the use of a chart.

To make a table easy to read, you should leave ample white space in and about it. If the table appears on a page on which there is also typed text, triple-space above and below the table. Leave a generous amount of space between columns and between the items in a column. The title and the table number should appear at the

TABLE II

CENTRIFUGAL FORCE CALIBRATION DATA

Element No.	Maximum g	Minimum g	Average g	Spread in g
1	2.21	1.65	1.93	0.56
	2.20	1.65	1.93	0.50
2	2.48	2.25	2.36	0.23
	2.50	2.22	2.36	0.28
3	3.08	2.59	2.84	0.49
	3.12	2.58	2.85	0.54
4	3.07	2.60	2.84	0.47
	3.10	2.60	2.85	0.50
		Average Spread		0.45

Figure 13 A table layout. *Source:* CAA Technical Development Report No. 48.

top. It is advisable to use Arabic numerals for the table number if Roman numerals have been used in the same report for numbering other kinds of graphic aids, or vice versa.

In separating parts of the table from one another—that is, one column from another, or one horizontal section from another—use single lines in most instances; but where you wish to give special emphasis to a division use a double line. Some people make it a practice to use a double line across the top of the table, under the title. Usually a single line should be drawn across the bottom of the table. The sides may be boxed or left open as seems most pleasing. But we'll add this caution: in case of doubt as to whether a ruling should be used at any given point in the table, leave it out. More harm will probably be done by too many lines than by too few, provided ample white space has been left.

Align columns of numerals on the decimal point, unless units of different quantities (like 2000, representing pounds, and 0.14, representing percentage) appear in the same column, in which case the column should be aligned on the right margin.

A heading should be written for each column, with the initial letter of important words capitalized. The headings should be written horizontally if possible, but if that would use too much space, they may be written vertically. Indicate units (like volts, Btu, cu ft) in the heading so that the units will not have to be noted in the column. If data is expressed in different systems of units, you should convert everything to the same system before entering the items in the table.

If the data in the table is not original, acknowledge the source in a footnote just below the bottom horizontal line of the table. Instead of using a superscript number in the table to refer to the footnote, use a superscript letter symbol (Roman), an asterisk, or some other convenient symbol. Tabular footnotes (that is, notes that refer to specific items in the table itself) should be placed between the bottom line of the table and the source note, if one is needed. Tabular footnotes may be keyed with the asterisk or a Roman superscript letter, whichever is not being used with the source note.

CONCLUSION

Graphic aids in their simpler forms are easy to prepare and easy to understand. In either simple or more complex forms, they often convey information or provide dramatic emphasis with an effectiveness that would be difficult or impossible to achieve in writing. On the other hand, if they are relied on too much, they may become a hindrance rather than a help. You should regard this chapter as merely a short introduction to the uses and techniques of one of the more valuable tools of science and technology.

Suggestions for Practice in Making Graphic Aids

Although the professional writer usually can make use of the services of drafting department personnel and technical illustrators in preparing graphs, charts, tables,

Federal Obligations for Research and Development (R&D)

In millions of dollars. For fiscal years.

Agency or function	1970	1974	1975	1976	1977	1978 est.	1979 est.
Total R&D[1]	15,340	17,415	18,988	20,723	23,929	26,420	27,972
Department of Defense	7,360	8,420	9,012	9,655	10,963	11,825	12,838
National Aero. and Space Admin.	3,800	3,002	3,064	3,447	3,703	3,876	4,192
Department of Energy[2]	1,346	1,489	2,047	2,464	3,536	4,196	4,175
Health, Education, Welfare	1,221	2,290	2,363	2,546	2,787	3,132	3,271
National Institute of Health	873	1,737	1,846	2,023	2,244	2,521	2,606
National Science Foundation	289	556	595	609	697	754	828
Department of Agriculture	281	379	420	462	547	632	636
National Defense	7,981	9,016	9,679	10,430	11,864	12,786	13,833
Space	3,510	2,478	2,511	2,863	3,066	3,141	3,383
Health	1,126	2,096	2,177	2,366	2,604	2,912	3,034
Energy level and conversion	425	665	1,186	1,439	2,302	2,863	2,827
Environment	322	659	795	847	954	1,066	1,082
Science and technology base	448	641	773	785	901	988	1,061
Transport and communications	590	703	641	636	705	829	837
Basic research	1,762	2,076	2,279	2,425	2,894	3,292	3,637
Applied research	3,455	4,435	4,798	5,448	5,947	6,565	6,846
Development	10,123	10,904	11,911	12,850	15,088	16,562	17,490
Total research	5,217	6,510	7,077	7,873	8,841	9,858	10,483
Life sciences	1,507	2,283	2,450	2,646	3,019	3,383	3,507
Psychological and social	321	434	448	530	602	708	757
Physical	893	990	1,072	1,221	1,511	1,696	1,819
Environmental sciences	476	634	733	770	921	1,012	1,110
Other	2,020	2,170	2,374	2,706	2,789	3,059	3,290

[1] Includes agencies and functions not shown separately.

[2] Prior to 1978, data represent expenditures of Energy Research and Development Administration.

Figure 14 Data for graphic presentation. *Source:* U.S. Bureau of the Census, *Pocket Data Book, USA 1979,* Washington, D.C., 1979, Table 556.

or drawings for reports and proposals, it is not a bad idea to try your hand at making a few — for fun and to gain some idea of the problems involved.

1. If you are in the process of preparing a research report, explore the possibilities of making use of some of the graphic techniques described in this chapter for complementing your presentation. Doubtless your source will contain many illustrations, but often they turn out to be difficult to reproduce. Sometimes it is possible to adapt and simplify an illustration from a source in such a way that it is more useful than the original (see Figure 14) would be in supplementing a portion of your own presentation.

2. Drawing on what you have learned from this chapter, devise at least two graphic means of presenting the data tabulated in Figure 14. How might color be effectively used?

SECTION VI

The Library Research Report

This section contains two chapters related to a single project: the writing of a library research report. The first of these chapters surveys the subject of how to find needed information in a library. The second discusses the writing of a library research report.

For professional people who hope to have more than minimal competence in their work, skill in locating published information is indispensable. Besides practice in acquiring this skill, the writing of a library research report provides a second important benefit. Such a report can be long enough to involve problems of form and organization that you have not encountered in the shorter exercises you have probably, up to this point, been writing.

For additional help in the preparation of the library research report, we encourage you to read the materials in Appendix D. Although these materials were prepared by the library staff specifically for use of students at the University of Texas at Austin, they should be useful in helping you select a subject for research, find background materials on the subject you choose, and identify primary sources, in both book and periodical form.

21

Finding Published Information

Introduction

This chapter will introduce you to various sources of information about what has been published on a given subject. There are many more sources than could be presented in this chapter, but once you have begun to make a deliberate study of library resources, one thing will lead to another, and your ability to locate books and articles will steadily improve. The materials described in this chapter should be sufficient to provide for your needs in the preparation of the library research report, which is discussed in Chapter 22, as well as to provide a basis for lifelong use of library resources.

Guides to the location of books will be considered first, then guides to the location of periodical articles, followed by a variety of other aids.[1]

Books

The most obvious source of information about books in the library is the card catalog. Many libraries have supplemented or replaced their card catalogs with

---

[1] We are indebted for some of the items listed in the following pages to Saul Herner, *A Brief Guide to Sources of Scientific and Technical Information,* 2nd ed. (Washington, D.C.: Information Resources Press, 1980), and to Saul Galin and Peter Spielberg, *Reference Books: How to Select and Use Them* (New York: Vintage Books, 1969).

computer printouts, microfilm or microfiche catalogs, or even online interactive catalogs. The holdings of most libraries are classified according to either the Dewey Decimal system or the Library of Congress system.

The Dewey Decimal system of classification divides the field of human knowledge into ten classes, with numbers designating the various classes. The ten classes of the Dewey Decimal system are:

000: General Works	500: Pure Science
100: Philosophy and Related Fields	600: Technology and Applied Sciences
200: Religion	700: The Arts
300: Social Sciences	800: Literature
400: Language	900: History, Geography, and Biography

The Library of Congress system groups materials into 20 basic categories, designating the groups by letters. The 20 basic categories are:

A: General Works
B: Philosophy, Psychology, Religion
C: History and Related Subjects
D: History and Topography (not including America)
E:
F: } American History
G: Geography, Anthropology
H: Social Science
J: Political Science
K: Law
L: Education
M: Music
N: Fine Arts
P: Languages and Literature
Q: Science
R: Medicine
S: Agriculture, Plant and Animal Husbandry
T: Technology
U: Military Science
V: Naval Science
Z: Bibliography and Library Science

The Library Catalog   Although no two libraries have catalogs that are exactly alike, these catalogs are likely to have one element in common: subject headings based on some kind of consistent terminology. In college libraries, the *Library of Congress Subject Headings* is the most widely used guide to the subject entries in the catalog. If you begin your search for books by looking in the subject headings list rather than directly in the catalog, you will save time and you will have an accurate view of what the particular library you are using actually owns or has access to. For example, if you went to a science library and looked for *scientific method* in the catalog, you would find nothing there. But if you looked up *scientific method* in the subject headings list, you would find a cross reference that would tell you to look for *science — methodology* instead, and you'd not be going home empty-handed. The subject headings list can also be of help in finding related subject headings that you might not have thought of on your own. For example, the entry in Figure 1 is related to *science — methodology*. In the subject headings list, more specific terms are preceded by *"sa"* (see also), whereas broader terms are indicated by *"xx"* (see from).

Once you have actually found a card (or "record" as an entry in a computerized catalog might be called), you can use information from it to gather further information (see Figure 2).

The *call number* helps you locate the book on the shelf of the library. Since books are usually shelved according to subject, you may find other books on the same subject near the book you intended to locate.

Next, the numbered words and phrases at the bottom of the card are called *tracings.* Most important are the *subject tracings,* indicated by Arabic numerals. These tracings indicate other subject headings that apply to this particular book. The book *Winning the Games Scientists Play* will be listed in the catalog under the term *scientists* as well as *science — methodology.* Many times these other subject headings will lead you to additional books on your research topic.

The *author entry* may also be important — you can look up the author's name in the catalog to find other books by that author. Frequently an author may write on a series of related subjects over the years.

Once you are on target as far as the subject you want, you'll need some basis for choosing among the various titles listed under a particular subject heading. Other parts of the card can be of help here. The date and publishing information will give you some idea as to how current the information is, and how scholarly or popular the author's style is likely to be. A book that is published by a university press or by a professional association in your field will likely approach your subject in a manner different from that of a book aimed at supermarket shoppers. A catalog entry will also usually tell you whether a book has a *bibliography* or *index.* A book with these features will often be more helpful for serious research than a book without them.

Science
   — Methodology

      sa Classification of sciences
         Communication in science
         Creative ability in science
         Experimental design
         Hypothesis
         Logic
         Stereology
         Scientific method
     xx Logic, Symbolic and mathematical

Figure 1   Subject heading entry.

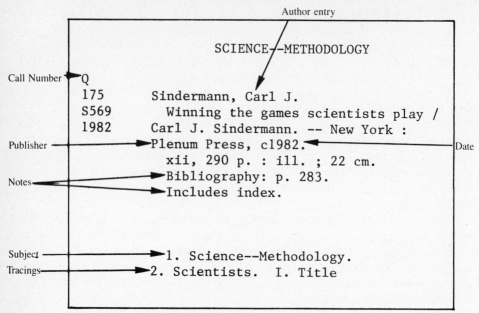

Figure 2   Library catalog card (subject card).

Other Guides to Books   Any one library is unlikely to have every book that has been published on a particular subject. Consequently some library users might like to be able to find lists to supplement the catalogs of the libraries they are using. Another reason to look for guides to books is to find out which libraries have a particular book, or what other books have been written by a particular author. The list below is highly selective; a reference librarian can refer you to other sources if the ones below do not meet your needs.

Subject Guides

Cumulative Book Index. A monthly publication. Cumulated every six months and biennially, listing alphabetically by subject, author, and title all books published in English.

Science Books and Films. A good source to find out what is currently being published in scientific and technical fields. Comes out five times each year; includes book reviews.

Subject Guide to Books in Print. Limited to books that are currently available. Annual list of books from more than 13,900 American publishers, indexed under 62,000 subject headings, with subject cross references. Updated by *Subject Guide to Forthcoming Books*.

Verification Guides

The National Union Catalog. This publication of the Library of Congress lists authors of works cataloged by the Library of Congress or by other contributing libraries. Although it is still published on a monthly basis, for most library users it is most valuable for its historical coverage.

OCLC (Online Computer Library Center). A national bibliographic database which allows users to search by author, title, or combination author-title. It provides complete catalog information, such as publisher, date, and subject headings, and in addition lists which libraries own a particular book or periodical. Many libraries have joined the OCLC network, or one of the other major networks.

Periodical Indexes and Abstracts

A research project seldom can be completed without the use of articles appearing in professional journals, as these publications contain information that is both recent and highly specific. Ordinarily, a researcher will begin the task of identifying periodical articles by using an indexing or an abstracting source. Both indexing and abstracting sources are used for the purpose of identifying published material on a particular subject or by a particular author. The technique for using these two kinds of sources, however, is slightly different.

Indexes are slightly easier to use, but they contain less information than abstracts. To use an index, turn directly to a subject of interest, such as "Information storage and retrieval systems." There you will find either a list of articles or a cross reference to another subject heading. A typical index entry appears in Figure 3. This entry, from the 1982 volume of *Applied Science and Technology Index,* tells us that an article entitled "Picture browsing system" appeared in a journal called *IEEE Transactions on Communications* in the December 1981 issue (Vol. 29) on pages 1968–1976. The article was written by A. N. Netravali and E. G. Bowen, and includes a bibliography, illustrations, and diagrams.

Unlike indexes where articles are listed in alphabetical order by subject, the articles in abstracts may appear in various orders. For example, many abstracts, such as *Computer & Control Abstracts,* establish a classification system and arrange articles in that way. A portion of the classification system of *Computer & Control Abstracts* appears in Figure 4.

If you are looking for an article on a specific topic such as "Data retrieval systems" rather than a general topic such as "Information science," you will save time by looking in the subject index of the abstracts for a reference to abstract or classification numbers dealing with the specific topic you are interested in. In the index entry shown in Figure 5, you are referred to classification section 7250L. You must turn from the subject index to the abstract section to find a full citation of the article, along with a description of its contents. As you can see in Figure 6, the abstract entry that you now reach gives you the same location information as you find in a periodical index. In addition, the abstract summarizes the contents of the article.

INFORMATION storage and retrieval systems

Picture browsing system. A. N. Netravali and
 E. G. Bowen, bibl il diags IEEE Trans Comm
 29:1968–76 D '81

Figure 3 Typical index entry.

7200	INFORMATION SCIENCE AND DOCUMENTATION
7210	Information services and centres
7220	Generation, dissemination, and use of information
7230	Publishing and reproduction
7240	Information analysis and indexing
7250	Information storage and retrieval
7250C	*Bibliographic systems*
7250L	*Other systems*
7290	Other aspects of information science and documentation

Figure 4 A portion of an abstract classification system.

Data handling	6130
Data preparation	55
Data reduction	73, 74, 78
<u>Data retrieval systems</u>	<u>7250L</u>
Data structures	6120
Data transmission	3370G, 5510
Database management systems	6160
DC motors	3260B, 3340H

Figure 5 Abstracts subject index entry.

7250L Other systems

16223 A picture browsing system. A. N. Netravali, E. G. Bowen (Bell Labs.,Holmdel, NJ, USA).
IEEE Trans. Commun. (USA), vol. COM-29, no. 12. p. 1968–76 (Dec. 1981).
The authors describe a system for storing pictures in a database and retrieving them from a remote location over a low bit-rate channel. The system improves the human interaction with the picture database by constructing an auxiliary text database containing a list of attributes of each picture, a hierarchical encoder-decoder and a light pen to select the areas of picture buildup. Picture searching (or browsing) takes place in two stages: in the first stage, knowing the required picture attributes, a user selects a subset of pictures by matching attributes to the text database: in the second stage, these selected pictures are displayed hierarchically so that a low resolution picture is reproduced first and made sharper gradually. A light pen allows the user to give priority to upgrade selected areas or reject a picture. Informal tests indicate that it is much easier to browse through a picture database using this sytem and that the time to retrieve a picture of given attributes decreases considerably compared to sequential picture presentation. (10 refs.)

Figure 6 An abstract entry.

Entries in both indexes and abstracts are highly abbreviated. If you cannot identify the name of the magazine by the abbreviation, you will find an alphabetical list of full names of magazines and their abbreviations in the prefatory pages of each volume of the *Index*. Note that volume and page numbers for a reference are given together, with a colon separating the volume from the pages. Some entries may show a + after a page number. This mark means that the article is continued in the back pages of the magazine.

Since many people are unaware of the range and variety of indexes and abstracting tools available in libraries, a selective list of indexes and abstracts that include articles relevant to scientific and technical subjects is shown below with some brief comments. This is not a comprehensive listing. Check with a librarian if none of these sources seems to cover your subject adequately.

Selected Indexes and Abstracts

General and Interdisciplinary

Indexes

New York Times Index (1913–   ). Published monthly and cumulated annually. Useful for finding newspaper articles.

Readers' Guide to Periodical Literature (1900–   ). Author-subject index to articles appearing in general interest magazines. Issued monthly, cumulated quarterly and annually. Valuable for the student researcher who needs articles written for a nontechnical audience.

Abstracts

Dissertation Abstracts International (1938–   ). Summaries of dissertations written at universities may be found in this index. Dissertations may be obtained from the school at which they were written or through University Microfilms, the publisher of this abstracting tool.

Humanities and Fine Arts

Indexes

Art Index (1929–   ). Cumulative author-subject index to selected list of fine arts periodicals and bulletins. Issued quarterly.

Humanities Index (1974–   ). Subject index to periodicals that cover literature, philosophy, history, linguistics, and other subjects in the humanities.

Business and Economics

Indexes

Business Periodicals Index (1958–   ). A monthly subject index to all fields of business.

Abstracts

The Wall Street Journal Index (1958–   ). An index to articles appearing in the WSJ.

Human Resources Abstracts (1975–   ). Includes summaries of articles dealing with employment, training, and related issues.

Social, Political, and Behavioral Sciences

Indexes

Public Affairs Information Service (P.A.I.S.). Bulletin (1915–   ). Subject index to articles, pamphlets, and government documents on economics, government, and public affairs.
Social Sciences Index (1974–   ). Covers broad range of social science journals.
Social Science Citation Index (1972–   ). An index that enables the user to find authors and articles most frequently cited by other authors.

Abstracts

Abstracts in Anthropology (1970–   )
Psychological Abstracts (1927–   )
Sociological Abstracts (1952–   )

Science and Technology — Interdisciplinary

Indexes

Applied Science and Technology Index (1958–   ). Formerly *Industrial Arts Index* (before 1958). Subject index to articles appearing in more than 200 periodicals dealing with engineering, trade, science, and technology. Issued monthly and cumulated quarterly and annually. Sometimes indexes books and pamphlets. Particularly useful to students because its general subject headings are broken down into subdivision classifications, thus often guiding the student to a satisfactory topic for research — one sufficiently limited for practical investigation.
Conference Papers Index (1973)
Science Citation Index (1961 –   ). Issued quarterly, with annual cumulations. International in scope.
General Science Index (1978–   ). A subject index to scientific fields such as biology, chemistry, nutrition, medicine, and psychology.

Abstracts

Corrosion Abstracts (1962–   )
Environment Abstracts (1971 –   )
Mathematical Reviews (1950–   )
Pollution Abstracts (1970–   )
STAR (Scientific and Technical Aerospace Reports) (1963–   )
Textile Technology Digest (1944–   )
World Textile Abstracts (1969–   )

Agriculture

Indexes

Bibliography of Agriculture (1942–   ). Cumulated annually. Cites references in fields of plant science, soils, forestry, animal husbandry, entomology, agricultural engineering products, economics, food and human nutrition.

Abstracts

Animal Breeding Abstracts (1933 –   )
Dairy Science Abstracts (1939–   )
Field Crop Abstracts (1948–   )

Food Science and Technology Abstracts (1969–)
Forestry Abstracts (1939–)
Horticultural Abstracts (1931–)
Plant Breeding Abstracts (1930–)

Computer Sciences

Indexes

Current Papers on Computers and Control (1969–). Monthly listing of titles in control engineering and automation.
Computer Literature Index (1971–). A quarterly index to basic articles in the field of computer science.

Abstracts

Computer & Control Abstracts (1966–)
Information Science Abstracts (1966–)

Engineering

Indexes

Current Papers in Electrical and Electronic Engineering (1969–). Monthly list of titles in subject fields.

Abstracts

Applied Mechanics Reviews (1948–)
Electrical and Electronic Abstracts (1898–)
Electronics and Communications Abstracts Journal (1967–)
Engineering Index Monthly and Author Index (1906–). Author-subject index to professional technical periodicals in all engineering fields, engineering college publications, government bureau publications, and research organization publications. Published annually. Gives bibliographical information; annotates publications, and reviews more than 1,200 publications, some quite advanced for undergraduate research projects.
International Aerospace Abstracts (published twice monthly by American Institute of Aeronautics and Astronautics) (1961–)
Solid State Abstract Journal (1957–)

Life Sciences

Indexes

Biological and Agricultural Index (1916–). Monthly with annual cumulations. Covers a wide range in the agricultural field.

Abstracts

Behavioral and Neural Biology (1979–)
Biological Abstracts (1927–)
Entomology Abstracts (1969–)
Genetics Abstracts (1968–)
International Abstracts of Biological Sciences (1954–)
Microbiology Abstracts (1966–)
Oceanic Abstracts (1964/66–)

Medical Sciences

Indexes
Cumulative Index to Nursing and Allied Health Literature (1961–). Broad coverage of nursing and allied health fields.
Index Veterinarius (1931–). Indexes major publications in this field.

Abstracts
Dental Abstracts (1956–)
Cumulated Index Medicus (1960–)
Virology Abstracts (1967–)
The Veterinary Bulletin (1931–)

Physical Sciences

Indexes
Current Papers in Physics (1966–). Semimonthly listing of titles of papers from more than 900 journals. Entries classified.

Abstracts
Astronomy and Astrophysics Abstracts (1969–)
Chemical Abstracts (1907–)
Gas Chromatography Literature Abstracts and Index (1952–)
Geo Abstracts (1960–)
INIS Atom Index (1970–).Incorporates nuclear science abstracts.
Metals Abstracts (1968–)
Meteorological and Geoastrophysical Abstracts (1950–)
Mineralogical Abstracts (1959–)
Petroleum Abstracts (1961–)
Physics Abstracts (1898–)
Theoretical Chemical Engineering Abstracts (1964–)

Governmental, Statistical, and Miscellaneous Information

Despite the great length of many card catalogs and the extensive variety of indexing and abstracting sources, many forms of information are not listed in these traditional sources. For example, publications of governmental bodies can be difficult to identify unless you are familiar with reference sources such as those listed below:

Monthly Catalog of United States Government Publications (1895–). Published by the U.S. Government Printing Office, this is the official catalog of U.S. documents. This index lists the most widely available U.S. documents.
Government Reports Announcements and Index (1946–). The U.S. government issues many grants and contracts to outside organizations to do work on its behalf. This publication, sometimes called "NTIS" because it is issued by the National Technical Information Service, abstracts and indexes the written results of these contracts.
U.S. Bureau of the Census. *Statistical Abstract of the United States* (1879–). This annual publication is a compilation of statistical information published by the U.S. government.
Congressional Information Service. *American Statistics Index* (1973–). This publication indexes and abstracts statistical publications of the U.S. government, providing greater

detail and more currency than *Statistical Abstract.* ASI has proven so useful to researchers that the publisher is now issuing two companion publications, described below.

Congressional Information Service. *Statistical Reference Index* (1981–). Statistical publications of state agencies, university research bureaus, and commercial research firms are among the items indexed and abstracted in this source.

Congressional Information Service. *Index to International Statistics* (1983–). The statistical publications of the United Nations and other international agencies are indexed in this publication.

Reference Works

Although encyclopedias and other special reference works are generally not basic sources of information for a student's research report, they are often extremely useful in the course of work on a subject. Encyclopedias may provide needed general information enabling the researcher to read books and articles with greater understanding. Special engineering and scientific reference works, as well as technical dictionaries and biographical reference books, may be just what is needed to find answers to difficult questions that arise during the course of the research.

Of particular use in determining what reference tools are available are Harold R. Malinowsky, *Science and Engineering Literature: A Guide to Current Reference Sources,* 3rd ed. (Littleton, Colorado: Libraries Unlimited, Inc., 1980); and for less specialized works, Eugene P. Sheehy, *Guide to Reference Books,* 9th ed. (Chicago: American Library Association, 1976). We'll list below, with brief comment, some of the principal — and most commonly available — works you ought to know about. You should realize, however, that there are numerous specialized encyclopedias and dictionaries for the various technical fields. There are encyclopedias for the fields of astronomy, physics, chemistry, chemical engineering, welding, petroleum, meteorology, horticulture, social sciences, and mathematics; and there are handbooks for nearly all specialized fields. In addition to the dictionaries we mention, there are many others for such fields as aviation, mathematics, physics, chemistry, geology, agriculture, electronics, petroleum, meteorology, forestry, and medicine.

Encyclopedias

General Encyclopedias

New Columbia Encyclopedia (4th ed., 1975). A good one-volume reference work with brief but simple and accurate entries.

Encyclopaedia Britannica (15th ed., 32 vols., 1985). This encyclopedia has three divisions: the Propaedia, or Outline of Knowledge; the Micropaedia, or Ready Reference; and the Macropaedia, or Knowledge in Depth. The encyclopedia covers all major fields of knowledge.

Encyclopedia Americana (30 vols., 1983). This encyclopedia contains tens of thousands of articles which help bridge the world of the specialist and the general reader. It is especially good on recent technological and scientific achievements.

Collier's Encyclopedia. Not as scholarly as *Encyclopaedia Britannica* or *Encyclopedia Americana,* but clear and simple and in readable type.

International Encyclopedia of the Social Sciences (1968). In 17 volumes; covers all disciplines of the social sciences.

International Encyclopedia of Statistics (1978). Lengthy explanations of statistical procedures and
   concepts are found in this two-volume set.

General Scientific Encyclopedias
McGraw-Hill Encyclopedia of Science and Technology. A basic source for every field of science and
   technology.
Van Nostrand's Scientific Encyclopedia (6th ed., 1983). Available in either one or two volumes,
   this encyclopedia provides an overview of topics in science and engineering.

Specialized Scientific/Technical Encyclopedias
Encyclopedia of Chemical Technology (24 volumes, 1978–84). Chemical substances and industrial processes are covered in this specialized encyclopedia.
Encyclopedia of Computer Science and Technology (2nd ed., 1983). All aspects of computer science,
   from integrated circuits to social implications, are covered in this one-volume source.
Prentice-Hall Encyclopedia of Mathematics (1982). This one-volume encyclopedia is written for
   the beginner in the field.
Encyclopedia of Polymer Science and Technology: Plastics, Resins, Rubbers, Fibers (1965–72). Sixteen
   volumes on all aspects of the field.
Encyclopedia of the Biological Sciences (2nd ed., 1970). More than 800 entries.
Encyclopedia of Physics (1981). Major concepts in physics are defined in this one-volume ency-
   clopedia.

Dictionaries

General Technical Dictionaries
Chamber's Dictionary of Science and Technology (1978). One volume; a good basic dictionary.
McGraw-Hill Dictionary of Scientific and Technical Terms (3rd ed., 1984). A useful and comprehen-
   sive technical dictionary.

Specialized Scientific/Technical Dictionaries
The Prentice-Hall Encyclopedia of Mathematics (1982)
Dictionary of Logical Terms and Symbols (1978)
Dictionary of the Environment (2nd ed., 1983)
Modern Dictionary of Electronics (1977)
Glossary of Geology (2nd ed., 1980)
Dictionary of Astronomy, Space, and Atmospheric Phenomena (1979)

Biographical References

American Men and Women of Science. This frequently updated publication contains brief biographical sketches of scientists in the United States. Entries include employment,
   honors, and publications.
Dictionary of Scientific Biography. This set contains lengthy biographies on the careers of noted
   scientists of the past, such as Albert Einstein and Isaac Newton.
McGraw-Hill Modern Scientists and Engineers. Essays about the careers and accomplishments of
   1140 engineers and scientists of the twentieth century appear in this three-volume set.

Leads to Trade Literature

In addition to the sources of information you may find in a library, there is a vast
amount of information published by industrial concerns in the form of bulletins,

catalogs, reports, special brochures, and the like, which can be of great usefulness. Most of these publications may be had free of charge by simply writing for them. The problem is to find out about them and to know where to write. Three books in the library may be of special value in solving this problem:

1. *Thomas' Register of American Manufacturers.* This multivolume work gives both an alphabetical and a classified list of American industrial organizations. Although it is designed as a purchasing guide, you can use it to find out where to write for information on particular products.
2. *Sweet's Engineering Catalogues.* Commonly referred to as *Sweet's File,* this compilation of catalogues can also guide you to sources for the purchase of various architectural and building items.
3. *The Gebbie Press House Magazine Directory.* This is a guide to magazines published by industrial organizations for circulation among employees, stockholders, and interested outsiders. Many of these house "organs" are valuable sources of information. One example is *The Lamp,* a magazine published by Exxon in a handsome format. It contains many articles of interest to the petroleum technologist.

Information Retrieval Systems

In recent years the process of searching computer databases to retrieve information has become an accepted part of the research process. Because you may need to use this kind of service occasionally while writing or researching in school or later when you're employed, we are providing a brief overview of information retrieval systems. There are many kinds of databases, but we'll focus primarily on nationally available bibliographic databases.

A database, in this context, is equivalent to a book or publication that is stored in a computer memory. When information from that "book" is needed, the researcher, rather than browsing through the entire item, can specify the particular fact or concept that is needed by conducting a "search" of that database. The results of the search may appear on the screen of a computer terminal, may be printed immediately in the library, or may be printed at a distant location and mailed to the researcher.

In libraries, databases are most widely used to obtain bibliographic information comparable to what is found in some of the indexes and abstracts listed earlier in this chapter. These citations may consist strictly of basic information such as author, title, journal name, date, and pages, or they may have additional information such as an abstract or an address from which to order the publication. In addition, some databases include the complete text of articles; such databases tend to be among the most expensive to search.

Besides supplying bibliographic information, nationally available databases can also be used to obtain statistical information, as, for example, the Consumer Price Index, or directory information such as the address of Bell Laboratories. In fact, any kind of information that is available in a machine-readable format can become a database. If you need to know whether a database on a particular subject is available, ask a librarian, or consult one of these reference books:

Carlos Cuadra, *Directory of Online Databases* (New York: Zoetrope, 1983), or Martha E. Williams, *Computer-Readable Databases: A Directory and Data Sourcebook* (Chicago: American Library Association, 1984).

Databases, however, do have limitations. They frequently provide only limited historical coverage of a subject. In addition, the success of a database search can be affected by the skill of the searcher or the quality of the database.

Libraries differ widely in the availability of particular databases and in their pricing of database search services. If you are interested in finding out whether a computer search would help you with your research assignments, we suggest you ask a librarian for assistance.

Conclusion

It is almost impossible to overestimate the value of a thorough acquaintance with library resources, and yet it is not difficult to learn your way around a library once you recognize that doing so calls for deliberate and methodical study. The aids that have been noted in this chapter are by no means sufficient in themselves to solve every one of your future library research problems, but with these aids as a nucleus and with a genuine interest in the subject, you will have no trouble in developing the knowledge you need.

Suggestions for Writing

One of the best ways of achieving a sense of professional confidence, aside from knowing your subject well, we believe, is knowing where to find information when it is needed. It will be to your advantage to become acquainted with the resources of your library, particularly in your own professional field. The suggestions below are intended to help you gain a knowledge of those resources — and enhance your sense of professional confidence.

1. Investigate the resources of your library in relation to your professional major. Find out which of the following are available: abstracts, biographical references, bibliographies, book review publications, directories, encyclopedias, dictionaries, handbooks, newspaper indexes, periodical guides, and indexes to government reports. You may have to ask the help of the reference librarian. Then write a brief report to a hypothetical student who is interested in specializing in your field, listing the basic information about each of the resources you find, together with information about the particular usefulness of the items, as well as information about where the books can be found.
2. Find out what professional periodicals relevant to your field of professional interest your library subscribes to. Examine an issue or two of each to determine what specialized sphere of interests each periodical appears to deal with (that is,

what sorts of articles the magazine publishes). Write a letter-report to a freshman who intends to major in this field, describing these periodicals and explaining in what ways they are useful.

3. Prepare a bibliography of your library's resources on a topic of interest to you (perhaps concerning the subject you will write a research report on). Include books, periodical articles, theses, dissertations, and government-sponsored research reports. Use the bibliographical form described in the following chapter (or the form prescribed by your instructor).

4. Find out where and in what publications you would find information about commercial products. Write a brief report explaining to your reader where to find information on a particular kind of product (say, high-fidelity speakers, oil additives for cars, transistors, laminated wood products, and so forth).

22

Writing the Library Research Report

Introduction

A study of technical writing that did not include the preparation of at least one report of four or five thousand words in length would be unrealistic. As you know, reports of this length, and indeed of much greater length, are common in industry and business. And there are some problems in writing a long report that are different from the problems of writing a short one, particularly in organizing and handling data. Furthermore, in an academic study of technical writing it is only in a fairly long report that you are likely to find a realistic synthesis or combination of the writing problems you have previously been studying more or less in isolation.

These are good reasons for writing a long report. Usually the most feasible way of preparing a long report in a course in technical writing is to write a library research report.

The library research method itself may seem somewhat unrealistic, however, consisting as it does of study and discussion of what other people have said about a subject. This method does not require you to wrestle with a mass of raw data, as was usually required of the workers whose books, articles, and reports you read in the library. This is a serious disadvantage. But in most respects the library research method is quite satisfactorily realistic. The problems of style, organization, format, handling of transitions, and so on are not different from what you will encounter in a report on original work. Moreover, the writing of library research reports is often required in science and industry — not to mention advanced courses in college. In government agencies, industry, and research organizations, one of the first steps when a new project is begun is often a "literature survey" of everything in print that may have some bearing on the subject. The results of such a survey are usually written up in the form of a report or used at the basis for a portion of a report.

Two other factors connected with the preparation of a library research report deserve mention. One is that it provides an excellent opportunity for increasing your knowledge of how to use the library. A second is that it is an unusual opportunity to study in detail some technical subject that you would like to know more about. Students sometimes find the preparation of such a report a first step toward mastery of a special subject; and such specialized knowledge is often helpful in securing a position. Numerous students, within our own knowledge, have shown prospective employers a report of this kind as part of the evidence of their fitness for a job.

Considered as a process, the library research report requires the following steps: (1) selecting a subject; (2) making an initial, tentative plan of procedure; (3) finding published materials on the chosen subject; (4) reading and taking notes; (5) completing the plan and writing the first draft; (6) documenting the text; and (7) revising the first draft and preparing the completed report. These steps will be discussed in the order stated, with the exception of step 3, to which the preceding chapter was devoted.

Selecting a Subject

In selecting a subject for your long report you should look for one that has the following qualifications: (1) the subject is interesting to you; (2) it is related to your major field; (3) it is a subject about which you already know enough so as to be able to read intelligently, but not one about which you have nothing to learn; (4) it is restricted enough in scope so that it can be treated adequately within about 5000 words; and (5) it is a subject on which sufficient printed material is available to you.

Most of these qualifications are self-evident and need no particular comment. As for the last of them, we would suggest this caution: don't assume that any library book or article or other document will be available to you until you have it in your own possession. You may find that a highly important article is at the bindery, or lost, or charged out to an uncooperative faculty member. One other point in the list above — the problem of scope — will be discussed at length later on. You may also wish to review the set of questions posed at the end of the chapter on proposals. See p. 279.

The five qualifications listed can, in general, be met by three different kinds of subjects: (1) subjects representing a project you are actively working on, like building a boat, designing a gas model airplane, or remodeling a room in a residence; (2) subjects concerned with the making of a practical decision, like choosing the best tape deck within a given price range, or the best outboard motor, or the best type of foundation for a given residence at a given location; (3) subjects that will add to your store of practical knowledge, like cross-wind landing gear for aircraft, offshore drilling platforms, or rammed-earth construction for small homes. Which one of these three types of subject matter is the best for you depends on your interests and your background.

You will find it helpful to make a list of possible choices under each of the three headings as one of the first steps in making your selection. As the next step, go to the library to find out whether sufficient material is available on the most attractive subject on your list. Right at this point, however, there is a strong possibility that you will run into the problem of limiting the scope of your subject. Chances are that your first formulation of the subject will prove too broad for a report of the length required. To make some suggestions about what to do, we'll consider an exaggerated example.

Suppose you are interested in the subject of television and decide that you will write a report on it. You consult the card catalogue and appropriate indexes to periodical literature in the library, and discover scores of articles and books dealing with television. Obviously, you cannot read them all. Even if you could, you would find from an examination of titles that they deal with so many different aspects of the subject that unification of the material would be next to impossible. Further examination would show that much of the material is superseded by later developments and so is of no value. But the main thing would be the impossibility of covering all the material available. Two courses are open: you may reject the subject of television entirely or you may limit your investigation to some particular phase of it.

Assuming that you are unwilling to give up the subject altogether because of your interest in it, you will find several opportunities for limiting the scope of the subject. Let's consider just a few. You might begin with a time limitation and see what has been published on television developments during the past few years. It would quickly become evident that, for this subject, this way of limiting the subject would be wholly inadequate. So you might try subdividing according to subject matter rather than time. Here the classifications in the card catalogue and indexes would be of service. For instance, you might find a number of articles devoted to various parts of television apparatus, such as antennas, amplifiers, or cathode-ray tubes. Or special broadcasting problems might prove of interest, such as the use of satellite relay stations. In short, examination of the titles of publications on the general subject in which you are interested should suggest any number of ways of limiting the subject to manageable proportions. Ultimately, of course, you must examine the publications themselves to be absolutely certain that they offer adequate, but not too abundant, material for a report of the length you are expected to write. But most of the work of limiting the scope of a subject can be done by careful thought initially and by careful study of available sources of information.

A second way of limiting the scope of a subject is through the choice of reader. As we remarked earlier, technical writing is almost always done with a specific reader or group of readers in mind, and the library research report should be no exception. Your instructor may ask you to direct the report to him or her, or to choose an imaginary reader — perhaps a student like yourself. In any case, as you can see, identification of a reader answers many questions about what material to include and what assumptions to make about the level of vocabulary and presentation.

In general, subjects of current interest are best, subjects that for the most part are treated in periodical articles. If whole books have been devoted to a subject, it is likely to be too broad for report treatment. This does not mean, of course, that books may not be used as sources of information. To return to our illustration above for a moment, you would be fairly certain of finding books on television in which chapters might be devoted to antennas, and these would be useful for a report on antennas. If you were to find, however, that an entire book or several books had been written on antennas, you would probably decide that the topic is too broad for adequate treatment in a 5000-word report.

Making an Initial Plan

Once a subject has been chosen and approved by your instructor, it is time to lay a few plans for general organization and coverage of the subject, so as to simplify and give direction to the task of reading and taking notes.

First, make a list of the things you want to find out about your subject. Add to this list those things you think your reader will want to know. Sometimes these items will be identical, sometimes not. This list will prove most useful to you when you read and take notes because you will have some idea of what to look for in the reading. Of course it will undoubtedly prove necessary to revise the list, perhaps a number of times. You may discover, for instance, that nothing has been published on some particular aspect of the subject you thought ought to be discussed. You may discover discussions of important aspects of the subject that had not occurred to you when you were making your tentative guidance outline. You may discover that certain aspects of a subject will have to be eliminated because of space limitations. Such a list or outline should not in any sense be regarded as final but merely as a general guide, something to give you a sense of direction.

If your knowledge of a chosen subject is so slight that you don't feel able to compile such a list, you should do some general reading on your subject first in order to acquire the necessary acquaintance with it. Then you can make a tentative outline. The importance of making a list for guidance will be made clearer in the discussion of note taking.

An example of the relation of an initial list of topics for guidance to a final report outline is shown below. The subject is the magnetic fluid clutch.

Initial Guidance List	*Final Outline*
1. What functioning parts does it have?	I. Introduction
	A. Definition of a clutch
2. How does it operate?	B. The need the magnetic clutch can
3. How much does it cost to make?	fill
4. How efficient is it?	C. Object of this report
5. Is it difficult to maintain and repair?	D. Scope and plan of this report
	II. Principle of operation
6. When was it developed and by whom?	III. Description of clutch
	A. Driving assembly

7. How does it compare with other   B. Magnetic fluid
   types?   C. Driven assembly
                                            D. Electric coil
                                       IV. Advantages and disadvantages
                                            A. Inertia
                                            B. Simplicity of design
                                            C. Leakage
                                            D. Ease of control
                                            E. Number of parts
                                            F. Smoothness of operation
                                            G. Fluid trouble
                                            H. Centrifugal trouble
                                        V. Applications
                                            A. Automotive
                                            B. Servomechanisms

Reading and Taking Notes

Once a preliminary list of sources (bibliography) has been compiled and a guidance list has been set down, it is time to begin reading and taking notes. In deciding what to read first from a list of sources, you should choose a book or article that promises to give a pretty general and complete treatment of the subject and which is simply and clearly presented. How a book or article rates in these qualities can be guessed at by examining titles and places of publication. For instance, an article entitled "Color Television Explained" appearing in a popular magazine is certain to be easy to understand and nontechnical in its treatment. By reading simply written articles covering your subject broadly, you will be better able to understand and use the information you find in books and specialized periodicals. It may happen that your judgment of a title and place of publication will turn out to be wrong, but in general you will simplify your job by following this procedure.

You can now begin reading and taking notes. This is a job that should be highly systematic from the start. The following paragraphs outline an efficient method.

Three basic requirements of any good system of reading and note taking are as follows: (1) the reading should be conducted according to a plan, not haphazardly; (2) the method of arranging the sequence of the notes should be highly flexible; and (3) the system should be economical of time.

The first of these three requirements has already been discussed. Its observance requires the preparation of an initial guidance list or outline so that pertinent materials can be selected and irrelevant materials ignored. We can go on, then, to discuss the second and third requirements.

Flexibility of arrangement of the notes is easily achieved by the use of cards. In theory, the method requires that only one note be written on a given card (4 by 6 cards are a convenient size). It is next to impossible to define "one note," of course. For our discussion, however, it will be sufficient to say that "one note" is any small unit of information that will not have to be broken up so that the parts can be placed at separate points in the report. When all note cards have been prepared and arranged in the proper order, it should be possible to write the report without ever turning forward or backward as you go through the pack of cards. Naturally, such perfection is scarcely to be expected in practice.

Several symbols or labels can be used to save time. The first of these is a heading, put at the top of the card (see Figure 1). This heading is useful in the process

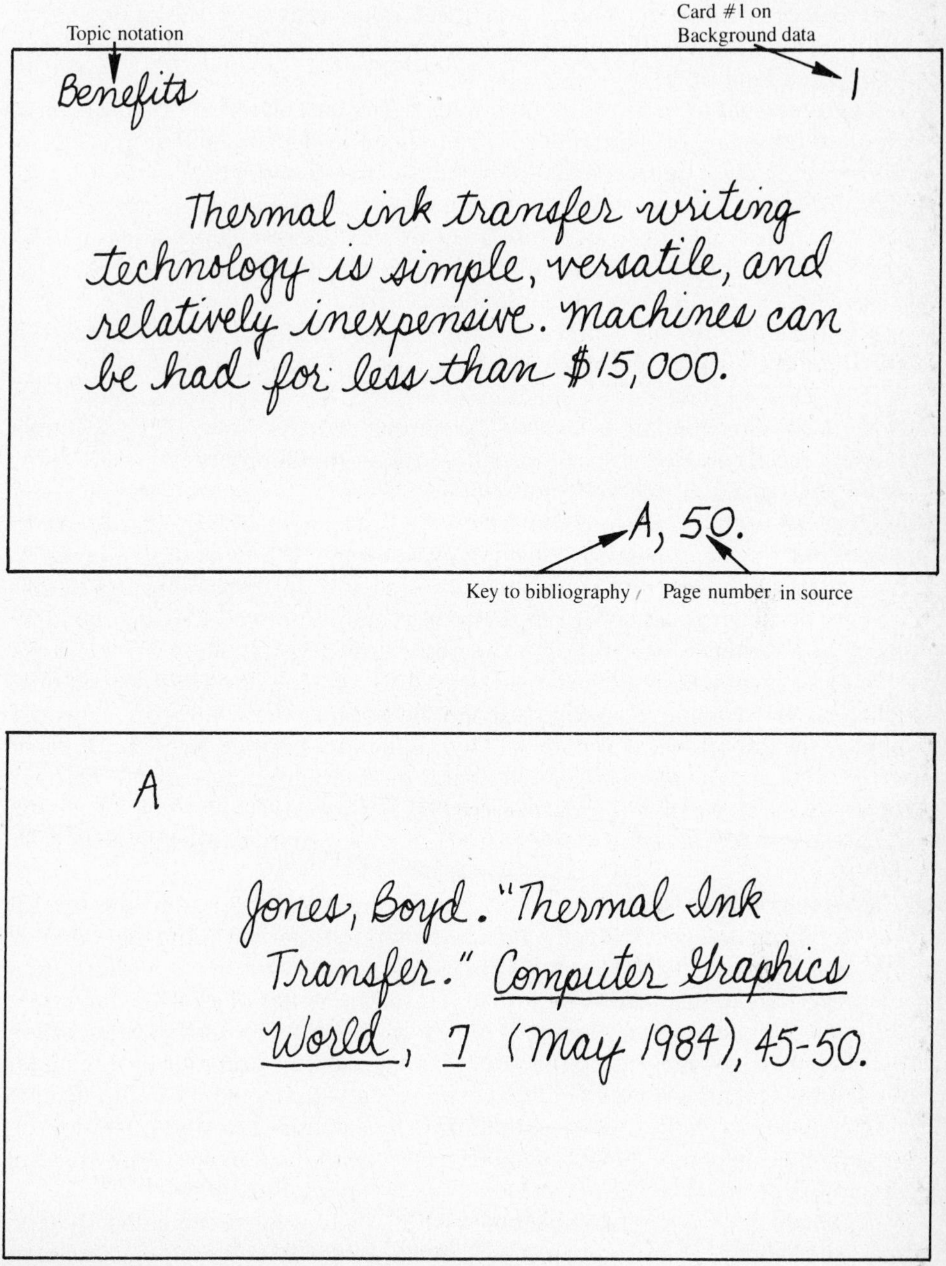

Topic notation

Card #1 on
Background data

Benefits   1

> *Thermal ink transfer writing technology is simple, versatile, and relatively inexpensive. Machines can be had for less than $15,000.*
>
> A, 50.

Key to bibliography   Page number in source

A

> *Jones, Boyd. "Thermal Ink Transfer." Computer Graphics World, 7 (May 1984), 45-50.*

Figure 1   A note card and a bibliography card.

of sorting and arranging the cards. It can be taken from the tentative outline. Some people like to add a symbol taken from the outline (like II A).

Secondly, it is convenient to use a symbol to indicate the source from which the note was taken. It is imperative that you indicate the source of every note so that the text can be documented. One way to keep a record of sources would be to write complete bibliographical data on each note card. This method would be inefficient, however, if you had more than two or three note cards from the same source (imagine writing eight or ten times the data for an article with a long title from a journal with a long title).

A better way is to write the bibliographical data on a blank card. You may find it helpful to use a card of a different size from the note card, or a different color, so that it is easy to distinguish the bibliographical cards from the note cards. On this bibliographical card, put a capital letter in the upper left-hand corner. Now, when you make a note from this source, instead of writing the complete bibliographical data on the bottom of the note card, all you need do is write the capital letter that will key the note card to the proper bibliographical card.

The last entry on the note card is the number of the page from which the information in the note was taken. This is shown in Figure 1.

One thing we have not yet mentioned is the nature of the notes themselves. What kind of note should one make — an outline, a series of words and phrases intended to recall complete discussions, full or almost word-for-word transcriptions, or what? Our advice is that your notes should first of all be entirely in your own words (except for quotations, about which we'll say more in a moment) and in summary form as far as possible. Secondly, we strongly recommend that notes be made full enough so that you will not be confused as to their meaning and significance later on when you come to use them. It is important to avoid using the same phrasing and sentence structure that the author of the article uses. To make the material your own, you should first read it carefully, making sure you understand it, and then put down what you want to use, briefly and in your own words. We would emphasize the importance of economy of words in note taking, to simplify the job of studying and using the notes later on. It is quite discouraging to read a note, perhaps some weeks after it was made, only to discover that it does not make sense to you and that you have to return to the source to find out what you tried to get down in your notes.

Direct quotations are not essential to a research report, but there are several reasons for using them occasionally. First, a quotation from a well-known authority may lend force to a passage. Second, it may be a courtesy to use an author's own words, especially on a controversial topic. Third, the writer of a report may use a quotation to emphasize that the report is merely giving an account of what others have said, and is not presenting the writer's own opinion. And fourth, it is sometimes desirable to quote a statement that expresses an idea in an unusually effective way. Whatever the reason for a quotation, it is essential that the quotation be absolutely exact. Quotation marks should be placed around it in your notes and the page number of the source noted.

In concluding this discussion of note taking, we urge you to remember that the system we have described is not a magic formula. You will have to use judgment at every step, and you will inevitably have to do a lot of work with the cards, discarding repetitious notes and filling in gaps. On the other hand, the system is efficient. It comes closer than any we know to satisfying the three basic requirements stated at the beginning of our discussion. Its advantages are attested by the fact that it is widely used.

Completing the Plan and Writing the First Draft

Once you have completed taking notes, it is time to prepare a final outline of your report. To do this properly, you will have to read through your notes carefully, perhaps a number of times. First of all, you will want to make yourself thoroughly familiar with the content of your notes. While doing so, you will be devoting some thought to the best order in which to present the topical divisions of your subject matter. It is here that the usefulness of the notetaking system just described becomes most apparent. With topic headings on each card, you can now rearrange the cards in a suitable order. This rearranging may require several experimental tries before you are satisfied.

For the Illustrative Material at the end of this chapter, we have reprinted some pages from a 5000-word report by a student. You might wish to study this section as a model for transforming your outline into the subhead system in your report. Models of documentary form will also be shown in a later section.

When the above preliminaries have been carried out, writing the first draft amounts to little more than transcribing your notes to paper in a connected, coherent discussion. They will not be transcribed verbatim, of course. Although your notes, if properly taken, will be in your own words rather than in those of the authors of your sources, you will do well to rephrase and reword many of the passages in your notes as an additional safeguard against reflecting the style of writing used in your sources. You will do a good deal of this anyway (that is, without making a conscious effort) if you have mastered the content of the notes thoroughly. Furthermore, you will be adding transitional statements, developing and clarifying some points of fact by supplementary discussion, making comments (evaluations and conclusions) about the facts from your sources, and the like. In short, the report itself will contain a good deal of writing that is all yours, and not merely transferred from a source to your paper.

Most inexperienced writers attempt too much when they undertake the first draft of a report. They try to devote attention not only to the subject matter itself, but also to style and correctness of expression. In writing the first draft, forget about style and correctness. Concentrate on the subject matter alone. Get down on paper what you want to say; there will be time later for smoothing out your sentences, correcting your spelling, punctuation, and choice of words. If you have a lot of inertia to overcome in getting started with your writing, do not conclude that you have no talent. Most writers are slow in getting started, even professional ones. Once a start is made, however bad, the task usually becomes easier. Awkward beginnings can be remedied later on.

Documenting the Text

Documentation is the recording of published source material used in a report. Of course unpublished materials, such as letters and dissertations, are also used occasionally in reports, but problems of documentation are concerned primarily with printed matter. Actually you will want to document your report (that is, put in footnotes) as you write your first draft; we are discussing documentation separately merely for convenience.

A bewildering variety of different methods are employed in documentation, preference differing from one editor to another, even within a given field. In a moment we will describe three widely used methods, but first we will comment on the purpose of documentation and where to include footnotes in your report. For convenience in explanation we will use "footnote," "footnoting," etc., to refer to the basic elements and methods of documentation in general even though these terms might not apply in the strictest sense in some instances. Until recently, footnoting was the most traditional method of documenting sources of information, in scientific as well as in other writing. Gradually, however, the footnote system of documentation has yielded to the economic pressures of higher printing costs. Even though it has now been abandoned by publishers of many scholarly journals, you will still find it useful to know, simply because it was used so widely when present library holdings were being put into print.

Placing Footnotes in the Text   Let's begin with a question. "When do you need to write a footnote, or a citation?" (For our immediate purpose, a footnote and a citation can be thought of as having the same meaning; we'll explain a difference between them when we describe the different systems of documentation.) For the library research report we are discussing in this chapter, the answer to the above question is simple. Every fact, idea, and opinion that you have obtained from your reading, whether you quoted it or paraphrased it, should be footnoted. One of the important values of writing such a report is the experience gained in keeping all the data involved under complete control.

Although the documenting of your report may seem like an alarmingly difficult task at first, it really isn't so very hard. Remember, your first draft has been largely a transcription of your notes, and your notes contain precise indications of the source and page number of each fact. You can, of course, put footnotes into your report during the process of writing the first draft. Or you may wish to wait until you have finished getting your discussion down on paper. It doesn't really matter when it is done. The important thing is to do it, and do it completely.

A number of questions naturally arise. Suppose one sentence contains information from two or more sources. Does this call for only one footnote, or for more than one? On the other hand, suppose several pages of discussion in the report are based on a single source. Does a footnote need to appear after each sentence, each paragraph, or at the end of the discussion?

Every unoriginal statement must be documented. That means two footnotes must be written if a single sentence contains data from two distinct sources. If a paragraph contains information from a dozen sources, a dozen footnotes appear at the bottom of the page. If several pages are based on one source, just one note is needed, at the end of the discussion. To put it another way, a footnote must appear at the end of each portion of discussion that is based on a particular source. The portion may be a phrase, a sentence, a paragraph, or a longer part of your composition. Let us repeat, your notes will have each fact identified as to source and page number. Except for the work involved, there will be no difficulty in documenting each fact.

"But suppose," you may say, "that in between paragraphs of information taken from sources appears a paragraph which is original, like an evaluative comment, or a transitional paragraph. Do I need a footnote for that?" The answer is no; you do not need to footnote yourself. "But," you may object, "how will the reader know that what I am saying is original and not taken from one of my sources? Or that I haven't simply forgotten to put in a note?" The answer to this question is that the reader can usually tell from the nature of the comment you are making — its content and style — that *you* are speaking, not one of your sources. Just remember to document all the facts you have secured in your research, all the information you acquired *after* beginning your investigation, and you will have done a satisfactory job of documenting your report.

Although the foregoing discussion may make it appear that you will have an extremely large number of footnotes in your report, it doesn't usually take nearly as many as you might think. Let's consider an actual case. One of our students wrote a report entitled *Tantalum as an Engineering Material.* He organized it according to the following main headings: Introduction, Occurrence, Extraction of Tantalum, The Working of Tantalum, Tantalum Alloys, Uses for Tantalum, and Costs of Tantalum. His bibliography contained 15 items. His report, which was 21 pages long, contained 32 footnotes.

Here's how they were distributed. None was necessary for the introduction because in that section he simply introduced the reader to tantalum as one of the rarer metals and explained what he proposed to discuss in the remainder of the report; he explained the purpose of the report, its plan of presentation, its limitations, and its point of view. Section II contained eight footnotes, five of them references to one source. It happened, you see, that most of his information on the occurrence of tantalum came from a United States Bureau of Mines article entitled "World Survey of Tantalum Ore." The other three articles referred to in this section had provided him with bits of information not contained in the above-mentioned item. He could have got by quite adequately with four footnotes instead of eight for this section in view of the fact that most of his data came from the one source. The third section of his report, on the extraction of tantalum, was based on information from one source, and he used two footnotes, one for each of the two paragraphs in this section. He could have used just one note. The fourth section, on the properties of the metal, contained seven footnotes, all but two of which were references to one source. This one source contained the most complete discussion of the properties of tantalum; the other two articles referred to gave him a few facts not contained in the chief source. Here, again, he could have reduced the number of footnotes from seven to three or four. His fifth section, on the working of tantalum, contained eight footnotes, six of them from three sources. In general, separate subdivisions of this section were based on different sources so that he found it necessary to put a footnote at the end of each topical subdivision of the section, plus an additional one for a direct quotation. The next to last section, on uses for tantalum, required four footnotes — he had found material on four uses in four different articles. The last section, on costs, contained two notes.

The ratio of report length to number of footnotes described above is fairly typical, but otherwise no particular significance should be attached to the example. Another report of the same length might contain twice as many footnotes, or half as many. It all depends on how many sources are used and the extent to which each is used. You should not use more footnotes than are needed, but you should use enough to make clear the source of all information secured during the process of investigation.

The method of footnoting just discussed provides for the kind of complete documentation that is most useful in a report written primarily for practice. In actual industrial and research reports and publications a somewhat different policy is followed because facts that are common knowledge to workers in the field do not need to be documented. What does need to be documented are principally those elements on which the validity of the report as a new piece of work depends, in addition to any elements concerning which a reader might be interested in seeking further information. More specifically, footnotes are used for (1) direct quotations; (2) controversial matters; (3) ideas of critical importance in the content of the report; (4) citation of well-known, authoritative writers; (5) acknowledgment of an author's originality in developing the idea presented; and (6) comments on additional material the reader might like to examine.

Although footnotes are primarily used for references to sources in a research report, they do have other uses. Definitions of technical terms used in the text may be put in the form of footnotes if it is felt that some readers may need the definitions. If a term is one of crucial importance, its definition should appear in the text. It may happen that a number of terms need to be used which may or may not be familiar to a reader; the writer will not want to interrupt his discussion repeatedly to supply definitions, and footnotes offer a satisfactory solution. Footnotes may also be used for other statements that do not properly belong in the discussion. Suppose, for instance, that you find all your sources but one in agreement on some point in your discussion. A footnote could be used to report this one exception to general agreement. Finally, let us say that footnotes should be kept to a minimum; although necessary for acknowledging sources of information and occasionally for supplementary discussion, they do interrupt the reader and certainly add nothing at all to the readability of reports. Do not put a lot of them into a report in the hope that they will make it more impressive.

Selecting a Documentation System   In addition to thinking about where to place footnotes in your report, you will need to consider how the footnotes themselves should be handled. From among many possibilities, we have selected three different types of documentation for consideration: a version of the author-date system, a version of the number system, and a version of the combined footnote-bibliography system. Each of these three systems has its own advantages and disadvantages.[1] The first two are probably the commonest systems used in science and industry.

We would not advise you to memorize these systems; in fact, we would not advise you to memorize even a single one of them. The reason for this advice is that in the course of your professional career you will probably have to employ a considerable variety of documentation systems, and there simply isn't much point in memorizing any particular one of them. What you do need to learn is, of course, to follow accurately the details of any model you are asked to use. We are presenting these three systems for your general information and to provide you with a model. Your instructor will probably want to designate one of the three for use in your library research paper.

Author-Date System. The name "author-date" is derived from the fact that the documentation within the text is made up of the name of the author being cited and the date of the work by that author, together with a specific page number if needed. All the publishing information the reader may require about the works noted is presented in a separate list at the end of the report. In this system, there is no footnote at the bottom of the page; hence, we are now using the term "cited" or "citation," instead of "footnote." Following are examples of citations in the text, using the author-date system.

"Spacetime, usually thought of as four-dimensional, may have as many as seven dimensions" (Friedman and van Nieuwenhuizen, 1985:74).

Seeking universal symmetry, physicists are trying to find as much antimatter as matter (Guillen, 1985:32).

IBM'S instructions include the discussion of fan-fold forms (IBM, 1975).

The evolution of television in the "information age" will result in larger spread sheets and wider computer-type displays with increased resolution (Nadan, 1985:135).

It is still possible, but not likely, that we will ever find out when and how the human race first acquired language (Simpson, 1968:14).

Lenard never forgave Röntgen for having discovered X-rays first, as he was convinced he had really discovered them himself (Keller, 1983:62).

One of the symptons of shock is dilated pupils (Shock, 1974).

Organized labor's recruiting efforts are bolstered by US study showing union workers are better paid (Labor Seeks Recruits, March 15, 1985).

The sailing ships in the Chesapeake Bay oyster fleet should be kept in operation for ecological as well as historical reasons (Doe, Sept. 10, 1986).

---

[1] For a detailed discussion of these and of other systems see the following books: *A Manual of Style,* 12th ed. (Chicago: University of Chicago Press, 1969); *Style Manual,* rev. ed. (Washington, D.C.: U.S. Government Printing Office, 1967); Council of Biology Editors, Committee on Form and Style, *CBE Style Manual,* 4th ed. (Washington, D.C.: American Institute of Biological Sciences, 1978); Margaret Nicholson, *A Practical Style Guide for Authors and Editors* (New York: Holt, Rinehart and Winston, 1971, paperback); *Publication Manual of the American Psychological Association,* 3rd ed. (Washington, D.C.: APA, 1983).

The first of the items above is a quotation from page 74 of an article by Friedman and van Nieuwenhuizen published in 1985. Although all the remaining items are paraphrases of the original text, not direct quotations, the same kinds of source information are given about them, with the exception that no page number is noted for the IBM entry; this entry does not need a page number because the fan-fold forms are discussed at points throughout the document.

A "Literature Cited" section accompanying the foregoing citations is illustrated below (admittedly, it would be a strange report indeed which included the variety of subjects indicated).

Doe, J. Sept. 10, 1986. Personal interview.
Friedman, D. Z. and P. van Nieuwenhuizen. 1985. The hidden dimension of spacetime. Sci. Amer. 252(3):74–81.
Guillen, M. 1985. The paradox of antimatter. Sci. Dig. 93(2)32–81.
International Business Machines Corporation. June 1975. IBM 3767 Communications terminal setup instructions. Fourth Ed. White Plains, New York, Loose-leaf pub. 18 pp.
Keller, A. 1983. The infancy of atomic physics: Hercules in his cradle. Clarendon Press, Oxford. 230 pp.
Labor Seeks Recruits in Right-to-Work States. March 15, 1985. Chri. Sci. Mon. 77(78):10.
Nadan, J. S. 1985. A glimpse into future television. Byte 10(1):135–150.
Shock, Physiological. 1974. Ency. Brit. Fifteenth ed. Helen Hemingway Benton, Chicago, Ill. 16:699.
Simpson, G. G. 1968. The biological nature of man. Pages 1–17 in S. L. Washburn and P. C. Jay, eds. Perspectives in human evolution. Holt, Rinehart and Winston, New York.

Let's consider the form of these entries, to observe certain features of the author-date system. Only the initials are used, rather than full first and middle names. The year of publication — even for a periodical article — follows the author's name. Only the first word — and proper nouns — of titles of articles and books are capitalized, except that when a title is the first element in a citation it is given normal capitalization. Titles of journals are abbreviated.[2] The first numeral at the end of the second entry above (252) is the volume number, the second is the number of the specific issue, and the last two are the pages on which the article begins and terminates, respectively. The Literature Cited entries, as a group, are in alphabetical order according to the authors' last names, or according to title if no author is given. If two or more articles by the same author are listed and were published in a single year, they would be distinguished in both the textual citation and the Literature Cited section by lowercase letters following the year, in chronological sequence, like this: 1986a, 1986b.

Number System. The number system is much like the author-date system but differs from it in these respects: (1) Only a superscript numeral, not the author's name and the date, appears in the text; (2) the entries in the Literature Cited section at the end of the report are listed in the order in which they appear in the text, not

---

[2] For a list of generally accepted abbreviations, see American National Standards Institute, 1969. American national standard for the abbreviation of titles of periodicals. ANSI Z39.5. American National Standards Institute, New York.

alphabetically; and (3) the specific page in the cited material on which the reference appears is indicated in the Literature Cited entry, not in the text entry.

Here are two examples of textual citations:

"Spacetime, usually thought of as four-dimensional, may have as many as seven extra dimensions."[1]

Lenard never forgave Röntgen for having discovered X-rays first, as he was convinced he had really discovered them himself.[2]

Instead of the superscript reference numerals shown above, some editors place the numeral on the same line as the text, enclosed in parentheses and preceded by the letter R (for Reference) and a dash, like this (R-1). This method is especially easy for the typist.

For either of the two ways of writing the numeral just shown, the entries in the Literature Cited section for the references above would be these:

1. Friedman, D. Z. and P. van Nieuwenhuizen, 1985. The hidden dimensions of spacetime. Sci. Amer. 252(3):74.
2. Keller, A., 1983. The infancy of atomic physics: Hercules in his cradle. Clarendon Press, Oxford, p. 62.

One advantage the number system has, as compared with the author-date system, is its adaptability for use in providing brief explanatory "footnotes," comments that may be of value to a reader but are unsuited for inclusion in the text. Such a comment can be signaled by a superscript numeral in the text, just as is done for a source reference.

Footnote-Bibliography System. The footnote-bibliography system provides much more information in the textual citation than do the systems already described. Informative footnotes are helpful to the reader, but are of course an added requirement for the writer and typist. Footnotes are placed either at the bottom of the page concerned or in a separate list at the end of the report. In students' library-research reports they are often placed at the bottom of the page because of an instructor's need to check the use of source materials, page by page, while reading.

There are three elements in the footnote-bibliography system: a superscript numeral in the text, a "footnote" either at the bottom of the page or in a list at the end of the report, and a bibliography at the end of the report.

The superscript numeral appears in the text exactly as it does in the number system, so we'll not illustrate it again.

The footnotes are brief indications of the character of what is being cited. A reader who wants to obtain a document cited will turn from the footnote to the bibliography to find the necessary publishing information.

Here are examples of footnotes (more than would likely appear on a single page of a typical report):

1. Friedman and van Nieuwenhuizen, "The Hidden Dimension of Spacetime," p. 74.
2. Guillen, "The Paradox of Antimatter," p. 32.
3. *IBM 3767 Communication Terminal Setup.*
4. Nadan, "A Glimpse into Future Television," p. 125.

5. Simpson, "The Biological Nature of Man," p. 14.
6. Keller, *The Infancy of Atomic Physics: Hercules in his Cradle,* p. 62.
7. Nadan, p. 137.
8. Simpson, p. 12.
9. *IBM,* p. 10.
10. "Shock, Physiological," p. 699.
11. "Labor Seeks Recruits," p. 10.
12. Personal interview with Dr. John Doe, Sept. 10, 1986.

[Note: No bibliographical entry is made for an interview, since a reader has no access to the interview.]

Figure 2 illustrates the form of footnotes at the bottom of a page. Note that a line is run part way across the page to separate footnotes from text. Aside from such obvious considerations as capitalization and punctuation, three principles in the footnotes deserve attention. The first is that the specific page (or pages) from which the cited information was obtained is indicated in the footnote itself. The second is that in the first citation of an article, the title of the article is shown, but not the title of the journal or the date. The third principle is that on all appearances after the first citation of a given document, the document may then be identified by no more than the author's last name, as in the example footnotes 7 and 8 above. However, if you have more than one document from a single author, then at least part of the title must be shown for each, in every citation after the first. If a document has no author, then an abbreviated title may be shown in the second and subsequent citations, as in footnote 9.

Bibliographical forms for the footnotes above are shown in the following list. As in the author-date system, the bibliography is in alphabetical order according to the author's last name.

Friedman, D. Z. and P. van Nieuwenhuizen. "The Hidden Dimension of Spacetime." *Scientific American, 252* (Mar. 1985), 74–81
Guillen, M. "The Paradox of Antimatter." *Science Digest, 93* (Feb. 1985), 32–81.
IBM 3767 Communication Terminal Setup Instructions, 4th ed. White Plains, N.Y.: International Business Machines Corporation, June, 1975. (Looseleaf)
"Labor Seeks Recruits in Right-to-Work States." *Christian Science Monitor, 77* (March 15, 1985), 10.
Nadan, J. S. "A Glimpse into Future Television." *Byte, 10* (Jan. 1985), 135–150.
Keller, A. *The Infancy of Atomic Physics: Hercules in his Cradle.* Oxford: Clarendon Press, 1983.
"Shock, Physiological." *Encyclopaedia Britannica* 15th ed. Chicago, Ill.: Helen Hemingway Benton, 1974. Vol. 16, 699.
Simpson, G. G. "The Biological Nature of Man," in S. L. Washburn and P. C. Jay (Eds.), *Perspectives on Human Evolution.* New York: Holt, Rinehart and Winston, 1968.

In this system, the number of the particular issue of the journal identified is not shown; it is replaced by the month, or month and day, of issue. The volume is shown. And, of course, the page numbers are those occupied by the entire article.

We hope the forms we have illustrated will be helpful, but we close these remarks on documentation with two cautionary suggestions. The first is that our brief treatment of the subject leaves a great many kinds of problems untouched; if you encounter a problem for which the models shown don't provide help, ask your

Dr. St. Clair reasons that:

These acoustic forces, arising from radiation
pressure, act to cause a concentration of the
suspended particles in the regions of
maximum displacement and to produce
attractive and repulsive forces between the
particles.[19]

Variables to Consider

Three variables must be considered in connection
with the precipitation of aerosols in large industrial
volumes: the sound field intensity in which the
aerosol is treated, the exposure time, and the frequency of the sound.[20]

Intensity. Although noticeable agglomeration is
caused at 140 decibels, an intensity of about 150
decibels is most efficient for industrial application.[21]
Sounds above 120 db, incidentally, are painful to the
human ear.[22]

Effective conversion of the energy of a generator
to sound depends upon the design of the generator and
treating chamber. A properly designed installation
may convert 40% to 60% of the compressed gas's
energy to sound energy.[23]

---

[19]St. Clair, "Agglomeration of Smoke, Fog, or Dust
Particles by Sonic Waves," p. 2439.

[20]Danser and Newman, "Industrial Sonic . . . ,"
p. 2440.

[21]Ibid., p. 2441.

[22]Jones, Sound, p. 245.

[23]Danser and Newman, p. 2440.

Figure 2   Illustration of a page with footnotes.

instructor for advice. The second suggestion is that there are really no universal rules about documentation. When you go to work for a company or agency, or when you want to publish an article in a certain journal, do them the courtesy of finding out what their custom is, and be guided by it.

Revising the Rough Draft and Preparing the Final Copy

After the rough draft has been completed, the next step is to revise the rough draft and prepare the report for submission. We'll assume that the rough draft has been documented. We suggest that you plan your work so that you will have plenty of time for the revision. It is an excellent idea to allow enough time so that you can lay your rough draft aside and forget it for several days, perhaps a week. The reason for this suggestion is that you will have difficulty spotting your mistakes if you undertake revision immediately after finishing the rough draft. (See the second example in Illustrative Material, Chapter 10, for an illustration of the kind of error too-hasty revision may produce.) You should read your rough draft objectively and critically, putting yourself as much as possible in the place of the person or persons who will read the finished report.

You can use the time between writing the rough draft and revising it to clean up other tasks incident to completion of the report: preparing the illustrations, preparing the cover and the title page, the letter of transmittal, the table of contents, the list of figures, and writing a first draft of the abstract.

As you start the final revision of the report, remember that you are making a revision, not a final copy. Making the final copy should be a purely mechanical operation, requiring no significant changes in the text. If you try to make revisions and final copy simultaneously, you will find yourself in such troubles as making a change on page 10, which in turn requires a change on page 5, which is already typed! The use of an electronic word processer can greatly reduce this problem, but doesn't eliminate it entirely. Sooner or later you must commit yourself to a final copy.

After you have completed the revision of the report, it is wise to go through it again several times, from cover to cover, deliberately checking each time for only one or two specific elements. Certainly the entire text should be checked once for grammar, with special attention to any specific kinds of errors that may tend to cause you trouble; once for transitions; and once for spelling if you have trouble with spelling.

Just what elements should be checked depends to a considerable extent upon the material and your own strengths and weaknesses as a writer. The technical writing checklist that we have included may be helpful. To this list we would strongly recommend that you add the following questions:

1. Are all the necessary functions performed by the introduction and the conclusion or summary?
2. Are the principles of the special techniques of technical writing, such as description of a process, made use of as needed?

Methodical use of a checklist is good insurance. If you can answer "Yes" to all the questions in it, you can turn in your report, confident that you have done your best and are reasonably sure that you have done well.

Report Appraisal

The list below is intended to assist you in planning, writing, and editing your own reports or in indicating to others the specific weaknesses of reports submitted to you for editing.

Before appraising a report, be sure to determine its exact purpose. What response is desired from the reader — or readers?

A Technical Writing Checklist

Before you begin, have you . . .
1. Defined the problem?
2. Compiled all the necessary information?
3. Checked the accuracy of all information to be presented?
4. Taken into account previous and related studies in the same field?
5. Learned all you can about who will read your presentation?
6. Determined why they will read it?
7. Tried to anticipate questions your readers will want answered?
8. Determined your readers' attitude toward the objective of the presentation?
9. Decided on the slant or angle you want to play up?
10. Checked the conformity of your approach with company policy and aims?

In making a plan, have you . . .
1. Planned an introduction that will introduce the subject matter and the presentation itself?
2. Arranged the parts of the presentation so that one part leads naturally and clearly into the next?
3. Included enough background information?
4. Excluded unnecessary and irrelevant detail?
5. Planned a strong, forceful conclusion?
6. Clearly determined the conclusions and recommendations, if any, that should be presented?
7. Settled on a functional format: headings, subheadings, illustrations, and other details?

In writing, have you . . .
1. Expressed yourself in language that conveys exactly what you want to say?
2. Used language that is adapted to the principal readers?
3. Used the fewest possible words consistent with clearness, completeness, and courtesy?
4. Achieved the tone calculated to bring about the desired response?

5. Tried to produce a style that is not only accurate, clear, and convincing, but also readable and interesting?
6. Presented all the pertinent facts and commented on their significance where necessary?
7. Made clear to the reader what action you recommend and why?
8. Correlated illustrations and art work closely with text?

In reviewing and revising, have you . . .
1. Fulfilled your purpose in terms of the readers' needs and desires?
2. Proofread painstakingly for errors in grammar, punctuation, and spelling?
3. Weeded out wordy phrases, useless words, overworked expressions?
4. Broken up unnecessarily long sentences?
5. Checked to see if headings serve as useful labels of the subject matter treated?
6. Deleted words and phrases that might be antagonistic?
7. Honestly judged whether your choice of words will be clear to the reader?
8. Checked whether transitions are clear?
9. Double-checked to see that the introduction sets forth clearly the purpose, scope, and plan of the presentation?

Finally, have you . . .
1. Finished the presentation on time?
2. Produced a piece of writing you can be proud of?

Illustrative Material

The following material is a series of pages taken from a library research report written by Curtis W. Jones, Jr., while a student, and reprinted here by his permission. The estimated length of the report was about 5000 words. To avoid confusion with the paging of our own text, we have deleted the page numbers from the table of contents. We are including only pages having some element of special interest as a possible model, such as the title page and a page showing how subheads were managed. The table of contents will indicate what has been omitted.

Report on

<u>FOUR-CHANNEL STEREO</u>

Submitted to
Dr. John Doe
for
English 317
University of Texas
Austin, Texas
April 23, 1976

by

Curtis W. Jones, Jr.

9008 Laurel Grove
Austin, Texas 78758
April 23, 1976

Dr. John Doe
Department of English
The University of Texas
Austin, Texas 78712

Dear Dr. Doe:

In accordance with your instructions, I have prepared the following report, entitled <u>Four-Channel Stereo</u>.

The purpose of this report is to explain the basic principles of four-channel stereo. It is assumed that the reader is familiar with stereo and that the reader has a general technical background. A short history and some general background are given. The report then explains the various systems used to reproduce four-channel sound. The applications of these systems in recorded material and broadcasting are then investigated. The main body of the report closes with a general summary.

I sincerely hope that this report will prove to be satisfactory.

Respectfully yours,

Curtis W. Jones, Jr.

Curtis W. Jones, Jr.

TABLE OF CONTENTS

ABSTRACT

Four-channel stereo was introduced in 1969. By 1970,
several major manufacturers were producing systems. Three
basic systems are used for four-channel sound. "Derived"
four-channel enhances stereo records. Matrix systems mix
the input channels to produce output. Discrete systems
produce four "discrete" channels.

Tapes were introduced quickly in open-reel and eight-
track formats, but cassette tapes have not been used because of a strict compatibility requirement. Records using
the matrix format were soon marketed, but producing discrete records was difficult.

Adapting the matrix system to FM broadcasting was
easy. Discrete broadcasting, however, involves FCC rule
changes and appears to be far in the future.

Report on

FOUR-CHANNEL STEREO

I. Introduction

A giant stride toward the goal of high quality sound
reproduction was taken in the 1950's when stereo was invented.[1] A big improvement over monophonic, or one-speaker
sound, stereo nevertheless did not satisfy everyone. Skeptics
pointed out that in live performances, at a concert, the
sound comes to the listener from all directions. Some
sounds reflect off the walls and reach the listener later
than other sounds. The desire to "surround" the listener
with sound led to the development of four-channel, or
quadraphonic stereo.

The purpose of this report is to explain the basic
principles of four-channel stereo. Discussion will be
centered on the basic systems and will be expanded to consider the applications of each system in reproducing
sound.

The five sections that follow consider (1) the history
of four-channel stereo, (2) some important background information, (3) the basic systems that have been developed, (4)
the applications of the basic systems in reproducing sound by
using recorded materials, and (5) the FM broadcasting of four-
channel sound.

V. RECORDED FOUR-CHANNEL MATERIAL

Four-channel would of course be useless without the availability of recorded material. As with stereo, tapes and records are the means of getting recorded material to the listener. In addition to these two methods of record- ing quadraphonic sound, headphones will be discussed in this section because they are a major part of the audio market and because it may be surprising that headphones can reproduce four-channel sound.

Tapes

Tape was the logical first choice for the repro- duction of quadraphonic sound. Tape with up to 24 tracks had been in use for years, and could easily produce four discrete channels.[19] Four-channel tapes can best be un- derstood by comparing them with their stereo counterparts.

The discussion that follows is divided into the three categories in which tapes are marketed. These are open-reel, eight-track, and cassette.

Open-Reel Tapes. Open-reel is the tape format used by professional recording studios. The 24-track machines men- tioned earlier are open-reel units.

<u>NOTES</u>

[1] Sessions, "Four-Channel Stereo From Source to Sound," p. 15.

[2] Petras, "Four Channels on a Disc. CBS-Sony SQ Matrix," p.25.

[3] Feldman, "Four-Channel Firms Up," p. 41.

[4] Petras, p. 25.

[5] Maynard, "Four-Channel Sound Today," p. 35.

[6] Fantel, "Four-Channel Stereo--Here at Last," p. 69.

[7] Maynard, "Experiment With Four-Channel Stereo," p. 34.

[8] Sessions, pp. 29-32.

[9] Maynard, "Four-Channel Sound Today," p. 34.

[10] Shorter, "Four-Channel Stereo; An Introduction to Matrixing," p.3.

[11] Petras, p. 27.

[12] Feldman, "The Evolution of Four-Channel Equipment," p. 26.

[13] Sessions, p. 96.

[14] Shorter, "Four-Channel Stereo; Some Commercial Quadraphonic Matrix Systems," p. 54.

[15] Skilling, "Electric Networks," p. 296.

[16] Sessions, p. 85.

[17] Fantel, p. 69.

[18] Sessions, p. 81.

[19] Feldman, "Four-Channel Firms Up," pp. 41-42.

[20] Feldman," Status Report: Four-Channel Tape Machines," p. 34.

[21] Ibid., p. 34.

[22] Ibid., p. 40.

[23] Ibid.,p. 41.

[24] Ibid., p. 35.

[25] Ibid., p. 35.

[26] Ibid., p. 40.

[27] Sessions, p. 96.

[28] Sessions, p. 65.

[29] Savon, "How It Works: Discrete Four-Channel With CD-4 Discs," p. 89.

[30] Sessions, p. 85.

[31] Jurgen, "Untangling the 'Quad' Confusion," p. 57.

[32] Sessions, p. 87.

[33] Jurgen, p. 57.

[34] Feldman, "Four-Channel Firms Up," p. 45.

[35] Salm, "New Four-Channel Stereo Techniques," p. 86.

[36] Sessions, p. 126.

[37] Haskett, "The Truth About FM," pp. 36-37.

[38] Haskett, p. 37.

[39] Haskett, p. 38.

[40] Salm, p. 35.

[41] "Four-Channel Stereo," p. 52.

[42] Haskett, p. 40.

[43] Tillett, "Quadraphonic Progress Report," p. 20.

[44] "Petition for Quadcasts Awaits FCC Decision," p. 34.

[45] Tillett, p. 21.

[46] Salm, p. 35.

[47] Salm, p. 35.

[48] Carey and Sager, "Quadraphonic Broadcasting--Current Proposals and the Way Ahead," p. 422.

[49] Meute, "Four-Channel Stereo FM--From One Station," p. 73.

APPENDIX

BIBLIOGRAPHY

Carey, Michael J., and John C. Sager. "Quadraphonic
    Broadcasting--Current Proposals and the Way Ahead."
    Wireless World, 80 (November, 1974), 422-425.

Fantel, Hans. "Four-Channel Stereo--Here at Last." Popular
    Mechanics, 136 (August, 1971), 66-70+.

Feldman, Leonard. "The Evolution of Four-Channel Equipment."
    Audio, 57 (July, 1973), 26+.

-----. "Four-Channel Firms Up." Popular Electronics, 4
    (December, 1973), 40-45.

-----. "Methods of Matrixing for Four-Channel Sound." Popular
    Electronics, 3 (January, 1973), 26-31.

-----. "On Matrix Quadraphonic Systems." Audio, 55 (October,
    1971), 20-22+.

-----. "Status Report: Four-Channel Tape Machines." Popular
    Electronics, 5 (July, 1974), 33-41.

-----. "Why the Four-Channel War Need Not Take Place." Audio,
    56 (July, 1972), 30+.

"Four-Channel Stereo." High Fidelity and Musical America,
    21 (January, 1971), 48-54.

Friedman, Herb. "What's Wrong With Four-Channel." Radio-
    Electronics, 47 (March, 1976), 76-77.

Haskett, Thomas R. "The Truth About FM." Electronics
    World, 82 (August, 1969), 37-40+.

Jurgen, Ronald K. "Untangling the 'Quad' Confusion." IEEE
    Spectrum, 9 (July, 1972), 55-62.

Maynard, Harry E. "Experiment With Four-Channel Stereo."
    Radio-Electronics, 42 (March, 1971), 33-38.

-----. "Four-Channel Sound Today." Radio-Electronics,
    42 (October, 1971), 33-36.

Meute, Jason P. "Four-Channel Stereo FM--From One Station."
    High Fidelity and Musical America, 20 (March, 1970), 72-73.

"Petition for Quadcasts Awaits FCC Decision." Electronics,
    44 (September 13, 1971), 34+.

Petras, Fred. "Four Channels on a Disc. CBS-Sony SQ Matrix."
    Radio-Electronics, 43 (March, 1972), 25-27.

Salm, Walter G. "New Four-Channel Stereo Techniques." Radio-
    Electronics, 41 (October, 1970), 33-35.

Savon, Karl. "How It Works: Discrete Four-Channel with CD-4
    Discs." Radio-Electronics, 45 (October, 1974), 36-38.

Sessions, Ken W. Jr. Four-Channel Stereo From Source to Sound.
    Blue Ridge Summitt, Pa.: Tab Books, 1973.

Shorter, Geoffrey. "Four-Channel Stereo. An Introduction
    to Matrixing." Wireless World, 78 (January, 1972), 2-5.

-----. "Four-Channel Stereo. Some Commercial Quadraphonic
    Matrix Systems." Wireless World, 78 (February, 1972), 54-57.

Skilling, Hugh H. Electric Networks. New York: John Wiley &
    Sons, 1974.

"SQ Eyes the 45 Market." High Fidelity and Musical America,
    24 (February, 1974), 31.

Tillett, George W. "Quadraphonic Progress Report." Audio,
    58 (July, 1974), 20-25.

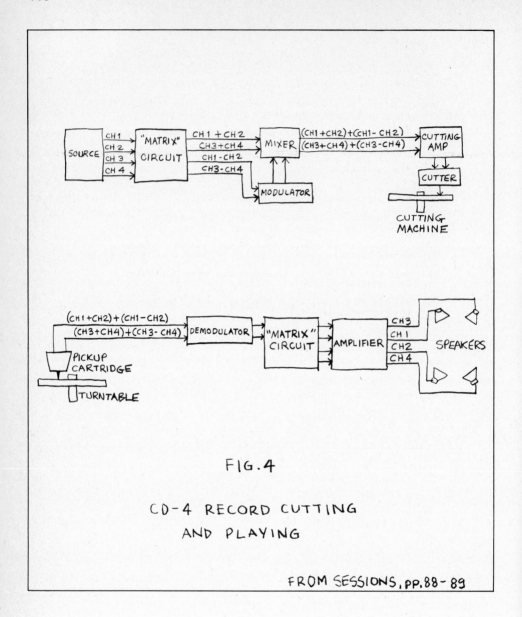

FIG. 4

CD-4 RECORD CUTTING
AND PLAYING

FROM SESSIONS, pp. 88-89

Suggestions for Writing

A choice of topic for a library research report should be made in accordance with principles discussed in this chapter, the resources of the library accessible to you, and the advice and approval of your instructor. The following list of possible topics for investigation is not intended to be limiting; we hope, rather, that it will be suggestive. We might also point out that many of the listed topics may need to be modified or limited so as to permit a full treatment of the subject in the length your instructor specifies. We have found, incidentally, that subjects of current and timely interest are often best, since most of the available information on them is likely to be found in periodicals (subjects about which entire books have been written are usually too broad to be dealt with adequately in a paper of the length normally specified).

Alcoholism and Low Blood Sugar
Problems of Air Traffic Control
Odor Control in Air Conditioning
Anti-Air Pollution Devices for Automobiles
New Developments in Electrically Powered Automobiles
Concrete Additives
Automobile Safety Devices and Standards
The Mechanical Heart: Current Developments
Marijuana: Physiological Effects
Microphotography
Miniature Television Cameras
Microminiature Integrated Circuits
Industrial Water Pollution by Radioactive Materials: Developments in Combating
Hallucinogenic Drugs
Forest Fire Prevention and Control
New Techniques in Desalination
Practical Applications of Time-Sharing Computer Systems
The Use of Lasers in Medicine
Ultrasonics in Machining
Teaching Machines
Food for Astronauts
Disc Brakes for Automobiles
New Techniques in Flow Metering
Cryogenics in Hydrocarbon Processing
Oyster Farming
Alluvial Gold Deposits: How to Discover
Torque Control in Gas Turbines
Experimental Living in the Sea
Clear-Cutting in Forests: Benefits and Hazards
Polymer-Impregnated Concrete
Designing an Insurance Program for an Individual
Word Processers

Environmental Effects of Salination of Soils by Irrigation
Laying Deep-Water Pipelines
Layout Technique in Advertising
Dolphin Research
High-Fidelity Component Compatibility
Tape Recorder Care and Maintenance
Computer Applications in _____ (select a field)
Government Specifications for Manual Writing
Information Retrieval by Machine

Appendixes

Appendix A

A Selected Bibliography

Grammar, Usage, and Style

Bernstein, Theodore M. *Miss Thistlebottom's Hobgoblins.* New York: Farrar, Straus, 1971.

Bryant, Margaret M. *Current American Usage.* New York: Funk & Wagnalls, 1962.

Condon, John C. *Semantics and Communication.* New York: Macmillan, 1966.

DeGeorge, James, Gary A. Olson, and Richard Ray. *Style and Readability in Technical Writing.* New York: Random House, 1984.

Estrin, Herman A., and Donald V. Mehus. *The American Language in the 70's.* San Francisco: Boyd & Fraser, 1974.

Evans, Bergen and Cornelia. *A Dictionary of Contemporary American Usage.* New York: Random House, 1957.

Flesch, Rudolf. *The ABC of Style — A Guide to Plain English.* New York: Harper & Row, 1964.

Follett, Wilson. *Modern American Usage.* New York: Hill & Wang, 1966.

Fowler, H. W. *A Dictionary of Modern English Usage,* 2nd ed. Rev. by Sir Ernest Gowers. New York: Oxford. 1965.

Graves, Robert, and Alan Hodge. *The Reader Over Your Shoulder: A Handbook for Writers of English Prose,* 2nd ed. New York: Vintage Books, 1979.

Lanham, Richard A. *Revising Prose.* New York: Charles Scribner's Sons, 1979.

Lanham, Richard A. *Style: An Anti-Textbook.* New Haven and London: Yale University Press, 1974.

Mager, N. H., and S. K. Mager. *Encyclopedia of English Usage.* Englewood Cliffs, NJ: Prentice-Hall, 1974.

McCartney, E. S. *Recurrent Maladies in Scholarly Writing.* Ann Arbor: University of Michigan Press, 1969.

Morris, William, and Mary Morris (Eds.). *Harper Dictionary of Contemporary Usage.* New York: Harper & Row, 1975.

Newman, Edwin. *Strictly Speaking.* New York: Bobbs-Merrill, 1974.

Nicholson, Margaret. *American English Usage.* London: Oxford, 1957. (Also New American Library paperback.)

Partridge, Eric. *Concise Usage and Abusage.* New York: Philosophical Library, 1955.

Perrin, Porter G., and W. R. Ebbitt. *Writer's Guide and Index to English.* 7th ed. Chicago: Scott, Foresman, 1981.

Shaw, Fran W. *30 Ways to Help You Write.* New York: Bantam Books, 1980.

Strunk, William, and E. B. White. *The Elements of Style,* 3rd ed. New York: Macmillan, 1979.

Trimble, John. *Writing with Style.* Englewood Cliffs, NJ: Prentice-Hall, 1975.

Turner, Rufus P. *Grammar Review for Technical Writers.* Reprint of 1964 ed. Melbourne, FL: Robert E. Krieger Publishing Co., 1981.

Williams, Joseph M. *Style.* Glenview, IL: Scott, Foresman, 1981.

Wood, Frederick T. *Current English Usage.* London: Macmillan, 1962.

Zinsser, William. *On Writing Well.* New York: Harper & Row, 1976.

On Simplifying Style

Bernstein, Theodore M. *The Careful Writer.* New York: Atheneum, 1965.

Flesch, Rudolf. *How to Say What You Mean in Plain English.* New York: Barnes & Noble, 1974.

Gowers, Sir Ernest. *The Complete Plain Words.* Baltimore: Penguin, 1975.

Gunning, Robert. *Technique of Clear Writing,* rev. ed. New York: McGraw-Hill, 1968.

Ryckman, W. G. *What Do You Mean By That? The Art of Speaking and Writing Clearly.* Homewood, IL: Dow Jones-Irwin, 1980.

Technical Publications Problems

Adamson, Rhomas A. *Inside Grant & Project Writing,* ed. Connie L. Pavlina. Salinas, CA: PAM Pubs., 1979.

Caird, Ken. *Cameraready.* Pasadena, CA: Cameraready Corporation, 1973.

Clarke, Emerson. *A Guide to Technical Literature Production.* River Forest, IL: TW Publishers, 1961.

Clarke, Emerson. *How to Prepare Effective Engineering Proposals.* River Forest, IL: TW Publishers, 1962.

Cremmins, Edward T. *The Art of Abstracting.* Philadelphia: ISI Press, 1982.

Cunningham, Donald H. *Creating Technical Manuals.* New York: McGraw-Hill, 1984.

Davis, Richard M. *Thesis Projects in Science and Engineering: A Complete Guide from Problem Selection to Final Presentation.* New York: St. Martin's Press, 1980.

Felker, Daniel B. (Ed.). *Document Design: A Review of the Relevant Research.* Washington: American Institute for Research, 1980.

Felker, Daniel B., et al. *Guidelines for Document Designers.* Washington: The Document Design Center, 1981.

Holtz, Herman, and Terry Schmidt. *The Winning Proposal: How to Write It.* New York: McGraw-Hill, 1981.

Jordan, Stello (Ed.). *Handbook of Technical Writing Practices,* 2 vols. New York: Wiley-Interscience, 1971.

Lehman, Maxwell (Ed.). *Communication Technologies and Information Flow.* Elmsford, NY: Pergamon Press, Inc., 1981.

Leiblich, Jerome H. (Ed.). *Instructions for the Preparation of Specifications, Standards & Technical Manuals.* Santa Ana, CA: Global Engineering, 1981.

Mandel, Siegfried, and D. L. Caldwell. *Proposal and Inquiry Writing.* New York: Macmillan, 1962.

Mandel, Siegfried, and D. L. Caldwell. *Writing for Science and Technology.* New York: Dell, 1970.

Melcher, Daniel, and Nancy Larrick. *Printing and Promotion Handbook,* 3d ed. New York: McGraw-Hill, 1966.

Reisman, S. J. (Ed.). *A Style Manual for Technical Writers and Editors.* New York: Macmillan, 1962.

Rogers, Raymond A. *How to Report Research & Development Findings to Management.* New York: Pilot Books, 1973.

Sachs, Harley. *How to Write the Technical Article & Get It Published.* Washington: Society for Technical Communication, 1976.

Teaching Technical Writing and Editing: In-House Programs That Work. Washington: Society for Technical Communication, 1976.

Technical Editing: Principles and Practices. Washington: Society for Technical Communication, 1975.

Walton, Thomas F. *Technical Manual Writing and Administration.* New York: McGraw-Hill, 1968.

Weil, B. H. (Ed.). *Technical Editing.* Westport, CN: Greenwood, 1975.

Whalen, Tim. *Preparing Contract Winning Proposals and Feasibility Studies.* New York: Pilot Books, 1982.

On Graphic Aids

Austin, Richard. *Report Graphics: Writing the Design Report.* New York: Van Nostrand Reinhold, 1983.

Beakley, George C., and Donald D. Autore. *Technical Illustration.* Indianapolis, IN: Bobbs-Merrill, 1983.

Beasly, David. *Design Illustration.* Exeter, NH: Heinemann Educational Books, 1979.

Bethune, James D. *Technical Illustration.* New York: Wiley, 1983.

Buil, Robert L. *Technical Illustration.* Los Altos, CA: W. Kaufman, 1984.

Cardamone, Tom. *Chart and Graph Preparation Skills.* New York: Van Nostrand Reinhold, 1981.

Dezart, Louis. *Drawing for Publication: A Manual for Technical Illustrators.* New York: State Mutual Book and Periodical Service, 1981.

Dicerto, J. J. *Planning and Preparing Data-Flow Diagrams.* New York: Hayden, 1964.

French, Thomas E., and Charles J. Vierck. *Engineering Drawing and Graphic Technology,* 12th ed. New York: McGraw-Hill, 1978.

Gibby, J. C. *Technical Illustration,* 3rd ed. Chicago: American Technical Society, 1969.

Gieseke, Frederick, et al. *Technical Drawing,* 5th ed. New York: Macmillan, 1967.

Gootschall, Edward M. *Graphic Communication '80's.* Englewood Cliffs, NJ: Prentice-Hall, 1981.

Graf, Rudolf F., and George J. Whalen. *How It Works, Illustrated.* New York: Popular Science/ Outdoor Life Book Division, Times Mirror Magazines, Inc., 1974.

Harvill, Lawrence R., and Thomas L. Kraft. *Technical Report Standards: How to Prepare and Write Effective Technical Reports.* Beaverton, OR: M/A Press, 1979.

Hicks, G. A. *Modern Technical Drawing,* 2 vols. New York: Pergamon, 1967–1968.

Hoelscher, R. P., C. H. Springer, and J. S. Dobrovolny. *Graphics for Engineers.* New York: Wiley, 1968.

Lefferts, Robert. *Elements of Graphics: How to Prepare Charts and Graphs for Effective Reports.* New York: Harper & Row, 1981.

On Technical and Scientific Reports

Blicq, Ron S. *Guidelines for Report Writers: A Complete Manual for on-the-job Report Writing.* Englewood Cliffs, NJ: Prentice-Hall, 1982.

Brown, John F. *A Student Guide to Engineering Report Writing.* Solona Beach, CA: John Fiske Brown Associates, 1982.

Brunner, Ingrid, J. C. Mathes, and Dwight Stevenson. *The Technician as Writer: Preparing Technical Reports.* Indianapolis: Bobbs-Merrill, 1980.

Damerst, William A. *Clear Technical Reports.* New York: Harcourt, 1972.

Gallegher, William J. *Writing the Business and Technical Report.* Boston: CBI Publishing Co., 1981.

Graves, Harold F., and Lyne S. S. Hoffman. *Report Writing,* 4th ed. Englewood Cliffs, NJ: Prentice-Hall, 1965.

Gray, Dwight E. *So You Have to Write a Technical Report.* Washington, DC: Information Resources Press, 1970.

Harkins, Craig, and Daniel Plung (Eds.). *A Guide for Writing Better Technical Papers.* New York: IEEE Press, 1982.

Holscher, Harry H. *How to Organize and Write a Technical Report.* Totowa, NJ: Littlefield, Adams, 1965.

Jones, W. Paul, and Michael Keene. *Writing Scientific Papers and Reports,* 8th ed. Dubuque, IW: Wm. C. Brown, 1981.

MacKenzie, Raymond N., and William E. Evans. *Technical Writing: Forms and Formats.* Dubuque, IW: Kendall/Hunt Publishing Co., 1982.

Mathes, J. C., and Dwight Stevenson. *Designing Technical Reports.* Indianapolis: Bobbs-Merrill, 1976.

Michaelson, Herbert B. *How to Write and Publish Engineering Papers and Reports.* Philadelphia: ISI Press, 1982.

Mount, Ellis (Ed.). *Technical Reports: Their Role in Sci-Tech Libraries.* New York: Haworth Press, 1981.

Pauley, Steven. *Technical Report Writing Today,* 2nd ed. Boston: Houghton Mifflin, 1979.

Rook, Fern. *How to Prepare a Science Project Report, Write a Research Paper, Format a Report.* Phoenix, AZ: Fern Rook, 1982.

Sawyer, T. S. *Specification and Engineering Writer's Manual.* Chicago: Nelson Hall, 1960.

Schmidt, Steven. *Creating the Technical Report.* Englewood Cliffs, NJ: Prentice-Hall, 1983.

Souther, James W. *Technical Report Writing,* 2nd ed. New York: Wiley, 1976.

Swanson, Richard. *For Your Information: A Guide to Writing Reports.* Englewood Cliffs, NJ: Prentice-Hall, 1974.

Ulman, Joseph N., and Jay R. Gould. *Technical Reporting,* 3rd ed. New York: Holt, Rinehart and Winston, 1972.

Weiss, Edmond H. *The Writing System for Engineers and Scientists.* Englewood Cliffs, NJ: Prentice-Hall, 1982.

Wilkinson, C. W., Peter B. Clarke, and Dorothy C. M. Wilkinson. *Communicating Through Letters and Reports,* 7th ed. Homewood, IL: Richard D. Irwin, Inc., 1980.

On Scientific and Technical Writing (Not restricted to reports)

Agnos, Thomas J., and Stanley Schatt. *The Practical Law Enforcement Guide to Writing Field Reports, Grant Proposals, Memos, and Resumes.* Springfield, IL: Charles C Thomas, 1980.

Alvarez, Joseph A. *The Elements of Technical Writing.* New York: Harcourt Brace Jovanovich, Inc., 1980.

Andrews, Clarence. *Technical and Business Writing.* New York: Houghton Mifflin, 1975.

Andrews, Deborah C., and Margaret D. Blickle. *Technical Writing: Principles and Forms,* 2nd ed. New York: Macmillan, 1982.

Barnett, Marva. *Elements of Technical Writing,* 2nd ed. Albany, NY: Delmar, 1982.

Barnett, Marva. *Writing for Technicians,* rev. ed. Albany, NY: Delmar, 1982.

Berry, Dorothea M., and Jordan P. Martin. *A Guide to Writing Research Papers.* New York: McGraw-Hill, 1971.

Bingham, Earl G. *Pocketbook for Technical and Professional Writers.* Belmont, CA: Wadsworth Publishing Co., 1982.

Blicq, Ron S. *Technically — Write! Communicating in a Technological Era,* 2nd ed. Englewood Cliffs, NJ: Prentice-Hall, 1981.

Bly, R. W., and G. Blake. *Technical Writing: Structure, Standards and Style.* New York: McGraw-Hill, 1982.

Brand, Norman, and John O. White. *Legal Writing.* New York: St. Martin's Press, 1976.

Brogan, John A. *Clear Technical Writing.* New York: McGraw-Hill, 1973.

Brusaw, Charles T., Gerald T. Alred, and Walter E. Olie. *Handbook of Technical Writing,* 2nd ed. New York: St. Martin's Press, 1982.

Burkett, W. D. *Writing Science News for the Mass Media,* 2nd ed. Houston: Gulf Publishing Co., 1974.

Campbell, John M., and G. L. Farrar. *Effective Communication for the Technical Man.* Tulsa: Petroleum Publishing Co., 1972.

Colmann H., and A. Barneas. *Writing Medical Papers.* London: Heinemann, 1974.

Cunningham, Donald H. *A Reading Approach to Professional Police Writing.* Springfield, IL: Charles C Thomas, 1972.

Davidson, H. A. *Guide to Medical Writing.* New York: Ronald, 1957.

Day, Robert A. *How to Write and Publish a Scientific Paper,* 2nd ed. Philadelphia: ISI Press, 1983.

Delaware Technical and Community College, English Department. *Writing Skill for Technical Students.* Englewood Cliffs, NJ: Prentice-Hall, 1982.

Dodds, Robert H. *Writing for Technical and Business Magazines.* Melbourne, FL: Robert E. Kreiger Publishing, 1982.

Durenberger, Robert W. *Geophysical Research and Writing.* New York: Crowell, 1971.

Ehrlich, Eugene, and Daniel Murphy. *The Art of Technical Writing.* New York: Bantam Books, 1969.

Eisenberg, Anne. *Effective Technical Communication.* New York: McGraw-Hill, 1982.

Estrin, Herman A. *Technical and Professional Writing.* New York: Preston, 1976.

Evans, John. *Beginners Guide to Technical Writing.* New York: Focal Press, 1983.

Ewer, J. R., and O. A. Latorre. *A Course in Basic Scientific English.* New York: Longman, 1975.

Ewing, David W. *Writing for Results,* 2nd ed. New York: Wiley, 1979.

Fallon, William K. (Ed.) *Effective Communication on the Job,* 3rd ed. New York: AMACOM, 1981.

Fear, David E. *Technical Writing,* 2nd ed. New York: Random House, 1978.

Fishbein, Morris. *Medical Writing, the Technic and the Art,* 4th ed. New York: C. C. Thomas, 1972.

Fox, Rodney. *Agricultural and Technical Journalism.* Westport, CN: Greenwood, 1952.

Freeman, Joanna M. *Basic Technical and Business Writing.* Ames, IA: Iowa State University Press, 1979.

Gensler, Walter J., and K. D. Gensler. *Writing Guide for Chemists.* New York: Harcourt, 1961.

Grantham, Donald J., et al. *Technical Communication.* Los Angeles: GSE Publications, 1975.

Grimm, Susan J. *How to Write Computer Manuals for Users.* Belmont, CA: Lifetime Learning, 1982.

Harris, John S., and Blake H. Reed. *Technical Writing for Social Scientists.* Chicago: Nelson-Hall, 1976.

Hays, Robert. *Principles of Technical Writing.* Reading, MA: Addison-Wesley, 1965.

Herbert, A. J. *The Structure of Technical English.* New York: Longman, 1975.

Higginson, Eric P. (Series ed.). *Writing Skills for Technical Students.* Englewood Cliffs, NJ: Prentice-Hall, 1982.

Hirschorn, Howard H. *Writing for Science, Industry and Technology.* New York: Van Nostrand Reinhold, 1980.

Hoover, Hardy. *Essentials for the Technical Writer,* 2nd ed. New York: Dover Publications, 1981.

Houp, K. W., and T. E. Pearsall. *Reporting Technical Information,* 5th ed. Encino, CA: Glencoe Publishing, 1984.

Huber, Jack T. *Report Writing in Psychology and Psychiatry.* New York: Harper & Row, 1961.

Kapp, Reginald O. *The Presentation of Technical Information.* New York: Macmillan, 1948.

King, Lester S., and Charles G. Roland. *Scientific Writing.* Chicago: American Medical Association, 1968.

King, Lester S. *Why Not Say It Clearly: A Guide to Scientific Writing.* Boston: Little, Brown, 1978.

Kolin, Philip C. *Successful Writing at Work.* Lexington, MA: D. C. Heath and Co., 1982.

Lannon, John M. *Technical Writing,* 3rd ed. Boston: Little, Brown, 1985.

Laster, Ann A., and Nell Ann Pickett. *Writing for Occupational Education.* San Francisco: Canfield Press, 1974.

Lawrence, Nelda R., and Elizabeth Tibeaux. *Writing Communications in Business and Industry,* 3rd ed. Englewood Cliffs, NJ: Prentice-Hall, 1982.

Levine, Norman. *Technical Writing.* New York: Harper & Row, 1974.

Long, Sandra Salser. *Transmission: Communication Skills for Technicians.* Reston, VA: Reston Publishing Co., 1980.

Lynch, Barbara S., and Charles F. Chapman. *Writing for Communication in Science and Medicine.* New York: Van Nostrand Reinhold, 1980.

Maimon, Elaine P., et al. *Writing in the Arts and Sciences.* Cambridge, MA: Winthrop Publishers, 1981.

Mandel, Siegfried. *Writing for Science and Technology.* New York: Dell Publishing Co., 1970.

A Manual for Authors of Mathematical Papers. Providence, RI: American Math Society, 1979.

Markel, Michael H. *Technical Writing: Situations and Strategies.* New York: St. Martin's Press, 1984.

Mehaffy, Robert E. *Writing for the Real World.* Glenview, IL: Scott, Foresman and Co., 1980.

Menzel, Donald H. *Writing a Technical Paper.* New York: McGraw-Hill, 1961.

Miles, J., D. Bush, and A. Kaplan. *Technical Writing: Principles and Practice.* Chicago: Science Research Association, 1982.

Mirin, Susan Kooperstein. *The Nurse's Guide to Writing for Publication.* Wakefield, MA: Nursing Resources, 1981.

Mitchell, John. *Handbook of Technical Communication.* Belmont, CA: Wadsworth, 1962.

Mitchell, John. *Writing for Technical and Professional Journals.* New York: Wiley, 1968.

Morris, Jackson E. *Principles of Scientific and Technical Writing.* New York: McGraw-Hill, 1966.

Morrisey, George L. *Effective Business and Technical Presentations,* 2nd ed. Reading, MA: Addison-Wesley, 1975.

Moyer, R., E. Stevens, and R. Switzer. *The Research and Report Handbook — For Business, Industry, and Government.* New York: Wiley, 1981.

Mullins, Carolyn J. *The Complete Writing Guide to Preparing Reports, Proposals, Memos, Etc.* Englewood Cliffs, NJ: Prentice-Hall, 1980.

Olsen, Leslie A., and Thomas N. Huckin. *Principles of Communication for Science and Technology.* New York: McGraw-Hill, 1983.

Pearsall, Thomas, and Donald Cunningham. *How to Write for the World of Work,* 2nd ed. New York: Holt, Rinehart and Winston, 1982.

Pickett, Nell Ann. *Practical Communication.* New York: Harper's College Press, 1975.

Pickett, Nell Ann, and Ann A. Laster. *Technical English: Writing, Reading, and Speaking.* New York: Harper & Row, 1984.

Rathbone, R. R. *Communicating Technical Information.* Reading, MA: Addison-Wesley, 1966.

Reeder, Robert C. *The Sourcebook of Medical Communications.* St. Louis, MO: C. V. Mosby Co., 1981.

Rickard, T. A. *Guide to Technical Writing.* San Francisco: Mining and Scientific Press, 1908.

Roman, Kenneth, and Joel Raphaelson. *Writing that Works: How to Write Memos, Letters, Reports, Resumes, and Other Papers that Say What You Mean, and Get Things Done.* New York: Harper & Row, 1981.

Roundy, Nancy L., with David Mair. *Strategies for Technical Communication.* Boston: Little, Brown, 1985.

Sherman, T. A., and Simon Johnson. *Modern Technical Writing,* 3rd ed. Englewood Cliffs, NJ: Prentice-Hall, 1975.

Smock, Winston. *Technical Writing for Beginners.* Englewood Cliffs, NJ: Prentice-Hall, 1983.

Spitzer, Michael, Michael Gamble, and Teri Kwal Gamble. *Writing and Speaking in Business.* New York: Random House-Alfred A. Knopf, Inc., 1984.

Stratton, Charles R. *Technical Writing: Process and Product.* New York: Holt, Rinehart and Winston, 1984.

Strong, Charles William, and Donald Edison. *A Technical Writer's Handbook.* New York: Holt, Rinehart and Winston, 1971.

Tichy, H. J. *Effective Writing for Engineers, Managers, Scientists.* New York: Wiley, 1966.

Tracy, C. R., and H. L. Jennings. *Writing for Industry.* Chicago: American Technical Society, 1974.

Trelease, Sam F. *Scientific and Technical Papers.* Cambridge, MA: M.I.T. Press, 1968.

Van Duyn, Julia. *The DP Professional's Guide to Writing Effective Technical Communication.* New York: Wiley, 1982.

Ward, Ritchie R. *Practical Technical Writing.* New York: Knopf, 1968.

Warren, Thomas L. *Technical Writing: Purpose, Process, and Form.* Belmont, CA: Wadsworth, 1985.

Weisman, H. M. *Basic Technical Writing,* 4th ed. Columbus, OH: Charles E. Merrill, 1980.

Woodford, F. Peter, ed. *Writing Scientific Papers in English.* New York: Elsevier, 1975.

Some Style Manuals

Keithley, Erwin M., and Philip J. Schreiner. *A Manual of Style for the Preparation of Papers and Reports: Business and Management Applications,* 3rd ed. Cincinnati, OH: Southwestern Publishing Co., 1980.

Kirkman, John. *Good Style: For Scientific and Engineering Writing.* London: Pitman Publishing, 1980.

Markus, John. *Electronics Style Manual.* New York: McGraw-Hill, 1978.

Messer, Ronald K. *Style in Technical Writing: A Text/Workbook.* Glenview, IL: Scott, Foresman, 1982.

Olsen, Gary, and James DeGeorge. *Style in Technical Writing,* New York: Random House, 1983.

Stockwell, Richard E. *The Stockwell Guide for Technical and Vocational Writing,* 2nd ed. Reading, MA: Addison-Wesley, 1982.

On Oral Presentations

Anastasi, Thomas E., Jr. *Communicating for Results.* Menlo Park, CA: Cummings, 1972.

Block, Jack, and Joe Labonville. *English Skills for Technicians.* New York: McGraw-Hill, 1971.

Hand, Harry E. *Effective Speaking for the Technical Man: Practical Views and Comments.* New York: Van Nostrand Reinhold, 1969.

Howell, W. S., and E. G. Barmann. *Presentational Speaking for Business and the Professions.* New York: Harper & Row, 1971.

Manko, Howard H. *Effective Technical Speeches and Sessions: A Guide for Speakers and Program Chairmen.* New York: McGraw-Hill, 1969.

Weiss, Harold, and J. B. McGrath, Jr. *Technically Speaking: Oral Communication for Engineers, Scientists, and Technical Personnel.* New York: McGraw-Hill, 1963.

Wilcox, Roger P. *Communication at Work: Writing and Speaking.* Boston: Houghton Mifflin, 1977.

On Business Letters and Reports

Barr, Doris W. *Communication for Business, Professional, and Technical Students,* 2nd ed. Belmont, CA: Wadsworth, 1980.

Berenson, Conrad, and Raymond Colton. *Research and Report Writing for Business and Economics.* New York: Random House, 1971.

Brown, Leland. *Effective Business Report Writing,* 3rd ed. Englewood Cliffs, NJ: Prentice-Hall, 1973.

Brusaw, Charles, et al. *The Business Writer's Handbook,* 2nd ed. New York: St. Martin's Press, 1982.

Dawson, Presley C. *Business Writing: A Situational Approach.* Encino, CA: Dickenson, 1964.

Himstreet, William C., and Wayne B. Baty. *Business Communications: Principles and Methods,* 7th ed. Belmont, CA: Wadsworth, 1983.

Lesikar, Raymond V. *How to Write a Report Your Boss Will Read and Remember.* Homewood, IL: Dow Jones-Irwin, 1974.

Lindauer, J. *Writing in Business.* New York: Macmillan, 1971.

Markel, Michael H., and R. J. Lucier. *Make Your Point: A Guide to Improving Your Business and Technical Writing.* Englewood Cliffs, NJ: Prentice-Hall, 1983.

Schutte, W. M., and E. R. Steinberg. *Communication in Business and Industry.* Huntington, NY: Kreiger, 1974.

Sheppard, M. *Plain Letters.* New York: Simon and Schuster, 1960.

Shurter, B. L. *Effective Letters in Business,* 2nd ed. New York: McGraw-Hill, 1954.

Swindle, Robert. *The Concise Business Correspondence Style Guide: Developing Writing Skills to Get the Results You Want.* Englewood Cliffs, NJ: Prentice-Hall, 1983.

Scientific and Technical Exposition with Instructional Comment

Anderson, W. Steve, and Don Richard Cox. *The Technical Reader: Readings in Technical, Business and Scientific Communication.* New York: Holt, Rinehart and Winston, 1980.

Blickle, Margaret M., and M. E. Passe. *Readings for Technical Writers.* New York: Ronald, 1963.

Bowen, Mary Elisabeth, and Joseph A. Mazzeo (Eds.). *Writing about Science.* New York: Oxford University Press, 1979.

Eisenberg, Ann. *Reading Technical Books.* Englewood Cliffs, NJ: Prentice-Hall, 1978.

Harty, Kevin J. *Strategies for Business and Technical Writing.* New York: Harcourt, Brace, Jovanovich, 1980.

Journet, Debra, and Julie Lipick Kling. *Readings for Technical Writers.* New York: Scott, Foresman, 1984.

Kolin, Philip C., and Janeen L. Kolin. *Models for Technical Writing.* New York: St. Martin's Press, 1985.

Lay, Mary M. *Strategies for Technical Writing: A Rhetoric with Readings.* New York: Holt, Rinehart and Winston, 1982.

Leonard, David C., and Peter J. McGuire. *Readings in Technical Writing.* New York: Macmillan, 1983.

Lynch, Robert E., and Thomas B. Swanzey (Eds.). *The Example of Science: An Anthology for College Composition.* Englewood Cliffs, NJ: Prentice-Hall, 1981.

Sparrow, Keats, and Don H. Cunningham. *The Practical Craft: Readings for Business and Technical Writers.* Boston: Houghton Mifflin, 1978.

Appendix B

Grammar and Usage

Introduction

The following materials are included as a convenience, for use when needed, and not in any sense as a substitute for a good grammar handbook. Everyone, even professional writers, occasionally needs help with problems of grammar and usage. On the other hand, in this book we cannot possibly provide the full range of assistance found in a thorough and systematic treatment of these problems. We have therefore selected materials on the basis of three limiting principles.

One of these principles is that of frequency. In an extensive survey of errors in technical writing, we discovered a curious fact. Technical writers do not characteristically make a great variety of significant errors in grammar and usage, but those errors they do make they tend to repeat over and over. Accordingly, we have included discussion of the errors we found most frequently. These errors involve certain subject-verb relationships, vague pronoun reference, coordination and subordination, dangling modifiers, and lack of parallel structure.

A second principle of inclusion is that of what might be called nuisance value. Certain kinds of errors are distinctly annoying to readers. Although we found such errors to be less frequent than those noted above, they are nevertheless fairly common. Two examples are the comma blunder and the run-on sentence.

The third and last principle on which we based a choice of material is clarification of the concept of grammar. Since the purpose of this appendix is to describe the "rules" of grammar and usage, it seems to us important, especially for scientifically trained readers, to have some understanding of where the rules come from. Therefore, it is with a brief discussion of this third topic that we begin.

This appendix is divided into three parts. Directly following the brief discussion of the meaning of grammar and usage is a section on common errors in usage, and then of certain special problems.

PART 1: THE MEANING OF GRAMMAR AND USAGE

To the linguist, the terms "grammar" and "usage" both mean the systematic way in which language functions to convey meaning. The linguist studies the problem of how a complex of structural patterns, governing the forms of words and sentences, operates so that we can communicate with one another. In a strict — and limited — sense, grammar has little to do with "correctness." It does have to do with what is possible and what isn't. To the linguist, a statement like "I seen him when he done it" is grammatical because it functions in accordance with a recognized and accepted organization and pattern of words. This does not mean that the linguist approves the statement, of course; any linguist would prefer "I saw him when he did it," since the latter sentence is in accord with accepted usage among educated users of English. In an important sense, the modern grammarian operates like the pure scientist, observing facts about the way language functions, and noting "rules" that stem directly from the observations, not from preconceived notions of how the language *should* function.

During the past several decades, "scientific" observation of the ways in which language operates has brought about the development of several new grammars. These grammars are at odds with much that has traditionally been taught in the schools. The traditional, or school grammar, together with its pronouncements on usage and style, had its origin in the eighteenth century with the publication in England of such books as Bishop Robert Lowth's *A Short Introduction to English Grammar* (1762), and in America of Lindley Murray's *English Grammar* (1802). As classical scholars, Lowth and his followers took as their model for English grammar the Latin grammar with which they were familiar, and they formulated English grammar to correspond with that of Latin, ignoring what present-day students of language take as an essential starting place for the discovery of a language's grammar — the ways in which a particular language actually is spoken and written.

Beginning roughly with the work of Leonard Bloomfield, whose *Language* was published in 1933, linguists have tried to discover how English actually functions. One result of their work has been the development of several new grammars, or methods of describing how the language functions. Among these new grammars, two are most important, the "structuralist" and the "generative-transformational."

The methodology of the structuralist grammar begins with identification of the smallest meaningful unit of speech, the phoneme, and moves on to the morpheme (a meaningful sequence of sound signals or phonemes), and then to the sentence — or, as one well-known text put it in its subtitle, "From Sound to Sentence in English.''[1] The second of the new approaches to the development of English grammar — and the one engaging the support and interest of most present day students of language — is the generative-transformational. Unlike the structuralists, the transformational grammarian distinguishes the deep structure of a sentence from its surface structure, formulates phrase structure rules that account for the constituents of deep structures and that specify their relationships to each other, and stipulates transformational rules that transform deep structures into surface structures.

We are keenly aware of the superficiality of the foregoing sketch of new developments in grammar. If you are interested in looking into the subject further, we suggest that you read Norman Stageberg's *An Introductory English Grammar* (4th ed. 1981), C. L. Baker's *Introduction to Generative-Transformational Syntax* (1978), Andrew Radford's *Transformational Syntax* (1981), and Herndon's *A Survey of Modern Grammars* (2d ed. 1976). The last named book attempts to survey developments of the past few decades for those with no linguistic background.

The important consideration for us here is that the inquiries into the nature of English grammar of the past several decades have been accompanied by a fresh look at the matter of ''correctness'' and acceptability in usage. From the point of view of a modern linguist, the criterion for good English is not to be found in the older grammarians' set of prescriptions or rules but in the observed practices of successful writers and speakers. As Robert Pooley said, ''Good English is that form of speech which is appropriate to the purpose of the speaker, true to the language as it is, and comfortable to speaker and listener. It is the product of custom, neither cramped by rule nor freed from all restraint; it is never fixed, but changes with the organic life of the language.''[2]

PART 2: COMMON ERRORS IN USAGE

Subject-Verb Relationships

You don't need to be reminded that the subject of a sentence must agree with the verb in number. Seeing to it that they do agree, however, is another matter. The following are particularly troublesome kinds of subject-verb relationships.

Indefinite Subjects   When the subject is an indefinite word, commonly a pronoun, the subject is usually identified by a following prepositional phrase, and the number of the object of the preposition normally determines the number of the verb. That is, the sense of the statement governs agreement. Let's look at some examples.

> 1. Both of these power supplies *are* satisfactory. [But notice that we would write, ''Either of these power supplies *is* satisfactory.'']

[1] A. A. Hill, *Introduction to Linguistic Structures* (New York: Harcourt Brace Jovanovich, 1958).
[2] Robert Pooley, *Teaching English Usage* (New York: Appleton-Century-Crofts, 1946), p. 14. By permission of the National Council of Teachers of English.

2. *Everyone* in the organization *makes* a weekly progress report. [Several people are obviously involved, but "everyone"—like everything, anybody, anyone—takes a singular verb.]

3. *Half* of the units *were* faulty. [But we would write "Half of the trouble is the fault of the drafting department."]

4. *Some* of the units *have* been in service for ten months. [But "Some of the material is no good."]

Collectives   Words in this category may take either a singular or a plural verb, depending on the sense of the statement. In other words, if the individuals that comprise the collective term are thought of separately, the verb should be plural; if they are thought of as a group, the verb should be singular. Pronouns referring to such terms must also agree. Study the following sentences:

1. The *number* of reports lost last year *was* large.
2. The *majority were* between 1.5 and 2.5 mm long.
3. A *pair* of workmen *were* taking turns inspecting the units.
4. This *pair is* not as good as that.
5. A *number* of the electrodes *were* burnt.

Subjects of Amount   As with indefinite subjects and collectives, subject-verb agreement with terms denoting amounts is governed by the sense of a statement, though terms denoting sums, rates, measurements, and quantities more commonly take a singular verb, despite their plural form. For example:

1. One hundred dollars per hour is high pay.
2. A thousand miles an hour is too fast.
3. Thirty-six inches is a yard.
4. Last year about forty hours was spent on that report. [One writer told us that he would think about each of those forty hours separately—and painfully—and would therefore use a plural verb!]
5. About eighty pounds of carbon is added to the mix. [Here we would choose a plural verb if the carbon is added pound by pound; for the example we have assumed that an 80-pound *sack* of carbon has been dumped into the mix—hence, the singular verb.]

Correlatives   When the parts of a compound subject are joined by such pairs as *whether/or, neither/nor, either/or,* the verb agrees with the nearer part of the subject, as in the sentence: "Either the mainspring or the connections are giving trouble." But note that *not only/but also* and *both/and* take a plural verb because "and" and "also" are clearly plus signs.

Relative Clauses   You will have no trouble in choosing the verb form in a relative clause if you simply remember that the verb must agree with the antecedent of the relative pronoun (which, that, who). Consider these examples:

1. This is one of those books that are worth studying. ["That" refers to "books" and thus "are" is required.]

2. This is one of those parts which are always giving trouble. ["Which" refers to "parts."]
3. One of the main errors which were involved was the post-computation check. [Note that "which" refers to "errors" and thus requires the plural verb in the relative clause; note also that the subject of the main clause is "one" and thus requires "was" as its verb.]
4. This is the one of those items which is faulty. [Note in this sentence that the presence of the word "the" before "one" leads us to use "is" after "which."]

Compound Subjects   Simple compound subjects in which the elements are joined by coordinating conjunctions or the correlatives normally present no problem. But compound subjects in which the initial item is singular take a singular verb form if the additional items that augment the subject are joined to it by *together with, no less than, as well as, along with,* and *in addition to.* Consider this example:

> The chief engineer, as well as the twenty engineers working with him, is of the opinion that the plan will work.

As a matter of fact, we should acknowledge that usage condones use of the singular verb with a compound subject in certain circumstances when (1) the elements forming the compound subject refer to one person, as in "Our Director — and friend — is sick"; (2) elements forming a compound subject are arranged in climatic order, as in "Our success, our growth, our survival depends on everyone of us working to capacity"; (3) the elements of a compound subject follow the verb, as in "There is promotion and money in this new effort of ours."

Vague Pronoun Reference

Since a pronoun conveys no information in itself but is meaningful only in reference to the word or phrase for which it stands, the reference should be unmistakably clear. Unfortunately, a good deal of ambiguity is found in technical writing, owing to careless use of "this," "which," and "it." We have found that "this" (and "it") is a particularly frequent offender when it is used as the subject of a follow-on sentence. Notice the lack of clearly defined reference in the following examples:

> Panels should be exposed at more than one test station on exterior racks and regular inspections should be made. This will require trained personnel. [Does "this" refer to exposing the panels, making inspection, or both? As the sentence stands, it is impossible to be sure. If inspections, the second sentence should begin "Inspections will. . . ."]

> The rotating scanning mirror is larger in effective diameter and must turn faster than the scanner. This will result in increased torque, requiring a more powerful drive motor. [Can the reader be immediately sure what "this" refers to?]

> This input is a prediction of cost, prices, taxes, and success based on history and present knowledge. It includes plans for when, where, and how much money will be devoted to each phase. [Can the reader be sure of what "it" refers to?]

The appended formulation for aluminum is designed to have fairly satisfactory self-cleaning properties, which makes it suitable for decorative purposes but not as good as white. [Here "which" probably refers to the fact that the formulation has self-cleaning properties. If reference is to "properties" the verb "makes" should be "make" to agree in number. A better version of the sentence is, "The self-cleaning properties of the appended formulation for aluminum make it suitable. . . ."]

Because these sentences have been taken out of context, their faults may appear so obvious that you would be inclined to say that any careful writer would avoid them. Yet errors like these are made over and over again in technical writing.

Coordination and Subordination

Most of us were taught that ideas of equal importance are expressed in independent or coordinate clauses and that ideas of less importance are expressed in subordinate or dependent clauses. Moreover, we were taught that certain conjunctions, like *and*, *but*, and *for*, are coordinating conjunctions and may be used in linking independent clause structures within a sentence; similarly, we were taught that certain adverbial subordinating words, such as *while, since, because, if,* and *when,* are used to introduce dependent, subordinate clauses that contain the lesser ideas or facts.

Although the validity of the "rule" that the main idea or most important fact should always be contained in the main or independent clause is highly questionable (judging from observation of the practice of accomplished writers), we can say that it is inefficient and wordy to express ideas of unequal importance by means of equal or coordinate structures. The practice of stringing together a series of facts by the addition of successive clauses joined by *and* and *but*, for example, can lead only to obscurity, monotony, and wordiness. Let's look at a few examples:

This value is best determined by actual test and it is 50 watts. [*Better:* This value, best determined by actual test, is 50 watts.]

Sand is the other important raw material and it is procured from an outside supplier. [Unless the writer wants to give equal stress to both facts expressed, it would be better to write: Sand, the other important raw material, is procured from an outside supplier.]

This estimate has been plotted in Fig. 3 and shows the likelihood that the meters will all fail at the same time. [*Better:* This estimate, plotted in Fig. 3, shows the likelihood that the meters will all fail at the same time.]

We believe it is important to recognize that coordination and subordination are formal, grammatical matters and that the structure of a sentence does not necessarily reveal semantic importance or impact. In other words, the most important idea of a sentence may — and often does — appear in a dependent structure (as in "Although your report is full of the grossest inaccuracies, obscurities of expression, and downright inanities, it is well typed"). Nevertheless, the use of subordinating structures is a useful way of achieving conciseness and of stressing what needs to be stressed. Let's take a look at an example of no subordination along with some examples of the same facts expressed in a variety of subordinated structures.

1. The chief engineer's report was a carefully written, brilliant analysis of the problem. It was about fifty typewritten pages in length. [No subordination.]
2. The chief engineer's report, which was about fifty typed pages, was a carefully written, brilliant analysis of the problem. [Subordination by clause.]
3. The chief engineer's report, covering about fifty typewritten pages, was a carefully written, brilliant analysis of the problem. [Subordination by participial phrase.]
4. The carefully written, brilliant analysis of the problem by the chief engineer covered about fifty typed pages. [Subordination by modifying phrase.]
5. The fifty-page report of the chief engineer was a carefully written, brilliant analysis of the problem. [Subordination by single word modifier.]
6. The chief engineer's report, about fifty typed pages, was a carefully written, brilliant analysis of the problem. [Subordination by apposition.]

These examples show opportunities for improving upon the version given in item 1.

Dangling Modifiers

A dangling modifier is one that has nothing to modify logically or grammatically, or one that seems to modify a word it cannot possibly modify. In technical writing, dangling phrases are very common, mainly because of the difficulties of describing action in the passive voice. Often — perhaps usually — these dangling phrases cause the reader no trouble, and many writers on the subject of usage take a lenient attitude toward their presence in sentences. Bergen and Cornelia Evans[3] say that

> The rule against the "dangling participle" is pernicious and no one who takes it as inviolable can write good English. In the first place, there are two types of participial phrases which must immediately be recognized as exceptions. (1) There are a great many participles that are used independently so much of the time that they might be classed as prepositions (or as conjunctions if they are followed by a clause). These include such words as *concerning, regarding, providing, owing to, excepting, failing.* (2) Frequently, an unattached participle is meant to apply indefinitely to anyone or everyone, as in *facing north, there is a large mountain on the right. . .*

And Wilson Follett[4] says that "Some participles have so far lost their obligation to serve nouns as adjectives that they have in effect become prepositions, parts of prepositional phrases, or adverbs."

Nevertheless in conservative, orthodox, formal English, you will surely escape criticism if you take care to relate action to a specific word that names the actor. Let's examine a few typical sentences:

1. *Dangling Verbal Modifiers*
After connecting this lead to pin 1 of the second tube, the other lead is connected to pin 2. [Who connects the lead to pin 1? It can't very well be "the other lead" that does so! Two correct possibilities suggest themselves. "After this lead has been

---

[3] From *A Dictionary of Contemporary American Usage,* by Bergen and Cornelia Evans. Copyright © 1957 by Bergen Evans and Cornelia Evans. Reprinted by permission of Random House, Inc.
[4] Wilson Follett, *Modern American Usage* (New York: Hill & Wang, 1966). p. 121.

connected to pin 1 of the second tube, the other lead is connected to pin 2." *Or:* "After connecting this lead to pin 1 of the second tube, the technician connects the other lead to pin 2." In this second sentence, the introductory phrase logically modifies the subject of the main clause, "The technician. . . ." The technician is the one who did the connecting. In the first sentence, the introductory active participial has been changed to passive to agree with the voice of the main clause.]

When starting the motor from rest in the forward direction, the main coil PEM is de-energized and the IR drop across PFN produces a flux to oppose the residual magnetism left by PFN. [The introductory phrase, "When starting the motor . . ." leads the reader to expect that the subject of the main clause will name the starter, but this expectation is unfulfilled. "Coil" is the subject of the main clause and it did not start the motor from rest. "When the motor is started from rest . . ." would solve the difficulty.]

In selecting the rectifier, current limiting resistors, and holdout coil, this hazard must be considered. [The introductory phrase may be kept if the main clause is made to read "The engineer must consider this hazard." Otherwise the introductory phrase must be changed.]

2. *Dangling Infinitive Modifiers*
To start the motor, the starter button must be depressed. ["To start the motor, the driver must depress the starter button" keeps the infinitive phrase from dangling because we now have "the driver" to relate the action to.]

To achieve a mix of the proper consistency, more sand must be added. [Main clause needs a subject like "you" or "the worker."]

Ordinarily, as we pointed out earlier, dangling modifiers are no real obstacle to understanding for the reader, but now and then, as in the following sentences, they cause amusement.

After drying for three days under hot sun, workers again spray the concrete with water.

After taking in a constant flow of oil for two days, the supervising engineer will note that the tanks are full.

As the Evans say, the trouble with such sentences as these is not so much that the verbal phrases dangle as it is that they don't: they are firmly attached to the subject of the main clause — and should not be.

Lack of Parallel Structure

Parallelism means the use of similar grammatical structure in writing clauses, phrases, or words expressing ideas or facts that are roughly equal in value. A failure to maintain parallelism results in what is called a "shifted construction." Parallelism is made clearer by the following illustrations:

1. *Parallelism of Word Form*
The report was both *accurate* and *readable*. ["Both" introduces two adjectives that describe the report. The parallelism would be lost if the sentence read. "The report was both accurate and it was easy to read."]

The process is completed by sanding, varnishing, and buffing the finish. [*Not:* "The process is completed by sanding, varnishing, and the buffing of the finish." The last item in the series is not parallel with the first two.]

2. *Parallelism of Phrases*
Preparing the soldering iron, making the joint, and applying the solder constitute the main steps in soldering an electrical connection. [All the initial terms of the phrases are participials to make the construction parallel. A failure of parallelism would give us something like this: "Preparation of the soldering iron, making the joint, and application of the solder. . . ."]

3. *Parallelism of Clauses*
That this machine is superior to the others and that this superiority has been demonstrated by adequate tests have been made clear in the report. [The introductory "that" of both clauses helps make the parallelism clear. A violation of this parallelism would exist if we had: "That this machine is superior to the others and this superiority is demonstrated by adequate tests have been made clear in the report."]

A shifted construction is sometimes caused by a change in point of view, as shown by the following examples:

A change from a personal style to an impersonal, objective one: "First I shall consider the points in favor of this program and second the disadvantages to the program will be considered."

A change from the indicative mood to the imperative: "First, the wires should be spliced. Next, take the soldering iron. . . ."

A change from the active to passive voice in the same sentence: "The maintenance crew wrap insulation around the joint before the repaired joint is replaced by them in the circuit."

The Sentence Fragment

A sentence fragment is a group of words, punctuated like a sentence but lacking some element without which it can't be a sentence. This error seems not to occur very often, but it does occur. And you will probably find that, on the job, no single kind of error is more annoying to your colleagues than a sentence fragment. Here are some examples.

1. Although few tropical diseases are caused by parasites. [*The problem:* This is a dependent clause and can't stand alone. *Corrections:* "Few tropical diseases are caused by parasites." *Or:* "Malaria is caused by a parasite, although few other tropical diseases are caused by parasites."]

2. The fire fighters were more and more affected by smoke inhalation. Finally succumbing altogether. [*The problem:* The second group of words has neither a clear subject nor a clear verb. *Corrections:* "The fire fighters were more and more affected by smoke inhalation, finally succumbing altogether." *Or:* "The fire fighters were more and more affected by smoke inhalation. They finally succumbed altogether."]

If you are having serious trouble with sentence fragments, the best thing to do is to study the concept of a sentence in a good grammar handbook (see Bibliography in Appendix A).

The Comma Blunder

A comma blunder is the separation of two independent clauses by a comma. It resembles a sentence fragment in that it is a result of failure to understand clearly what a sentence is. Examples follow:

1. The weather has mostly been clear, it rained on Monday. [*The problem:* There are two independent clauses, or two complete sentences, with a comma between them. The comma falsely indicates that one of the clauses is dependent. *Corrections:* "The weather has mostly been clear, but it rained on Monday." *Or:* "The weather has mostly been clear; it rained on Monday." *Or:* "The weather has mostly been clear. It rained on Monday."]

2. He finished his work Tuesday, however, he left town this morning. [*The problem:* It's impossible to tell which clause "however" modifies. *Corrections:* "He finished his work Tuesday, however; he left town this morning." *Or:* "He finished his work Tuesday. However, he left town this morning." *Or:* "He finished his work Tuesday; however, he left town this morning."]

The Run-on Sentence

The run-on or fused sentence occurs when two or more independent clauses are joined without any punctuation or connecting words between them. It is a third type of error arising from uncertainty about what a sentence is. Here is an example.

She worked a split shift as a nurse she didn't like it. [*The problem:* Because of the lack of any punctuation or connecting words after "nurse," the reader expects the last four words in the sentence to modify the preceding part. Therefore the reader has to correct this misunderstanding. *Corrections:* "She worked a split shift as a nurse, but she didn't like it." *Or:* "She worked a split shift as a nurse; she didn't like it." *Or:* "She worked a split shift as a nurse. She didn't like it."]

Inconsistency of Verb Tenses

All that is needed to avoid inconsistency in verb tenses is a moment's thought. For example, consider the following: "The experimental apparatus was set up and the laboratory assistant cleans up after the experiment was completed." Obviously, "cleans" is inconsistent with the tense of the other verbs — and the facts.

Who and Whom

The rule for "who" and "whom" is simple enough: *who* is the subjective form and *whom* the objective. The trouble comes, of course, in applying the rule. Here are some examples of correct ussage:

1. This task will be assigned to whoever can do it best. [*Comment:* "Whoever" is the subject of the clause "whoever can do it best." This entire clause is the object of "to."]

2. Whom do you think this report was intended for? [*Comment:* "Whom" is the object of "for."]
3. It was the chief engineer upon whom we relied for advice. [*Comment:* "Whom" is the object of "upon."]
4. The president, whom the critics denounced, acknowledged that he had made some errors. [*Comment:* "Whom" is the object of "denounced."]
5. The forest ranger was the one who we all thought should receive the commendation for bravery. [*Comment:* "Who" is the subject of the clause "who should receive the commendation for bravery."]

When in doubt about "who" and "whom," you might try substituting a personal pronoun in an equivalent construction. If "he," "she," or "they" works, use "who." If "him," "her," or "them" works, use "whom."

Placing Modifiers

A word, a phrase, or a clause may serve as a modifier. To avoid ambiguity and confusion for your reader, you must take great care to locate modifiers in such a way that their function within the sentence is clear. Consider the following sentence.

The man who spoke seldom met with success.

Does this sentence mean that the man who seldom spoke met with success, or that the man who spoke didn't meet with success very often?

Or consider the changes in meaning that can occur with shifts in the position of "only" in these sentences:

Only Sarah turned in her report on time.

Sarah only turned in her report on time.

Sarah turned in only her report on time.

Sarah turned in the only report on time.

Sarah turned in the report on time only.

Sarah turned in the report only on time.

Finally, here is an example of a poorly located phrase.

The group leader praised those who had continued their research because of these results. [*Clearer:* "Because of these results, the group leader praised those who had continued their research," *Or:* "The group leader praised those who, because of these results, had continued their research."]

A Note on the Split Infinitive

Most older textbooks on English grammar contain a rule against splitting an infinitive. That is, they say it is an error to place an adverb between the word "to" and the present-tense form of a verb. An example of a split infinitive is "to slowly go." The rule against the split infinitive had its origin in the fact that the earliest English grammars were modeled on Latin grammars, and in Latin the infinitive form is a single word. It surely would make no sense at all to "split" a single word. But in English an infinitive is two words.

Recent grammars often say that some split infinitives are acceptable, others not. We cannot ourselves see anything greatly wrong with, for example, "He likes to really run" when the emphasis is on "really." In fact, we believe such a statement has a different meaning from "He really likes to run," or "He likes really to run," or "He likes to run really." Of course we do not encourage splitting an infinitive with an adverb that can itself function as a verb (for example, "to better prepare the surface for painting"); the phrase "to better" is momentarily ambiguous. Moreover, we don't particularly like statements such as "He told us to not start yet." We prefer "He told us not to start yet." But our preference concerning these last problems has to do with style and clarity, not the splitting or nonsplitting of the infinitive. In other words, we prefer the accurate statement put as gracefully and naturally as possible. Nevertheless, many, many people have been taught to avoid all split infinitives, and there are many, many readers who disapprove of all split infinitives.

Shall and Will

In all but the most formal writing, Americans have just about given up using "shall," even though many can recall having been taught that "shall" should be used in the first person, both singular and plural, to express simple futurity and that "will" should be used in both the second and third persons. You may also recall having been taught that to express determination, threat, command, promise, inevitability, or willingness, you should use "will" in the first persons and "shall" in the second and third, and that in questions the form chosen is that expected in the answer. In speech, we suspect that most of you (and we) have simply used the contractions I'll, you'll, and we'll, and have not bothered at all whether "shall" or "will" was intended — or appropriate. And many people use these contractions in their writing as well, as we have in this book.

We wouldn't even bother to bring up this subject were it not for the fact that in some technical documents, such as contracts and other documents subject to legal interpretation, the choice of "shall" or "will" may have very significant implications, significant enough, indeed, to be a deciding factor in a case at law. If at some time or another you find yourself writing a document with such legal dimensions, you might find it advisable to consult the company legal staff before committing yourself to a final draft. Pending that time, here are some guidelines from the most reliable sources we know. We believe the following practices are common:

1. Use *shall* in all persons in contracts containing stipulations to be carried out by a party to the contract.
2. Use *shall* in formal resolutions.
3. Follow traditional usage in all documents subject to legal interpretation.

But we must close these remarks on "shall" and "will" by stressing that what we have said is tentative and incomplete. It might interest you to know that Wilson Follett's *Modern American Usage* has 22 pages on the subject, that H. W. Fowler and his brother (regarded as authorities on British usage) devoted 21 pages of their *The King's English* to this problem, and that the famous grammarian, Otto Jesperson, spent 117 double-column pages of his monumental *A Modern English Grammar on Historical Principles* on this complex subject.

PART 3: SPECIAL PROBLEMS OF USAGE

Special problems of usage that need attention are the use of abbreviations, numbers, symbols, word forms (particularly compounds), capitals, italics, punctuation, and spelling. Form, layout, and bibliographical forms also require consideration, but these are discussed elsewhere.

Since usage in these matters is not standardized throughout the country, we can lay no claim to final authority. You may discover, for instance, that some of the suggestions we make are not followed in the organization you work for. If so, you should certainly follow the rules of your own group. The rules below, however, are based on those accepted by the most widely recognized authorities and may be used with confidence.

Abbreviations

Abbreviations should be used only when they are certain to be understood by the reader. Otherwise, the term should be spelled out. Certain terms, of course, are commonly abbreviated everywhere — Dr., Mr., Ms., No., and the like.

The best authority we know for the use of abbreviations of scientific and engineering terms is the list approved and published by the Society for Technical Communication in May 1975. (The authoritative publication on symbols is ANSI/IEEE Std 260-1978, *IEEE Standard Letter Symbols for Units of Measurement,* published by the American Society of Mechanical Engineers. This publication, it is important to emphasize, lists approved *symbols* of quantities and units, not abbreviations of *words.* For instance, the approved abbreviation of the word "ampere" is amp, but the symbol for ampere is A. In many cases the abbreviation for a word and the symbol for a unit are identical, as atm for atmosphere, Btu for British thermal unit, g for gram, and so on; it is nevertheless important to differentiate between the abbreviation of a word and the symbol for a quantity or unit because the latter is independent of language.) The following rules are in agreement with the above-named publication (a list of the more common, approved abbreviations may be found in Appendix F):

1. In general, use abbreviations sparingly in the text of reports, and never when there is a chance the reader will not be familiar with them.

2. Abbreviations for units of measurement may be used, but only when preceded by an exact number. Thus, write "several inches," but "12 in."; or "several centimeters," but "12 cm." Do not use an abbreviation of a term that is the subject of discussion; thus, do not write, "The bp was quickly reached." Write "The boiling point was quickly reached." Abbreviations may be justified in tables, diagrams, maps, and drawings where space needs to be saved.

3. Spell out short words (four letters or less) like ton, mile, day.

4. Do not use periods after abbreviations unless the omission would cause confusion, as where the abbreviation is identical to a word. Thus, write "in." rather than "in" because the latter might be mistaken for the preposition. Some exceptions are "cot" for cotangent, "sin" for sine, "log" for logarithm. These abbreviations could scarcely be confused with the words.

5. Do not add an "s" to form the plural of an abbreviation. The number preceding an abbreviation of a unit of measurement sufficiently marks the expression as plural. Thus, write "128 bbl" rather than "128 bbls." Exceptions are "Nos." for Numbers, "Figs." for Figures, "Vols." for Volumes. In footnotes, the plural of pages is given as "pp."

6. Write abbreviations in lower-case letters rather than capitals unless the term abbreviated is a proper noun. Thus, write "hp" rather than "H.P." or "HP" for horsepower, but write "Btu" for British thermal unit. Exceptions are terms used in illustrations or bibliographical forms, as shown above.

7. Abbreviate titles only when they precede a proper name that is prefaced by initials or given names. Write "Professor Jones" rather than "Prof. Jones." "Prof. J. K. Jones" is acceptable.

8. Do not space between the letters of an alphabetical designation of an organization. Write "ANSI" for the American National Standards Institute, "ASEE" for American Society of Engineering Education, "ASME" for American Society of Mechanical Engineers, and so forth.

9. Use abbreviations that are more readily recognized than the spelled-out form. Thus, in reports, "FM" is as acceptable as "frequency modulation."

10. In reports where a term is used repeatedly, use the accepted abbreviation, but give a spelled-out parenthetical explanation upon first using it. Thus, you could write ". . . 1200 Hz (hertz) . . ." and thereafter use "Hz."

Symbols

Symbols are generally to be avoided in text. Custom may permit the use of certain symbols in particular organizations, however, and our recommendation is that you observe closely what local practice is and follow it. But while symbols are generally to be avoided in text, they are justifiable in tables, diagrams, and the like because of the need to conserve space. You are probably familiar with most of the commonly accepted symbols, such as " for inches, ' for feet, × for by, # for number, / for per, & for and. A few symbols, like % for percent and ° for degree are so commonly used in text that most readers are as familiar with the symbol as with the spelled-out term.

Numbers

The following rules represent commonly accepted practice in the use of figures:

1. Use figures for exact numbers for ten and above and spell out numbers below ten. Where several numbers, some above and some below ten, appear in the same passage, use figures exclusively. Thus, write:
   10 days
   eight resistors
   five tubes
   27 motors
   11 condensors, 8 tubes, and 27 feet of wire
2. Use figures in giving a number of technical units, as with units of measurement, whether below or above ten:
   8 kHz
   2500 hp
   28,000 Btu
   3 bbl
3. Spell out either the shorter or the first number in writing compound number adjectives:
   thirty 12-in. bolts
   8 six-cylinder engines
4. To avoid possible confusion in reading, place a zero before the decimal point in writing numbers with no integer:
   0.789
   0.0002
   Do not place zeros to the right of the last figure greater than zero unless you wish to show that accuracy exists to a certain decimal; thus you might write 6.7000 if accuracy to the fourth decimal exists.
5. Spell out fractions standing alone, as "three-fourths of the staff members." But with technical units, use figures:
   3-1/2 gpm
   5-1/4 sec
   Note the form used; 3½ and 5¼ are not desirable in typed copy because the fractions tend to blur, especially on carbon copies, and because typewriters do not have all fractions.
6. Omit the comma in four-digit numbers (practice is not uniform on this point, but the trend is toward omission):
   7865
   98,663
7. Follow conventional usage in writing street addresses, dates, and sums of money:
   4516 Spring Lane
   3600 Fifty-fourth Street
   March 11, 1986
   $8,000,000 or 8 million dollars or $8 million

8. Do not use numerals at the beginning of a sentence; numerals may be used for round-number estimates or approximations:
   Twenty-seven seconds elapsed (*Not:* "27 seconds elapsed").
   about 30 a minute
   nearly 500 arrived
9. Do not use two numerals in succession where confusion may occur:
   On August 12, eleven transformers burned out.
10. Use numerals for the numbers of pages, figures, diagrams, units, and the like:

    Fig. 8, stage 4, page 6, unit No. 5, Circuit Diagram 14.

Hyphenation of Compounds

Usage is rather uncertain in the handling of hyphenation — as illustrated in reports that are quoted earlier — but the following practices are generally approved:

1. Hyphenate compound adjectives which precede the term they modify:
   alternating-current motor
   ball-and-socket joint
   4-cycle engine
   2-ton trucks
2. In general, hyphenate compound verbs such as "heat-treat," "direct-connect."
3. Do not hyphenate adverb-adjective combinations, such as "newly installed," "readily seen."
4. In general, do not hyphenate compound nouns (such as boiling point, building site, bevel gear, circuit breaker) except those composed of distinct engineering units of measurement (such as foot-candle, gram-calorie, volt-ampere, kilogram-meter). Many compounds are, of course, written as one word (such as setscrew, flywheel, overflow).
5. In specific cases, try to observe and follow the practice of careful writers.

Capitalization

In general, technical writing style calls for no departure from the conventional rules for the use of capital letters. You have learned to capitalize proper names, names of cities and states, official titles of organizations, and so on. Any reputable dictionary or handbook of English can guide you as to conventional usage (and most of them contain a prefatory section stating the "rules"). We should like to call attention to two practices common to reports:

1. Capitalize all important words in titles, division headings, side headings, and captions. By "important" is meant all words except articles, prepositions, and conjunctions.
2. Capitalize Figure, Table, Volume, Number as part of titles. Thus reference would be made to Figure 4, Table 2.

When in doubt, do not capitalize.

Punctuation

The sole purpose of punctuation is, of course, to clarify thought, to make reading easy. Punctuation that does not contribute to this purpose should be avoided. Most of your difficulties with punctuation are likely to arise in the use of the comma, the semicolon, and the colon. For information on other punctuation marks, see any good handbook of grammar.

The Comma   The principal uses of the comma are:

1. Between independent clauses connected by a coordinating conjunction (and, but, for, or, nor, yet). But if commas are used within any of the independent clauses constituting a sentence (in accord with one or more of the rules below) a semicolon must be used between the clauses. Study these two sentences:

    The fixed coil is permanently connected across the line, and the movable coil is connected across the motor armature.

    The fixed coil, providing a unidirectional magnetic field in which the moving coil acts, is permanently connected across the line; and the movable coil, which operates to close the indicated contact, is connected across the motor armature.

2. After introductory clauses or phrases preceding the main clause of the sentence:

    After workers had completed the first part of the job, they immediately began the second.

    By jumping on the instant of the explosion, they avoided injury.

3. Between items of a series:

    The power supplies, the amplifiers, and the resistors are to be considered now.

    The engine was efficient, cheap, and light in weight.

4. Around parenthetical, interrupting expressions, appositives, and nonrestrictive modifiers:

    This plan, unless completely misjudged, will bring great success.

    This circuit breaker must, obviously, be kept in repair.

    They approved, for the most part, of our research plans.

    Ms. Jackson, chief technical adviser, returned yesterday.

    The chief project engineer, who used to work on the west coast, is responsible for the new procedure.

But not around restrictive modifiers:

    The generator that was tested yesterday is the one needed in this installation. (Restrictive modifiers, like "that was tested yesterday," cannot be left out without destroying the meaning of the sentence.)

The Semicolon   The semicolon is a stronger mark of separation than the comma, almost as strong as the period. It is chiefly used between independent clauses not connected with one of the coordinating conjunctions and between clauses connected with a coordinating conjunction that are quite long, or unrelated, or contain commas. Study these sentences in which semicolons appear:

> The first of these devices has failed after one year's use; the second has lasted five years.
>
> One of these instruments has never had to be replaced; however, it is showing signs of wear.
>
> Even after months of study, they failed to solve the problem; but, in some ways at least, they made a great deal of progress.

The Colon   The colon signals that something is to follow, usually something explanatory, as shown in the following examples:

> A few tools were available: a lathe, a power hack saw, and a drill press.
>
> Operation was becoming uneconomical: both labor and fuel costs were more than had been anticipated.
>
> There are three steps in the process: cutting, grinding, and polishing.

The colon is also used in certain special ways, as in the salutation in a business letter (Dear Sir:), in separating hours and minutes in a statement of time (10:30 A.M.), or in separating volume and pages in a bibliographical entry (17:43 – 50).

Parentheses   Parentheses (. . .) are used to enclose interpolated matter, whether a single word, a phrase, a clause (perhaps even a complete sentence), or numbers and letters in sequence. Parentheses are often used to set off matter that constitutes an interruption to the main statement, particularly when the interruption is more marked than that which would be shown by commas (such as an appositive). Usually, the interpolated matter is explanatory, illustrative, or supplementary; sometimes, however, it is only loosely related to the main thought being expressed. Study the following illustrative sentences:

> Three old typewriters (all out of working order) will be junked.
> This book (I believe you've seen it before) will be used as our text.
> The illustration (Fig. 4) shows all working parts of the device.
> The process has three main steps: (1) selecting a topic, (2) collecting information on the topic, and (3) writing a report on that information.
> Materials for the project will probably cost you about thirty dollars ($30).
> We will have to set forth in our report
>   (a) Immediate objectives
>   (b) Long-term objectives
> Athens (Georgia) will be the site of the meeting.

The Dash   The dash (in typescript it is a double hyphen, as --) is similar in its function to the comma and the colon, and a pair of dashes serves about the same purpose as parentheses, but is considered less formal. Because of this informality, dashes are sparingly used in technical writing. The principal uses of the dash are listed below:

1. To set off discontinuities, or sudden breaks in the thought being expressed. For example, "The first talk — by the way, have you finished preparing yours? — was on environmental protection in small urban areas."
2. To introduce a list (less formally than with a colon). For example, "Several items need further study — the apostrophe, brackets, and exclamation points."
3. To introduce further explanation. For example, "Most dashes are equivalent to commas — that is, they separate elements within a sentence."
4. To indicate a pause before an unexpected word or phrase. For example, "Scientists occasionally create verbs of doubtful validity — 'pessimize,' for instance."
5. To set off (less formally than with parentheses) parenthetical expressions. For example, "The effect of this style upon readers — most of them at any rate — is most unfortunate."
6. Before final summarizing statements. For example, "Word choice, sentence structure, paragraph development, format, and organization are all elements of an important problem — style, in short."
7. Before the name of the author of quoted matter. For example, "Give me liberty or give me death" — Patrick Henry.
8. To indicate omission of a word or letters. For example, "My boss said my report wasn't worth a d —." This use of the long dash is not very common in these days of frankness.

The Apostrophe   The principal uses of the apostrophe (') are as follows:

1. The apostrophe and *s* are usually added to form the possessive form of singular nouns and plural indefinite pronouns that do not end in an *s* or *z* sound (such as everyone, one, everybody, and nobody). Plural nouns already ending in an *s* or *z* sound require only the apostrophe (trustees' decision, students' dormitory). Singular nouns ending in an *s* or *z* sound form the possessive either with or without an *s* following the apostrophe. The final *s* is omitted if following the basic rule produces a form difficult to pronounce, such as Jones' report, Professor Adams' lecture, and similar forms.
2. The apostrophe is used to show the omission of a letter in contractions. Examples include it's (it is), can't (cannot), and wouldn't (would not).
3. The apostrophe is used to show the plural of words used as words, letters as letters, figures as figures, and so on. Look at the following sentences:

     His report contained too many *utilize's.*

     He always wrote his 7's with a mark through them, like 7.

     The department hired twenty-five T.A.'s. (A T.A. is a teaching assistant.)

Note: Some organizations now approve of omitting the apostrophe in writing the plural of letters or numbers, unless a word results from the omission. For example, "The Ds outnumbered the As in this class." But the letters a, u, and i could not, obviously, be so written because *as, us,* and *is* are words.

4. The apostrophe is used only with the second (or last) term in forming the possessive of a compound showing unit possession. For example, "United States and Canada's agreement." This practice applies to multiple word terms, such as Director of Personnel, son-in-law, Chairman of the Board, etc.

5. Remember that the apostrophe is not used to form the possessive of personal pronouns like it (its), who (whose), or they (their). The apostrophe is commonly omitted in formal titles of organizations, such as State Teachers Association and Technical Writers Association.

6. The apostrophe is often used to show the omission of digits in writing a year, as in "Back in '64" rather than "Back in 1964."

Quotation Marks   Quotation marks are used in the following ways:

1. To enclose direct quotations, either from a speaker or a text. One exception to this practice occurs when you quote extensively from a printed source. A common practice is to show that the matter is quoted by setting it off by extra margin and, in a double-spaced manuscript, by single-spacing the quoted matter. No firm rule exists, but if a quotation runs four or more lines, it is customary to set it off in this way — without enclosing it in quotation marks.

2. When a long quotation consisting of more than one paragraph is presented, quotation marks are used at the beginning of the first paragraph, at the beginning of each successive paragraph, and at the end of the last paragraph — but not at the end of intervening paragraphs. (Obviously, this convention does not apply if the practice mentioned above in rule 1 is followed.)

3. Single quotation marks are used for a quotation within another quotation. For example, "I must admit that I heard another student say 'I hate writing reports'; but I heard only one such remark."

4. Quotation marks are used to enclose titles of articles, titles of chapters in books, titles of poems, paintings, lectures — in short, they are used with titles of items appearing within the covers of a larger publication, such as the name of a magazine or the title of a book. Titles of the whole publication (periodicals, books, newspapers) are usually italicized, or underscored in typescript.

5. Quotation marks are customarily used to enclose words or phrases you wish to call special attention to. Examples might be technical terms you suspect are unfamiliar to your readers, a word used in a special sense, a coined term, a trade term, or such shoptalk terms and slang words for which you may feel slightly apologetic. Look over the following sentences:

> A "diaper" in architecture is not the baby garment you might first think of; it is geometric design used as surface decoration in art and architecture.

> He said he would come up with a "ballpark" estimate of the cost of the proposed research.

> We must really "get it all together" before undertaking this project.

> "Outta sight" appears to be a complimentary term.

> A "leader" is a printing term for a series of periods to indicate omission of words.

6. Quotation marks and punctuation: When a word or phrase within a sentence is placed in quotation marks and must be followed by a comma, the comma must be placed inside the final quotation marks:

> One of the newer coinages is the term "pessimize," but it has not yet won acceptance.

But if a quoted word or phrase appears before a semicolon, the semicolon appears after the final quotation marks:

> A new coinage is the word "pessimize"; it has not yet won acceptance.

If a quoted word or phrase appears at the end of a sentence, the period must appear ahead of, or inside, the final quotation marks:

> One of the newer coinages is the word "pessimize."

A final caution about using quotation marks to enclose slang terms, trade terms, shoptalk, or any term you are uncertain enough about to put in quotes: use such terms sparingly. It's really far better to use a term for which no apology is needed, or one your reader is almost certain to understand.

Other Marks of Punctuation We've lumped certain other marks of punctuation together here at the end of our brief treatment of punctuation, primarily because we believe their use is already quite familiar to you. These marks include the period, the question mark, the exclamation mark, and brackets. Let's just very briefly review their functions.

1. *The period* is used at the end of all sentences except, of course, those that ask questions or those expressing strong emotion (for which the question mark and the exclamation point are used). Periods are also used after some abbreviations, mostly those in common use such as Dr. (for doctor), Mr. (for mister), Mrs. or Ms. (for mistress), A.M. (for ante meridian), P.M. (for post meridian), and U.S. (for United States). But see the section on abbreviations for more information. As you know, periods are not used with the initial letters of the name of an organization used as its title: UN, UNESCO, NATO, TVA (among the few exceptions is M.I.T.). Periods may also be used to show that words have been omitted from a quotation; it's customary to use three such periods, or four if the omitted words occur at the end of the quoted sentence or passage. Finally, periods are used in showing decimal fractions (as in 12.4 percent) and in showing dollars and cents amounts (as in $126.10). It is not customary to use periods after page numbers, after capital letters used as designations, abbreviated symbols of the names of chemical elements, or such terms as 1st, 2d, 5th.

2. *Question marks* are used, naturally, at the end of an interrogatory sentence. They are not used at the end of indirect questions, such as "He asked whether we were ready to make our report." Sometimes a question mark may be used to express uncertainty about, say, a date, as in "His earliest published work — in 1910? — was by no means extraordinary."

3. *Exclamation points* are used at the end of an emphatic statement, or at the end of a sentence expressing surprise, lack of belief, or strong emotion. Sometimes an

exclamation point is used after an emphatic interjection, as in "Oh! What have you done?" Exclamation points should not be used for emphasis, especially in technical writing. If you want to emphasize the importance of a statement, try to find another way of pointing to its significance — like just coming right out and telling the reader how important it is.

4. *Brackets* are rarely used in technical writing and then, usually, only in printed texts. When used, they enclose matter extraneous or incidental to the context. The most common instance is editorial interpolation or comment, as in quoted matter in which a misspelled word appears, or in an explanatory note provided by an editor (for example: He said, "That year [1986] was our most productive yet.").

Spelling

Misspelled words rarely obscure meaning, but they do attract attention and distract the reader. It is well to remember that many readers, who couldn't spot an unattached modifier or an example of faulty parallelism if their lives depended on it, can spot a misspelled word with embarrassing ease.

How does one learn to spell correctly, then? By memorizing rules? By carefully "sounding out" the syllables of a word — by correct pronunciation? Perhaps, but we are dubious. For one thing, there are many, many exceptions to the rules. For another, many words are pronounced in a way that is not at all closely related to the way they are spelled. Many changes in sound that are not reflected in changed spellings have occurred over the years. In short, spelling is often not phonetic. Examples are well known to us all — words such as "gnat," "heifer," "zinc," and "cough." Fortunately, many of the nonphonetically spelled words are those most people have learned early in their schooling.

Despite our skepticism about the value that a knowledge of spelling rules may have in transforming a poor speller into an expert one, you may find some use — and comfort — in reviewing some of the more frequently applicable ones. In any case, we'll set them down. But before we do, let us urge you to make a habit of memorizing the correct spelling of any word you discover you've been in the habit of misspelling. Even if you are a very poor speller, it is most unlikely that you misspell a very large number of the words in your working vocabulary. When there's any doubt at all in your mind about the way to spell a word, look it up. And always make it a practice to check spelling carefully when you revise what you've written. We know a person who told us he was such a poor speller that he always read through every manuscript he'd written *backward* — starting with the very last word of the last sentence and working back to the beginning. This seems pretty drastic to us but maybe he knew what was necessary for him. We do know that in proofreading it's very easy to see an imaginary text, the text you think you wrote, rather than what is actually on the page.

For what they are worth, then, five of the principal rules of spelling are listed and illustrated in the discussion that follows.

1. When *ie* or *ei* is written, write *i* before *e* when the sound of the combination is

long *e*, except when these letters come after the letter *c,* or when the sound is long *a.* You probably remember hearing the doggerel that goes

> *I* before *e*
> Except after *c*
> Or when sounded as *a*
> As in *neighbor* and *weigh.*

There are, we're afraid, many exceptions to this rule. Seize, leisure, either, neither, and species are just a few.

2. When adding an ending (called a suffix) beginning with a vowel to a word ending in silent *e*, the final *e* is dropped, but when the suffix begins with a consonant, the final *e* is retained. Examples of dropping the final silent *e* are familiar: writing, and hoping. Examples of the retention of *e*'s are also familiar: movement, improvement, hopeful, likely, and hopeless.

Exceptions to both spelling practices abound: duly, truly, and awful drop the final *e* in the root word; words ending in soft *-ce* and *-ge* retain the final *e* when *-able* or *-ous* are added, as in courteous and changeable. You may also recall that a final *e* may be retained when a vowel suffix is added if the word could be confused with another: dyeing/dying, for instance.

3. When a vowel suffix is added to a one-syllable word ending in a single consonant preceded by a single vowel, or is added to a polysyllabic word accented on the last syllable, the final consonant is doubled. Familiar examples include stopped, dropped, occurred, lapping, etc.

4. When a suffix is added to a word ending in *y* and preceded by a consonant, the *y* is changed to *i* unless a double *i* would result. Here are some examples: worried, cried, merciful, cities, and accompaniment.

5. When forming the plural, simply add *s* unless the plural adds a syllable, in which case *-es* is added. Examples:

> book — books
> report — reports
> batch — batches
> match — matches

But remember rule 4 about nouns ending in *y*. Also remember that the plural of letters, signs, and figures is formed by adding *s* (as explained in the section on punctuation).

The five rules just listed are the principal ones — at least in the sense that they cover a large proportion of the words writers have trouble with, but many others could be listed, as indeed they are in readily available references. We suggest that you might be interested in the much more complete treatment to be found in a reputable dictionary. *Webster's Ninth New Collegiate* (9th ed., Merriam-Webster Inc., 1984), for instance, devotes a sizable section ("A Handbook of Style") to the subject.

One additional effort on your part may help you with whatever spelling problems you may have, and that is the study of a list of frequently misspelled words. It can certainly do no harm to examine such a list to see whether you are in the habit of spelling the words as they are given. We would repeat, however, that the surest

path to successful spelling lies in carefully and meticulously learning to spell correctly those words you discover you've been misspelling (keeping a list is helpful to
many people) and in forming the habit of careful proofreading. There follows a list
of frequently misspelled words. In some cases, we've listed along with the alphabetical entry those words commonly confused with the word listed.

Commonly Misspelled Words

accelerate
accept (confused with except)
accidentally
accommodate
accompanied
accumulate
achievement
acknowledgment
acquainted
across
address
advice
advise
affect (confused with effect)
aggravate
all ready
already
all right (not alright)
all together
altogether
always
amateur
among
amount
apparent
appearance
appropriate
approximately
arctic
argument
around
arrangement
article

beginning
believed
benefited
born
borne

boundary
breath
breathe
brilliant
break
broke
bulletin
buried
business

capital
capitol
carrying
changeable
choose
chose
chosen
cite (confused with site, sight)
coming
committee
comparative
compatible
competition
complement
compliment
conceive
conquer
conscientious
continually
coolly
copies
course
criticism
curiosity

dealt
decided
decision
definite

definition

dependent

description

desirable

despair

destroy

develop

different

disappeared

disappointed

disastrous

discipline

dissipation

divided

division

doesn't

drought

dual

duel

dyeing

dying

efficiency

eight

eliminated

embarrassed

emigration (immigration)

eminent (imminent)

emphasize

environment

equipped

especially

exaggerated

excellent

excitement

exhausted

exhilaration

existence

experience

explanation

familiar

fascinating

finally

foreign

formally

formerly

forth

forty

fourth

generally

genius

government

grammar

grievance

guard

handle

hangar

height

hindrance

hurriedly

imagination

immediately

incidence

incidentally

incidents

independent

indispensable

intelligence

interesting

interfere

interpreted

interrupted

irresistible

its

it's

itself

knowledge

laboratory

laid

lead

led

leisure

lightning

loose

lose

losing

maintenance

manufactures

manufacturers

mathematics
meant
medicine
miniature
minute
mysterious

naturally
necessary
nevertheless
nickel
ninety
ninth
noisily
noticeable
nowadays

obstacle
occasion
occasionally
occurred
occurrence
off
omission
omitted
operate
opinion
opportunity
optimistic
original
outrageous

paid
parallel
paralyzed
particularly
partner
passed
past
pastime
perform
perhaps
permissible
perseverance
persistent
personal
personnel

persuade
phenomena
phenomenon
physically
piece
pleasant
portrayed
possess
practically
precedence
precedents
preceding
preference
preferred
prejudice
preparations
prescription
principal
principle
privilege
probably
procedure
proceeded
professional
prominent
propeller
psychology
pursue

quantity
quiet
quite
quitting

realize
really
received
recognize
recommend
referred
relieve
remembrance
repetition
resource
rhythm
ridiculous

sacrifice
safety
satisfactorily
scarcely
schedule
seize
sense
separate
severely
shining
similar
sincerely
source
specimen
speech
stationary
stationery
stopped
strength
strenuously
stretched
studying
succeed
successful
superintendent
supersede
suppress
surely
surprise
synonym

temperament
temperature
their

there
together
too, to, two
toward
transferring
tremendous
tries
truly
twelfth

undoubtedly
unnecessary
until
usually

valleys
valuable
varieties
view
vigorous

weather
whether
whole
wholly
who's
whose
worrying
writing
written

your
you're

Sentences for Revision

The numbered sentences below are divided into two groups of 25 sentences each. The first group is concerned with kinds of errors discussed in Part 1 of this appendix, and the second group with kinds of errors discussed in Part 3.

These sentences can be used diagnostically. That is, through discovering which of the sentences give you trouble, you can identify particular problems you have. You can then go back to the appropriate part of the preceding discussion and study the principle involved. You will need your instructor's help in this process.

Some of the sentences in the two groups below are correct and some contain errors. Revise those that contain errors.

Group I

1. The new component is likely to be difficult to install but promises better performance. This will have to be checked out.
2. After connecting the high-pressure side of the line to the gauge, the reading startled everybody.
3. Either the ratchet or the bearings is ruined.
4. This is one of those projects which attracts attention.
5. Whoever made these repairs was ingenious.
6. The last worker on the assembly line locks the part into place and the inspection tag is attached.
7. The first completed machine, together with its accessories, are being shipped.
8. A number of the units was defective.
9. The only known way to cure the malfunction is to actually hit the side of the cylinder with a mallet.
10. If the operator doesn't forget to turn off the compressor.
11. The results are known and their meaning was clear.
12. These parts are long-lived and are manufactured by the John Doe Company.
13. The remaining three problems, which are less difficult, can be postponed.
14. Although data are available in large quantities, the committee refuses to make a decision.
15. Whom do you think you are?
16. The final shipment is not here yet, it is badly needed.
17. Upon finishing that operation, the surgeon heaved a sigh of relief.
18. It was too early to tell, the curing process was quite complex.
19. When the annealing oven started the fire the automatic sprinkler system came on everybody got wet.
20. The computation was incomplete. This oversight must be corrected.
21. This is one of those jobs that demand concentration.
22. The designer to whom we felt the credit belonged was quite young.
23. Neither the window frames nor the door has been painted.
24. Having seen with his own eyes the water pouring over the spillway.
25. The first steps are the clearing of the ground, the leveling of the ground, and removing large rocks.

Group II

1. This antique radio contained 5 tubes.
2. The container leaked 300 British thermal units per hour.
3. We need a 2 ton hoist.
4. Two alarming conditions appeared: a drop in voltage, and a fluctuating current.
5. On December 17 14 units were shipped.
6. This component weighed 400 lb. without accessories.

7. The exact reading was .748.
8. This technique, although expensive solved the problem.
9. Prof. Doe owned a car with a 510 H.P. engine.
10. The motor, which was rebuilt, will be acceptable.
11. The handle is several cm long.
12. Because they liked the foreman better public relations resulted.
13. The fullback was satisfied with his thirty nine yd runs.
14. You will need three tools; a saw, a hammer, and a plane.
15. The connecting rod is 20 centimeters long.
16. Please look at table 20.
17. The tower was 40 ft. high.
18. Over 2/3 of the investment was lost.
19. The company granted a 21-day extension.
20. A pressure of 10 lb per sq in. was inadequate.
21. The pointer read 0.632.
22. The saw was binding in the kerf, the teeth needed resetting.
23. The tank's capacity of 150 bbl was sufficient.
24. This newly installed generator is defective.
25. The results are found in Vol III.

Appendix C

Excerpts from *Human Factors in the Design and Utilization of Electronics Maintenance Information*

Taken altogether, the following pages contain the most illuminating insight into the practical problems of technical writing we have ever seen in print. From the annoyance of the schematic electronic circuit diagram that has too many folds in it, to the analysis of a reader's probable vocabulary, here are the problems faced by the technical writer.

We are presenting this material for two purposes. One is to make it available for leisurely examination by interested readers. The other is to use it as a source of helpful examples: at several points in this book, reference is made to these pages to provide practical illustrations of principles being discussed.

The information in this report was gathered through interviews with people responsible for maintenance of naval electronic equipment, both on board ship and at installations ashore. As indicated in a section of the report not reprinted here, one of the principal objectives of the study was to learn how to produce technical materials that would meet the needs of maintenance personnel in a combat zone who might have no aid or advice from outside the ship itself. The interviews on which the report is based were conducted by a team of psychologists and experienced navy technicians.

Only a small portion of the report is reproduced here. Included are the statement of the problem with which the report is concerned, and a section devoted to comments made by the people who were interviewed. Deleted portions are indicated by asterisks.

Research Report
Report 782

Human Factors in the Design and Utilization of Electronics Maintenance Information

J. H. Stroessler, J. M. Clarke, P. A. Martin (Soc, USN), and F. T. Grimm (Soc, USN)

U.S. Navy Electronics Laboratory, San Diego, California
A Bureau of Ships Laboratory

THE PROBLEM

Determine who uses technical manuals for Navy electronic equipments, how they are used and for what purposes, the extent to which present content is adequate for present users, and the functional relationship between technical-manual and other types of maintenance information.

* * * * * *

The problem was broken down into more specific components to arrive at questions which might be effectively answered by a field study. These are:

1. What are the tasks for which technical manuals and related materials are used?
2. What are the characteristics of the personnel performing these tasks?
3. How effective are the technical manuals and related materials as tools for the performance of each major task when used by particular categories of personnel?
4. To what extent are various parts of the technical manuals and related materials used in connection with each major task by each category of personnel?
5. How appropriate are the technical manuals and related materials to the needs and capabilities of the various user groups?
6. How are the technical manuals and other maintenance publications related in subject matter coverage, format, use, etc.?

* * * * * *

COMMENTS ON INDIVIDUAL TECHNICAL MANUALS

The following are significant observations made by personnel interviewed of the technical manuals suggested for particular study by the problem assignment.

AN/SPS-8

HELPFUL CHARACTERISTICS
The cable layout is outstanding as to function, origin, and termination of each line.

ADVERSE CHARACTERISTICS
1. Volume I is too big and bulky.
2. Tracing a single circuit involves use of too many schematics.
3. The schematics are folded too many times.
4. It is hard to maintain continuity in going from one schematic to the next.
5. The extra schematics in the folder are too large to be practical.
6. The theory of operation "snows" both Navy electronic technicians and the technical representatives of commercial companies.
7. The manual lacks a description of the purpose of the equipment.
8. The checks called for are too complicated.

9. Information on particular components and circuits is scattered through too many sections (antenna information cited as an example).
10. There is no antenna alignment procedure; no gear tolerances are given.
11. Mechanical parts are not identified well enough.

SRT Series
(Note: This manual was very difficult to obtain. The only copy the Task Group could get for study was in possession of a systems design group at NEL. This was an "Advance Form Instruction Book." Comments should be interpreted with the latter fact especially in mind since maintenance personnel occasionally must work with just such incompletely developed books.)

ADVERSE CHARACTERISTICS
1. Many photographs and schematics are omitted.
2. The manual is very confusing; there are inconsistencies between schematics.
3. Schematics contain errors.
4. Corrective Maintenance sections are too vague.
5. The circuitry is very difficult to follow.*

SPA-4

ADVERSE CHARACTERISTICS
1. Too many corrections to be made in pen-and-ink.
2. Many difficulties arise due to differences in terms and their meanings between manuals for BUORD and BUSHIPS portions of the system.

SPA-8A

HELPFUL CHARACTERISTICS
1. Pages of waveforms are useful; time base relationship diagrams and inclusions of distortion components in idealized waveforms are particularly helpful. Waveforms would be even more useful if more information on timing within the cycle were included. It would also be helpful if the origin of the signal resulting in each particular waveform were indicated. Waveforms should be in sequence. Wealth of details helpful — it has *all* the useful information.
2. Theory of Operation section is outstanding for understandability.
3. The folder of separate schematics is helpful.
4. Listing of inputs is very useful.

ADVERSE CHARACTERISTICS
1. Repeater alignment procedure is so detailed that the continuity of steps is lost. Technical representatives have a simpler, better alignment procedure.
2. Adjustment sequence needs improving.
3. Lacks enough data to enable personnel to substitute a different type of oscilloscope for the one recommended.

UQN-1B

HELPFUL CHARACTERISTICS
1. Clearly written and helpful as a whole.
2. Simplified schematics very good

---

* Expert personnel find this equipment difficult to maintain due, in part, to crowding. Certain difficulties in using the manual probably originate in the poor maintainability features of the equipment itself, which are reflected in the manual in the form of hard-to-trace circuit diagrams, vagueness, etc.

ADVERSE CHARACTERISTICS
1. Has many inaccuracies in schematics.
2. Some simplified schematics in Theory of Operation section which originally
   appeared in the manual for UQN-1 are inaccurate in reference to UQN-1B.
3. Corrections are difficult to obtain.

Miscellaneous Manuals
The following list contains desirable features of various current technical manuals
as pointed out by personnel interviewed.

1. TBS: Uses color well, especially in showing individual signal paths on block
   diagrams. Uses color effectively on schematics, but not so extensively as in
   Sangamo books.
2. VF: Overlays very effective in showing mechanical construction. A series of
   photographs and exploded views would probably accomplish the same results at
   less cost. Use of color to differentiate mechanical parts involved in some operations is very effective. Servicing block diagram is very helpful.
3. AN/CP6: Very good layouts on test equipments and test hookups.
4. AN/SPS-8A: Cabling diagrams are very good, but manual is too bulky.
5. AN/SQS-4: Good layout. Test set-up for checking preamplifier stages, etc., with
   oscilloscope are helpful. Pictures of waveforms on schematics would be an
   improvement.
6. AN/SQS-12: Display of variable pulse length in color is helpful. In general a good
   manual, but lacks waveforms.
7. BLR: Has two extremely helpful tables. One lists test equipments, their characteristics, and use data in reference to particular units and sections of units. The
   other is a guide by unit numbers to corrective maintenance data in the manual itself.

RESULTS AND DISCUSSION

Effectiveness of Technical Manuals

USE AND FUNCTIONS
The various sections in current technical manuals are so different in content and
purpose that they must be evaluated individually in terms of their functions and
users. In doing this, there are three main questions to be answered: (1) Does a given
section provide material that is useful for the function it is intended to serve? (2)
Can the necessary information be readily found and assembled (in the sense that
each piece joins with and supports others) for application to particular maintenance
problems? (3) Is the information presented—through text and visual aids—in a
manner that makes it readily understood and usable? The first question may be
answered in a very general way, by examining the extent to which each section is
used by major groups. Table 11 presents such a breakdown.

One of the most important differences between civilian and military maintenance
personnel lies in their use of the Installation section. Among the civilians interviewed, this section was the one most frequently reported as receiving extensive use.
Most of these users are in the planning, design, and installation groups in Naval
Shipyards. Civilian personnel differ from the military also in that their responses
indicate little or no use of the Operation, Operators Maintenance, and Preventive
Maintenance sections.

TABLE 11 Use* of Technical Manual Sections by 107 Military Personnel and 44 Civilians

	Frequently Used		Seldom Used	
Section	Mil.	Civ.	Mil.	Civ.
1. General description	3	0	8	0
2. Theory of operation	76	14	0	4
3. Installation	2	23	76	4
4. Operation	7	0	6	1
5. Operators maintenance	3	0	8	0
6. Preventive maintenance	13	0	3	0
7. Corrective maintenance	101	16	0	0
8. Parts list	82	10	10	0

* Infrequent mention of a section may reasonably be interpreted as lack of use.

INFORMATION LOCATION

Fifty-seven percent of 132 subjects interviewed concerning corrective maintenance information stated that necessary data were often hard to locate, and that even when they could be found, this was at the cost of considerable "flipping back-and-forth" among the pages. Highly trained and experienced men, both civilians and Navy personnel, reported these experiences more frequently than did inexperienced men. The probable causes of this difference are: (1) inexperienced men try to solve fewer difficult trouble-shooting problems; (2) inexperienced men are inclined to blame their difficulties on themselves rather than on the technical manuals. The following direct quotation from an interview subject typifies the attitude of inexperienced personnel: "It's not for a third-class to question the book."

Further reasons for experiencing difficulty in locating corrective maintenance information were also indicated. First, the information needed for trouble-shooting is, in many cases, so scattered that the technician frequently makes several frustrated attempts before finding what he needs. Second, there appears to be no standard organizational procedure that tells the reader where to look for the information he requires. A member of the Task Group in collaboration with a highly trained engineer who had been head of a technical manual writing group made a limited test of the time required to find needed information. The technical manual selected was typical of recent publications, NAVSHIPS 92501 (A), for the AN/FRT-27. The output from the oscillator buffer stage in the transmitter was selected at random as the information to be sought. The first step was to try to find the voltage reading on the simplified schematic, where it did not appear. The second step was to try to find it on the detailed schematic; it was not there either. Nor could it be found in the Theory of Operation section, which was examined as the third step. The fourth step was to look into the trouble chart where the voltage reading was finally found after a search through three pages of the chart. Total time required to find this information was about 20 minutes. Graduate engineers have reported spending days in off-and-on search before locating needed data.

The third main cause of difficulty in locating information was said to be the indexing system, which is quite different from that used in standard civilian references. The index in a Navy technical manual, when it has one, consists of the Table of Contents arranged alphabetically with clusters of heterogeneous topics under main headings. Many current technical manuals are not indexed at all. Of 38 subjects who discussed indexing, all agreed that a better index would be of help to them.

INFORMATION ARRANGEMENT

Closely related to the need for less hunting of information and less "flipping back-and-forth" are certain problems of information arrangement. The trouble-shooter finds it necessary not only to locate information, but to assemble it, mentally, for use. This is difficult with the present arrangement of technical manuals. Information useful in locating circuits and stages is found in block diagrams. Information required to gain a functional understanding of circuits is for the most part in Principles of Operation, but these data must be supplemented with the graphic representations and component values found on schematics. Part of the information required for identifying parts and components is obtained from photographs, and the rest from schematics. Data from tables as well as data from schematics are required in order to trace signals by comparing a sequence of measurements with values given in the manual. The significance of these measurements may be derived from the Principles of Operation, from a trouble-chart, or even from the General Description.

One experimental arrangement having the object of integrating trouble-shooting data was shown to personnel interviewed. Eighty-eight percent thought it helpful. This arrangement was developed by the RCA Service Company for use in their Inter-Level Electronics Training program. In this presentation a circuit is laid out on a single fold-out showing a block diagram, a simplified schematic, and an actual schematic in left-to-right order. The block diagram includes a short statement of the purpose of each stage, and a resume of the stage's function. On the same page with the simplified schematic, there are condensed analyses of the circuit and possible component failures. All components are included in the complete schematic with a brief description of each stage.

INFORMATION PRESENTATION

Block Diagrams

Functional block diagrams are useful to most maintenance men interviewed during this survey. They are most frequently used in gaining an over-all understanding of the relationship between circuits and components and the manner in which each major unit or stage contributes to the transformation of inputs into outputs. It appears necessary to visualize these relationships, and a good block diagram appears to be the best, if not the only, device which can aid in developing an adequate mental picture at the proper level of generality. It is important to bear in mind that trouble-shooting (except for "Easter-egg hunting") means going from the general to the particular, from the whole to a part, eliminating as one goes until the faulty segment is reached. As one 20-year chief put it, "No equipment is complicated once you get to the place where the fault is."

Block diagrams represent the general, gross aspect of the equipment. Thus they often serve as a starting point in trouble-shooting, helping the maintenance man to identify the area and/or circuit where the fault is located, and enabling him to begin reasoning as to what may have caused the unwanted effects. A block diagram may also serve as a bridge between the equipment itself and the schematics. Maintenance men — particularly the inexperienced — will frequently go from a part they have identified as faulty to the block diagram and thence to the schematic. Or they may reverse the process, going from the circuit segment or component they have been studying on a schematic, to a block diagram and thence to the equipment.

These aspects of the trouble-shooting process emphasize the need for close integration between the Theory of Operation section — where block diagrams are generally located — and the Corrective Maintenance section. They also make it apparent that locating useful block diagrams in the General Description section may cause the maintenance man needless flipping back-and-forth between widely separated aids.

Simplified Schematics

Simplified schematics are also found very useful, particularly at the stage when the maintenance man is trying to understand how a circuit works. They aid in tracing the *course* of a circuit simply because they are not encumbered with all the data required for signal tracing.

"Circuit tracing" and "signal tracing" are often used interchangeably. It is useful, however, to think of circuit tracing as the following of a path, or course, without regard to the characteristics of the energy which follows this course. Signal tracing may be defined as the effort to follow the flow of energy along a course, determining the segments in which it is normal and those in which it is not normal.

Photographs and Drawings

Seventy percent of the 124 men interviewed on the subject considered the photographs and pictorial drawings in manuals helpful. The main uses of these visual aids are: (1) in gaining a concrete understanding of the way an equipment functions, i.e., circuit and component interrelationships, and their contribution to outputs; (2) in identifying parts; (3) in locating parts. Pictures should be clear with sharp definition to make individual parts stand out. Callouts and captions should be large enough, and spaced far enough apart, to be easily read. It appears best — when a choice must be made — to sacrifice completeness of coverage to legibility. Good, clear drawings are as helpful as photographs, in the opinion of maintenance personnel interviewed.

Pictorial block diagrams of complex equipments and systems are especially effective in giving an over-all understanding. These may be made up of photographs, or drawings, or a combination of both. Their usual location — the General Description section — is not the best, however, because these diagrams are most often used in conjunction with the Theory of Operation section. It is held by most maintenance personnel that visual aids and the text to which they are related should be as close together as possible.

"Sams Approach"

The above discussion leads logically to a consideration of the frequently-mentioned "Sams approach" since the Sams Company maintenance manuals make use of a large number of photographs with callouts. Further characteristics of this company's manuals are (1) schematics having standardized format, symbols, and location of components; (2) waveforms and voltages on the schematics themselves; and (3) necessarily limited text.

Although many Navy maintenance personnel who have used Sams Company books have found them effective, most of these men doubt whether they would prove satisfactory for Navy equipments without considerable adaptation. There are several reasons behind this doubt. First, most Navy equipments are much more complex than the civilian equipments for which Sams books are written; thus, Sams type books for Navy equipments would necessarily have to be more complex and bulky. Further, it has yet to be proved that the standardized schematics used by the Sams Company are easier to read out than the best of the schematics in Navy technical manuals. Finally, there are reasons to believe that the Sams books do not contain *all* the information required in Navy maintenance situations.

SECTION ANALYSES

Corrective Maintenance Section
The extensive use made of this section (see table 11) makes it profitable to consider each of its separate sub-sections individually.

Schematics. All available evidence shows that schematics are the heart of the trouble-shooting process.* They are the maps showing current and signal flow — charting their courses between input and output and indicating the modifying points through which they pass. Schematics also provide a large part of the data required to understand a circuit and measure its performance.

In general, dependence upon schematics increases with specialization and experience. Experienced Navy personnel (chiefs and first class) tend to rely on schematics more heavily than do the less experienced rates. Some shipyard and tender repair specialists use very little of the manuals beside the schematics. The reasons seem to be: (1) specialists require less complete information on a given equipment than other technicians because a majority of their problems are repetitions of — or similar to — problems they have solved before; (2) highly experienced men generally have a reasonably complete understanding of the way each equipment under their care functions. Consequently, good schematics provide both these categories of personnel with most of the information they require for corrective maintenance, except when faced with a completely unfamiliar equipment or a very unusual malfunction. In such cases, experienced men may use all parts of the Corrective Maintenance section, and parts of other sections as well.

Most experienced maintenance personnel feel that wherever the waveform has significance it should be shown on the schematic. The oscilloscope on which the waveform was obtained should be indicated, and time base and peak voltage data should be given.

The complaints most commonly made against schematics by personnel interviewed were:

1. Lines were spaced too close together.
2. Too many lines crossed one another.
3. The signal path was not emphasized graphically.
4. The course of a current flow was highly intricate.

---

* The "color test" is the shipboard technician's favorite way of showing how much more the schematics are used than other portions of the manual. The edges of the schematics soon take on dark hues ranging from a greasy gray to black-brown. The rest of the pages have edges ranging from pale gray to virgin white.

5. Too many symbols were crowded onto a page.
6. Symbol numbers were too small to read easily.

Conversely, a sizeable majority of the respondents found schematics relatively easy to read when the following conditions obtained:

1. Color was used to trace a signal path.
2. The "in-line" convention was used to simplify the pathway.
3. Signal paths were emphasized with heavy lines.
4. Schematics were relatively uncrowded.
5. Symbols and numbers were readily visible.
6. "Roadmapping" was used to facilitate part location.
7. Realistic waveforms were shown instead of idealized ones.

According to 86 percent of the comments, the indication of inputs from, and outputs to, related equipments is desirable. Ninety percent of maintenance personnel would like to have a voltage tolerance range given on the schematic, and a special marking to indicate each point at which the voltage level is critical and no tolerance is permitted. Most maintenance personnel reported finding related voltage and resistance (V/R) tables very useful, particularly if they were of the pullout type and were not hidden when the manual was closed.

* * * * * *

Theory of Operation Section
Next to the Corrective Maintenance section, the most used part of technical manuals is the Theory of Operation (or Principles of Operation) section. Practically all maintenance personnel — civilian and military — study this section when they are performing corrective maintenance on an equipment that is relatively new to them, or when they are trying to solve a particularly baffling maintenance problem on any equipment. It should be borne in mind that, to the inexperienced maintenance man, all equipments are unfamiliar and most maintenance problems are baffling. The reason for studying this section is to gain an understanding of how the equipment is supposed to work.

Understandability and Readability. The understandability of the language used in the text is as important in the Theory of Operation section as any other place in the manual. Ninety percent of the instructors discussing the problem stated that the Theory of Operation section is the most difficult for their students. Consequently, the matter of difficulty level or readability of the writing (actually, level of comprehensibility) is of great significance.

The most realistic way to describe the appropriate level of comprehensibility is in terms of reader experience. The Theory of Operation section (and other textual material) should be written for personnel with a mean experience of about one year and ranging roughly from six months to two years. Text written at this level will meet the needs of men who have been converted from other rates and normally rated men whose actual experience with maintenance is limited but whose learning ability is high. It allows time for the average man to acquire the "real" (sensory) referents necessary for comprehension of practical — as contrasted to academic — electronics. To make the comprehensibility level higher would mean cutting off a large proportion of low-rated maintenance personnel who have responsible work to perform. To set it lower would mean that technical manuals would have to be written primarily for educational purposes since, on the average, the maintenance man with less than a year's experience does more learning than actual maintenance.

* * * * * *

The average Navy maintenance man has at least 12 years of civilian schooling. "Readability" in the technical sense popularized by Rudolf Flesch is, therefore, not a pressing consideration, although improvement would doubtlessly result if technical manual writers held average sentence length down to 25 or 30 words and used a minimum of polysyllabic words (like "minimum" and "polysyllabic"). As for the terminology use, most "A" school graduates have acquired sufficient technical vocabulary to read the words.* Most of them consult more experienced men and reference books to obtain the meanings of words they do not know.

Slightly more than half of the personnel discussing this matter felt that the Theory of Operation section was written clearly with easily understood words and sentences. The remainder believed that understandability of the section needed improvement because it was usually too technical and assumed too much reader background knowledge. This was frequently expressed in the statement, "The Theory of Operation section is written from an 'engineer's point of view.'"

Explanations. The explanation of concepts in the Theory of Operation section was frequently mentioned by the personnel interviewed as a characteristic in need of improvement. Here is the point at which the maintenance man's training in basic theory meshes with the more specific information in the manual. Some maintenance people feel that the less experienced men would be benefited if Theory of Operation sections were written in such a way as to recall the applicable principles to mind.

In contrast to this, other maintenance personnel feel that some Theory of Operation sections already contain too much engineering logic. This was expressed by the statement that the manual is "not written from the maintenance point of view."

Although opinion is divided as to whether or not the Theory of Operation sections currently are written from the "maintenance point of view," certain conclusions seem tenable on the basis of comments made: (1) the maintenance man should be required to make as few mathematical computations as possible. (2) The text should not attempt to justify the circuitry or explain its particular virtues. (3) The text should enable the maintenance man to bring his basic knowledge to bear in deriving a functional understanding of cause-and-effect relationships. The maintenance man is apparently more interested in what takes place along the course of a circuit than in why it happens; that is, he is more concerned with the direct physical causes of effects than with the scientific principles underlying the occurrences. In short, as one man put it, "maintenance personnel want maintenance information, not design engineering information."

About 40 percent of the maintenance personnel commenting on the matter feel that explanations in the Theory (Principles) of Operation section are not complete (detailed) enough. They believe that references are not cited frequently enough, and that while simple, obvious stages are often given in elaborate detail, unusual and

---

* In general, "A" schools give basic electronic training, "B" schools give more advanced training, and "C" schools give specialized training on individual equipments.

complex circuits are often given cursory treatment. Elaborating the obvious wastes the time of experienced men, and misleads the inexperienced by making them think that simple stages are difficult — thus causing them to spend time hunting for mysteries that are not there. All circuits should be described rather than being merely labelled, even though it is unnecessary to explain them in detail. Statements such as "From this point on, the sweep circuit functions as an ordinary DC amplifier circuit" leave too much for the maintenance man to interpret for himself.

It has been suggested by some of the experienced civilians interviewed that the underlying cause of this imbalance in the explanations is the lack of understanding on the part of some technical writers. It would be natural for a writer confronted with an already-developed equipment, finding that the engineers who had developed it lacked time to explain everything to him, to elaborate the stages he understood best and treat the less understood parts obscurely or in a very general manner.

Standardization. The technical vocabulary of "A" school graduates is generally adequate so long as only standard terms are used. Maintenance personnel are, however, frequently puzzled by manufacturers' unique designations and the variations of engineering terms which have not become standardized. Over 80 percent of the comments made on this subject indicated a strong feeling that the terms and symbols used should be standardized, including the standardization of inputs and outputs, and the "A" and "B" sides of tube symbols.

It is evident that standardization should be carried as far as possible. However, standardization is an ideal impossible of complete realization because engineering language is growing constantly. Several words having the same meaning may be in current use for a time before one becomes the standard term. The inclusion of a glossary containing any unique or unusual terms seems to be the most logical solution for the problem. Seventy-five percent of the maintenance men commenting on the matter believed that such a glossary would be helpful. Obviously, the glossary ought to cover all the text, but, as a few men indicated , it might be best placed at either the beginning or end of the Theory of Operation section.

Appendix D

Library Research Materials for Technical Writing at the University of Texas at Austin

The materials presented in this appendix were prepared by librarians at the University of Texas at Austin for the use of students of technical writing. More specifically, they were designed to be used in conjunction with Chapters 14, 21, and 22 of the present text, the chapters dealing with proposals, finding published information, and writing the library research report.

The materials handed out to students include the following: (1) a discussion of selecting a topic of research, (2) a technical writing research worksheet (including one especially designed for students of business), and (3) special bibliographical guides for students majoring in engineering, life sciences, computer science, nursing, and business. Most of the students who take technical writing major in these disciplines.

Not reproduced here are additional handouts designed to help students find periodicals in the various campus libraries. Call numbers of the reference works cited in these materials (as well as other items of purely local relevance) have been omitted since these numbers will not be the same in all college and university libraries. Since the introductions to the handouts for the various disciplines are the same, we have omitted all but the first. We have also omitted instructions for the location of specific libraries and special collections at the Austin campus.

We are grateful to the General Libraries of the University of Texas for permission to present these materials and we wish to acknowledge the contributions of those individuals who prepared them (listed alphabetically): Elizabeth Airth, Susan

Ardis, John Burlingham, Nancy Elder, Suellen Fortine, Ann Neville, Kay Nichols, Barbara Schwartz, and Wanda Smith. We are especially grateful to Ann Neville, Kay Nichols, and Barbara Schwartz, who conferred with us about these materials and were responsible for developing and editing them.

SUBJECTS FOR TECHNICAL WRITING RESEARCH

For this assignment you will do preliminary research on a topic based on one of the subjects listed on this handout. The information you collect will be used to prepare a Research Proposal. The proposal will describe your Library Research Report.

The primary disciplines for this project are engineering, the life sciences, nursing, the computer sciences, and business. The areas suggested on the next few pages have been selected not only to provide you with a wide range of possible subjects, but also to limit your choices to subjects on which adequate information can easily be found. If you choose a subject not listed here, plan to spend extra time doing your research and talk to a librarian before you begin. A librarian can help you develop a workable topic and can recommend useful sources.

Each of the areas listed on the next few pages is too broad for an effective search; to save time and energy later, you need to think of ways to restrict the focus of the topic you choose. A specific focus makes it easier to develop your topic statement and facilitates your research by helping you to eliminate whole areas that are not pertinent to your topic.

Selecting a Topic in Engineering

As you look at the topics listed below, think about the kinds of *problems* they present. A *problem* and its *solution(s)* are the primary focus of your Research Proposal and Report. One way to narrow your focus to a particular problem is to ask questions about the subject. For example, suppose you are considering *water pollution.* What pollutes water? Chemicals? Oil? How does oil get into the water? Spills from ships. Leaks from wells. What kinds of problems does spilled oil cause? Fish killed. Damage to shores. Are you interested in the effect of shoreline damage on wildlife? Or the ways damage is cleaned up? Direct your attention to a particular problem and plan to present, in your Research Proposal, an explanation of the problem and how you intend to research its solution(s).

Possible Topics for Research

Fiber optics; flood control; flywheels; food engineering; foundations; geothermal energy; human factors engineering; hydroelectric power; industrial robots; industrial safety; industrial waste disposal; insulation; integrated circuits; internal combustion engines; manganese nodules.

Wind energy; weather forecasting; water supply; water purification; traffic engineering; television antennas; tall buildings; synthetic fuel; solar heating; sewage disposal; water pollution; short takeoff and landing aircraft; radioactive waste disposal; prefabricated building; electric vehicles; daylighting.

Air pollution; artificial satellites; automatic control systems; aviation accidents; batteries; biomass; bridges; building materials; coal liquefaction; cogeneration; concrete construction; corrosion; dams.

Power transmission; polymers; pilotless planes; petroleum transportation; oil well drilling; petroleum refining; mass transit; microprocessors; natural gas; noise pollution; nuclear reactor safety; oil reclamation; oil sands; oil shales; petroleum exploration.

Selecting a Topic in the Life Sciences

As you look at the areas listed below, think about the *problems* they preent. A *problem* and its *solution(s)* are the primary focus of your Research Proposal and Report. One way to narrow your focus to a particular problem is to ask questions about the subject. For example, suppose you are interested in *species population*. What species? Is the population increasing or decreasing? Why? Does this present a problem? To whom? Why? What can be done about it? What cannot be done? Is there a solution that has not been tried? How could it be implemented? Direct your attention to a particular problem and plan to present, in your Research Proposal, an explanation of the problem and how you intent to research its solution(s).

Possible Areas of Research

Genetics — evolution; natural selection; speciation; heredity; mutagenesis; genetic engineering; adaptation; DNA.

Behavior — animal behavior; animal communication; animal migration; biorhythms; biological clocks; territoriality; niche.

Natural history and ecology — tropical forests; wildlife conservation; marine ecology; desert fauna and flora; species population; symbiosis; pesticides; endangered species.

Body systems or phenomena — hormones; enzymes; immunity; viruses; blood; animal pheromones; homeostasis; neural transmitters; fever; thermoregulation; death; prenatal development.

Health related — chemotherapy; interferon; antibiotics; vitamins; food additives; biomedical engineering; zoonoses; psychosomatic disorders; stress; microscopy; laboratory animals.

Selecting a Topic in Nursing

The subjects listed below are very general; you will need to select a particular approach. Each subject presents a variety of *problems*. Think about a specific problem in terms of *health-care planning, health-care delivery,* or *ethics.* You will probably need to restrict your focus further by concentrating on a particular kind of situation. Suppose you are interested in *nursing homes*. Are you concerned with whom they serve? How they are staffed? Patient neglect? Recreational activities provided? Direct your attention to a particular problem and plan to present, in your Research Proposal, an explanation of the problem and how you intend to research its solution(s).

Possible Areas of Research

Public health; nursing homes; facilities (surgicenters, day care centers); delivery systems; training (of health professionals such as dieticians, physical therapists); specialized units (hemodialysis, coronary care); gerontology or geriatrics; patient rights; patient education (self-care, nutrition, breast-feeding); psychosomatic or psychophysiological disorders; stress; drug use; abortion; mental health; death and dying; euthanasia; genetic counseling; human rights and human experimentation.

Selecting a Topic in the Computer Sciences

Your Research Proposal and Report will focus on a *problem* in the computer sciences and your recommended *solution(s)* to that problem. For this assignment, you will choose one of the broad subjects listed below, narrow your perspective by identifying a current problem that is related to your chosen subject, and present solution(s) to that problem. One way to narrow your focus to a particular problem is to ask questions about the subject. For example, suppose you are considering *computer crime.* What types of crime can be committed with a computer? Theft. Fraud. Embezzlement. What are the targets of computer theft? Databases. Computer programs. Computer equipment. Who commits these thefts? Industrial competitors. Employees of a victimized company. Vandals. How can database thefts by competitors be foiled? "Lock and Key" terminals. Data encryption. User recognition schemes. Which is the best solution? Direct your attention to a particular problem and plan to present, in your Research Proposal, an explanation of the problem and how you intend to research its solution(s).

Possible Areas of Research

Computer Software — programming languages; database systems; programming methodology; operating systems; software engineering; natural language processing; macroinstruction.

   Computer Hardware/Systems/Technology — distributed processing; computer architecture; computer graphics; memory technology; simulation and models; computer networks; digital systems design; computer performance evaluation; microcomputers and microprocessors; data communications; computer-assisted design.

   Computer Applications — government and law; business; education; medicine and health care; humanities; social sciences.

   Computers and Society — computer crime; personal computing; privacy and the computer; the future of computing; computer personnel; automation and its effects on the workplace; computer law.

   Artificial Intelligence — pattern recognition; speech recognition; robotics; learning systems; expert systems; language translation; problem solving.

Selecting a Topic in Business

Business research is usually undertaken with very specific objectives in mind. The purpose of the research is often to find information that can be used to make a decision or to back up a proposed change in the way something is done. Your choice for this assignment is limited to a specific *company* and a particular *problem*. Choose a company from the list of "Selected 'Fortune 500' Companies." You should imagine yourself to be an insider in this company, which is considering one of the following changes:

introduction of a new product line
a new marketing strategy
a new advertising campaign
expansion of its operations overseas
a merger
diversification
introduction of new technology
use of new equipment
a new management technique or techniques
a price change
opening a new plant
new personnel policies
a new budgetary system
reorganization

To make a decision about a proposed change, you will need to examine and critique the past performance of both your *company* and the *industry* of which it is a part, finding as much relevant information as possible. For example, if you are considering a new advertising campaign, you need to find out what types of advertising have been done before, which have been successful, what your competitors have done, etc. Your company needs a detailed study that supports the change you recommend.

RESEARCH STRATEGY FOR TECHNICAL WRITING

Many jobs in scientific, technical, and health-related fields require the preparation of research proposals and research reports. The strategy you follow to complete this worksheet is a process that you can use successfully in other assignments, both in school and in your career. The search strategy takes you from the general to the specific, from older materials to recent developments. As you go through the process, you will become familiar both with the library that contains the information you need and with standard research sources in your field.

By following the steps in the worksheet, you will be gathering the information and compiling the bibliography you will need to write a Research Proposal. When the Proposal has been accepted, you will need to build upon the basic information you have already found to write the Library Research Report.

Your immediate objective is to plan and implement a search strategy on a topic. The steps you will take to complete this process are listed below.

Step I. Get started.
  1. Select a preliminary topic.
  2. Review note-taking procedures.
  3. Become familiar with the campus library you will be using.

Step II. Find background information.
  1. Revise your preliminary topic based on background reading.
  2. List important terms relating to your topic.
  3. Identify additional sources you would like to check later.

Step III. Find primary sources and recent material.
  1. Identify useful materials.
  2. Locate materials in the library.
  3. Evaluate materials.

Step IV. Prepare your Research Proposal.

Step V. Complete your research for your Research Report.

Step I. Getting Started

1. Use the list of "Subjects for Technical Writing Research" as a starting point. Each subject mentioned is too broad to use for a research assignment. Select some problem to focus on.

   Write your preliminary topic here:

2. For the research strategy to be most useful, you need to *read* the materials you list as you progress through each step of your search. This enables you to choose the kinds of information you need and avoid spending time listing items that are not

directly related to your topic. Plan to keep a record of your research for these reasons:

You will save time. You will be able to avoid retracing all your steps when you want to go back to an item you have already used.

You will have an accurate record to use for the bibliography for your Proposal and Report.

Others will be able to find the same information you have found.

One way to keep a record is with *bibliography cards* and *note cards.*

(a) A *bibliography card* has a *complete citation.* FOR BOOKS, a complete citation includes author or editor, title, publisher, and place and date of publication; FOR ARTICLES, author, title of the article, title of the publication, volume number, date, and the page numbers on which the article appears.

A bibliography card for a technical report, government document, or conference proceeding volume should include the complete name of the group or agency that issued it, the title, and the date and place of publication. Because reports and other publications that are neither books nor articles can be difficult to find, you should also write down any report number, series number, or classification number listed with the title. This extra information will give you a better chance of finding the item.

(b) A *note card* has the author's last name, page numbers, and notes.

3. Use the list of sources for your discipline (handed out in class). It tells you which library has most of the materials you will need. To locate materials efficiently, you must become familiar with the library you will be using most often. As you begin your research, think about these questions:

(a) Is there a handout explaining the layout of the library? (If so, it may provide answers to some of the questions below.)

(b) Is the card catalog divided into Name/Title and Subject sections, or is everything in one catalog?

(c) Once you have a call number, is there a chart, a sign, or a handout to help you locate the item?

(d) Are the periodicals arranged by call number with the books or alphabetically by title in a separate area of the library?

(e) Does the library have collections of special materials such as technical reports or government documents?

(f) Is there a handout or sign that tells you the hours when reference assistance is available?

Use the space below to write down any information about the library you may need to consult later.

Step II. Finding Background Information

Background information is material that provides a broad overview of a particular topic, thus giving you a framework in which to structure your search. Choose a background source from the annotated bibliography for your subject to complete the tasks described below. Read this entire section before you begin working.

1. *Revise preliminary topic.* Review the preliminary topic you chose in Step I. Think about this topic as you read a relevant article or chapter from the background source you have chosen. Then clarify, refocus, expand, or limit your original topic based on the following criteria:

   (a) *Audience.* Imagine your audience to be the board of directors, the president, or a vice president of a company, or another audience as specified by your teacher.

   (b) *Purpose of assignment.* Your objective is to identify a problem, explain it to people who know little about it, and give them enough information to decide whether to approve your research proposal.

   (c) *Researchability.* Because this particular assignment requires the use of library materials, you need to be sure that you have chosen a topic for which printed information is available. If you are unable to locate background information in any of the sources listed on the bibliography, discuss your topic with a librarian before you spend any more time on research for this assignment.

   Use the space below to state your topic in one or two sentences. Your topic statement should clearly define the problem you will be researching.

2. *List important terms.* As you read a background source, notice key words, terms, and phrases that are used to describe your topic. Make a note below of terms that you will want to look up when you look for primary sources and recent materials in Step III of this worksheet. If some terms are unfamiliar to you, check one of the dictionaries listed in the annotated bibliography you got in class.

3. *Identify additional information sources.* As you read the background sources you have located, notice any references to other sources. Sometimes these sources are cited in the article itself, but more often you will find a list of sources at the end of the article. The sources cited are generally key works in that subject area. If you think any of the items will be useful in your research, be sure to make a bibliography card for each. You will need a complete citation.

Step III. Finding Primary Sources and Recent Material

Once you have gained an overview of your topic and have identified a problem concerning it, you will be ready to search for possible solutions to the problem. The three parts of this process are described below:

1. Identify useful materials:
   (a) Turn to the section labeled "Finding Primary Sources and Recent Material" in the annotated bibliography for your field. Read the introduction, then select one of the indexes listed there and look up your topic in the index. Check several volumes. Make bibliography cards for the items that seem useful.
   (b) If you need additional information, check one of the other indexes listed on your bibliography, or ask a librarian to suggest other sources.
2. Locate in the library the items you have listed on bibliography cards:
   (a) Categorize the items: are they articles, books, government publications, technical reports?
   (b) To find *articles,* follow the suggestions in the handout on finding periodicals you were given in class. HINT: Your teacher will probably prefer that you use more articles from *journals* than from *magazines.* Journals are considered to be scholarly and often have footnotes and bibliographies; magazines are thought to be more popular — *Time* and *Newsweek* are examples of magazines. However, in a library both magazines and journals are called *periodicals,* and are found through the same methods. You will also encounter the term *serial,* as in SERIALS LIST. The term *serial* includes magazines, journals, and certain other items as well. Frequently these terms are used interchangeably.
   (c) To find a *book,* look in the card catalog for the name of the author or for the organization that issued the book. Ask a librarian if you don't find a card for the item you want.
   (d) To find *government publications, technical reports,* or other material you can't identify, ask a librarian for help.
3. Evaluate materials. As you look at various articles, books, and reports, you should consider the following criteria to determine which items will be the most useful for your research:
   (a) *Timeliness.* Is the item recent? Does it cover recent developments? Does it give a historical perspective?
   (b) *Coverage and relevance.* Does the item cover your topic in depth or only briefly? Does the item add to the information you already have? Will you use this item in your Research Proposal or Library Research Report?
   (c) *Nature of research.* Is the item a report of the writer's own research *(a primary source)* or does the item discuss the work of others *(a secondary source)?* What special knowledge or qualifications does the author possess? Does the article include footnotes? Is there a list of sources consulted? Will any of these sources be useful in your research?

Optional assignment: Use the criteria discussed above to write a critical evaluation and an abstract of one of the articles you have found.

Step IV. Preparing Your Research Paper

By the time you are ready to write your Research Proposal, your topic should be narrowed to a specific problem, and you should know what kinds of solutions are possible. In your Research Proposal, then, you explain the problem and how you intend to research the solutions to that problem.

Step V. Completing Your Research

After you have written your Research Proposal and have studied your teacher's comments on it, you will need to begin work on your Library Research Report, in which you will suggest solutions to the problems you have identified. For your Research Report, you may need to return to Step II of this worksheet, depending on how satisfied you and your teacher are with your topic statement and Research Proposal. You will definitely need to gather additional primary sources and recent material (Step III). As you continue your research, you should consider using additional volumes of indexes you have already used, trying other indexes, and checking different subject headings. In addition, feel free to ask a librarian to suggest other sources; many specialized indexes are not listed on your bibliography.

RESEARCH STRATEGY FOR TECHNICAL WRITING [BUSINESS]

Many jobs in business-related fields require the preparation of research proposals and research reports. The strategy you follow to complete this worksheet is a process that you can use successfully in other assignments, both in school and in your career. The search strategy takes you from the general to the specific, from older materials to recent developments. As you go through the process, you will become familiar with both the library that contains the information you need and the standard sources basic to business research.

By following the steps in the worksheet, you will be gathering the information and compiling the bibliography you will need to write a Research Proposal. When the Proposal has been accepted, you will need to build upon the basic information you have found to write the Library Research Report.

Your immediate objective is to plan and implement a search strategy on a topic. The steps you will take to complete this process are listed below.

Step I. Get Started.
1. Choose a company from the list of "Fortune 500" companies that was handed out in class.
2. Review note-taking procedures.
3. Learn the location and arrangement of the campus library that contains business materials.

Step II. Find background information.
1. Locate background information on your company and the industry in which it operates.
2. Select one aspect of that company's activities to focus on.

Step III. Find periodical articles.
1. Identify useful materials.
2. Locate materials in the library.
3. Evaluate materials.

Step IV. Prepare your Research Proposal.

Step V. Complete your research for your Research Report.

Step I. Getting Started

1. For this assignment, you are researching some aspect of a particular company. Use the handout listing selected "Fortune 500" companies to choose one for your project.

List your company:

[Directions 2 and 3 are omitted here because they are the same as those on pp. 506–507.]

Step II. Finding Background Information

To complete this section, you need to use the annotated bibliography of business information sources (handed out in class).

This step will help you to narrow the scope of your research and formulate a statement which defines the precise topic you will research and report on.

1. You need to locate some basic background information on the company you have chosen and on the industry within which it operates. The sources listed under "Corporate Information" and "Industry Information" on the annotated bibliography will be useful.

   (a) For your company, you should look for such information as a brief corporate history, a list of major lines of business, the location of the corporate headquarters, a list of the major subsidiaries, the number of employees, financial information such as operating revenues, and any major changes in the company's financial position in recent years.

   List the source(s) you found most useful for information on your company:

   (b) For your company's industry, you should look for such information as the structure of the industry (including a list of the major segments of that industry); names and market share of the major companies in the industry; important trends affecting the industry; and political, economic, or other events likely to affect the industry.

   List the source(s) you found most useful for information on your industry:

2. Now that you have some basic background facts on your company and its industry, you need to take a closer look at your company's recent activities to help you decide on the particular aspect of the company you will research. For this part of the assignment, you will need to use corporate annual reports which are described in the annotated bibliography.

   [Directions for finding material in UT—Austin libraries were given here.]

   (a) Locate and read through the most recent annual report for your company.
   (b) While you still have the annual report in hand, look at the suggestions under "Business" on the list of "Subjects for Technical Writing Research" (handed out in class). Using the information you have gathered about your company,

you should be able to choose an aspect of its activities on which to concentrate.

(c) Write a brief description of the specific area of your company's activities you plan to examine.

Step III. Finding Periodical Articles

A library research report should be based on a variety of sources. The material you have found so far has provided basic information. You now need more specific material on your company's past performance in the area you plan to examine. Some of the questions you may need to answer are: What has my company done in this area in the past five or ten years? Which activities or changes have been most successful? Why? What has been done by other companies in the same industry? Is my proposal new or has someone else already tried it?

An important source of business information is the articles published in various newspapers, business periodicals, and trade journals. To locate articles on a particular industry or company, you need to use the appropriate indexes.

1. Using several years of at least two of the indexes listed on the annotated bibliography, locate and make bibliography cards for four items that pertain to your topic. If you do not find that many articles, you may need to change the terms you are using or select another index, or you may need to change the focus of your paper. Talk to a librarian for assistance.
2. Use ''Periodicals: How to Find Them'' (handed out in class) to learn how to locate the periodicals you have identified. If you have any problems ask a librarian for help.
3. Select *one* of the articles that you locate. On a separate sheet, write an *abstract* and a *critical evaluation* of the article. The article you choose should be fairly long and relate to your topic. Although you are writing only one evaluation, the points listed below can help you to determine the value of other articles as well.

Points to consider: Is the article recent? Does it cover your topic in depth? Does it deal only briefly with your topic? Does it cover recent developments? Does it give an historical perspective? Does it add to the information you already have? Is there a bibliography? Are there footnotes? Are any of the sources listed items that you have already included in your bibliography? Will you use this article in preparing your Proposal on your topic?

Step IV. Using Your Research

When you have completed this worksheet, you should have enough information to write a Research Proposal. Focus on the *company* you have chosen and the particular *problem* you propose to research.

Step V. Finding Additional Information

To support your recommendations in the Library Research Report, you will need more information than you have collected so far. You can find additional articles by going back to the indexes listed on the annotated bibliography. Often you will need more statistical information to support a statement, to provide a basis for comparison and contrast, to develop a theory, or to show trends. The annotated bibliography lists useful sources for statistics. If you need other information, a librarian in the Reference Services Department can help you to identify additional sources which are appropriate for your topic.

In your Library Research Report, you will explain your company's situation and support the change that you recommend.

ENGINEERING

This is a list of some basic research sources in engineering. Beneath each title listed is a description of the subject areas it covers and an explanation of why it is useful.

The guide is arranged in *search strategy* order. A search strategy is a method of doing research by progressing in a logical manner from one type of information to another, from general sources to highly specialized ones. By developing and using a search strategy, you can find materials you need without missing important sources and without wasting time on irrelevant ones.

Finding Background Information

When you are beginning research in any unfamiliar subject area, you first need an overview of the subject. Listed in this section are a variety of sources for background information. Choose the source that seems most appropriate for your topic and your level of expertise in this field.

Specialized Encyclopedias

Use one of the science encyclopedias listed below if you have little or no technical background in engineering.

McGraw-Hill Encyclopedia of Science and Technology. 5th ed. New York: McGraw-Hill, 1982. This fifteen-volume encyclopedia covers most branches of science and technology. Use the index in Volume 15 to determine which volumes include articles on your topic. Bibliographies at the ends of articles refer you to other sources of information.

Van Nostrand's Scientific Encyclopedia. 6th ed. New York: Van Nostrand Reinhold, 1983. In two volumes, this encyclopedia covers such fields as earth and atmospheric sciences, energy technology, life sciences, materials sciences, mathematics, information sciences, physics, and chemistry.

Ralston, Anthony (Ed.). *Encyclopedia of Computer Science and Engineering.* New York: Van Nostrand Reinhold, 1983. Use the index at the back of this one-volume encyclopedia to locate background articles on topics such as microprocessors, integrated circuits, robotics, and human factors in computing.

Handbooks

If you have some technical knowledge of engineering, use one of the following handbooks to locate background information. *Handbooks* are compact reference books which provide charts, graphs, tables, formulas, and other specific data often needed by engineers.

Standard Handbook for Mechanical Engineers. 8th ed. New York: McGraw-Hill, 1978. This basic handbook for mechanical engineering covers the design of machines and of processes used to generate power and apply it to useful purposes. The detailed index at the back of the volume will help you locate articles on subjects such as corrosion, internal combustion engines, flywheels, and solid waste disposal. The short bibliographies at the beginning of each section will give you ideas of other sources to check.

Considine, Douglas M. *Energy Technology Handbook.* New York: McGraw-Hill, 1977. Illustrations, graphs, and bibliographies are included in this handbook. Each major chapter covers a broad subject, such as solar, coal, or petroleum technology, and has its own detailed table of contents. To find information on a specific topic, check the index at the back of the volume. *The United States Energy Atlas* is another good source of information on this subject.

Gaylord, Edwin H., and Charles N. Gaylord (Eds.). *Structural Engineering Handbook.* 2nd ed. New York: McGraw-Hill, 1979. Information about the design, planning, and construction of various kinds of structures is in this handbook. Use the table of contents or the index to find material on subjects such as bridges or tall buildings. Some chapters have bibliographies.

Dictionaries

The following dictionaries may be helpful if you encounter unfamiliar terms in the materials you find.

McGraw-Hill Dictionary of Scientific and Technical Terms. 3rd ed. New York: McGraw-Hill, 1984. This dictionary provides clear definitions of terms used in various scientific and technical fields.

Sippl, Charles J., and Roger J. Sippl. *Computer Dictionary.* 3rd ed. Indianapolis: Howard W. Sams, 1980. This dictionary briefly defines basic terms in computer science, and includes some electronics terms as well.

Finding Primary Sources and Recent Material

Unlike *Readers' Guide* and other indexes you may have used in the past, indexes that cover technical fields often have additional features that provide extra information for the researcher. Such extra features can make these indexes somewhat more difficult to use. Two types of specialized indexes are described below.

An ABSTRACTING INDEX usually requires a two-step approach. Whereas in *Readers' Guide* you can find a full citation to an article by looking directly under a subject heading, in an abstracting index you must look first in a subject index (an alphabetical sequence) to find an abstract number, then in an abstract section or volume (a numerical or classified sequence) to find a full citation. *Pollution Abstracts,* described below, is an example of an abstracting index.

An abstracting index provides a summary of an article, but a CITATION INDEX does not. Instead a citation index focuses on footnotes, although it does have subject access. A citation index tells you how often and by whom a particular article has been used as a source or footnote. If an article has been cited (referred to) frequently, it may be assumed to be important or influential. If an article is cited in important journals in your field or by authors who have written articles you have found useful, you may find the article relevant to your current area of research. *Science Citation Index,* described below, is an example of a citation index.

In the indexes listed here, you may find references to sources such as conference proceedings, technical reports, and government documents, as well as to
journal articles. This list is only a sample of the many specialized indexes available;
check with a librarian if you need help using one of the indexes listed below or if you
need a more specialized index.

Indexes and Abstracts

Applied Science and Technology Index. New York: H. W. Wilson, 1958 to the present. Because of
its broad coverage of technical fields, this index is a good place to begin looking for
periodical articles on almost any topic related to engineering.

Engineering Index. New York: Engineering Index, Inc., 1982 to the present. Journal articles,
reports, books, and other technical materials are listed in this index. This is a classified
index, so items on the same subject will be listed together, and you can find both the
abstract number and the actual citation by looking directly under your subject. If you
do not find your subject at first, try looking for synonyms or broader terms in this index,
or ask a librarian for help.

Energy Research Abstracts. Oak Ridge, TN: United States Department of Energy, Technical
Information Center, 1975 to the present. Use this classified abstracting index to find
material on coal, petroleum, nuclear or geothermal energy, and related topics. Check
the subject contents in the front of any issue to identify the classification number for
your topic. This number will be the same in each issue.

Pollution Abstracts. Louisville, KY: Data Courier, 1970 to the present. This index with abstracts
is arranged by broad categories, such as radiation, noise, and sewage and wastewater
treatment. The index in the back will help you find specific subjects, such as names of
particular pollutants.

Computer and Control Abstracts. London: Institution of Electrical Engineers and the Institute of
Electrical and Electronics Engineers, 1973 to the present. The articles and technical
papers in this monthly index range from basic to highly advanced in level. Use the
"Classification and Contents" page of each monthly issue to locate articles on your
subject. A more detailed "Author-subject" index is published twice every year. Check
with a librarian if you cannot locate articles on your topic.

Science Citation Index. Philadelphia: Institute for Scientific Information, 1961 to present. There
are several ways to use this index. You can look up your *subject* to find articles. Or, if you
have found a particularly good article, you can use this index to locate other articles by
the same *author,* or articles that *cite* that article.

This index can also be used to *evaluate* your sources. An item that has been cited by
many writers is likely to be one that has made a valuable contribution in its field, and it
may be useful for your research as well.

This index is very different from other indexes, so read the instructions for using it,
located with the volumes, before you begin. Ask a librarian if you need additional help.

When you have followed the steps described in this bibliography, you should have
located the materials you need for your research. If you have any problems, ask a
librarian for help.

LIFE SCIENCES

Finding Background Information

When you are beginning research in any unfamiliar subject area, you first need an overview of the subject. Listed in this section are several sources for background information.

The first two sources listed are specialized encyclopedias. They cover a wide range of subjects and are particularly useful when you have little or no familiarity with the subject. The other sources are varied. Read the descriptions closely to decide which items may be most useful for your topic and your level of expertise in the field.

McGraw-Hill Encyclopedia of Science and Technology. 5th ed. New York: McGraw-Hill, 1982. This fifteen-volume encyclopedia covers most branches of science, technology, and health-related subjects. Because of its arrangement, you must use the index volume to locate articles on your topic.

Gray, Peter (Ed.). *Encyclopedia of the Biological Sciences.* 2nd ed. New York: Van Nostrand Reinhold, 1970. This encyclopedia covers all aspects of the life sciences, with discussions of such subjects as viruses, microscopy, symbiosis, antibiotics, and genetics. The articles are somewhat shorter than those in McGraw-Hill. Look in the index for page numbers in boldface type; they refer to the major articles on a subject in the encyclopedia.

Scientific American Cumulative Index 1948–1978. New York: Scientific American, 1979. With this cumulative index to 30 years of *Scientific American,* you can use the bound volumes of this magazine as you would a general scientific encyclopedia. It is especially useful for information on body systems and health-related subjects. Each topic entry in the index lists related terms, briefly summarizes the article, and tells the issue in which the article is located. Short bibliographies appear at the end of each magazine issue.

Ricklefs, Robert E. *Ecology.* Newton, MA: Chiron Press, 1973. Begin here for a good introduction to the study of ecology. This textbook includes articles on the physiology and behavior of organisms, on genetics, and on evolutionary biology. Territorial behavior, natural selection, adaptation, symbiosis, and evolution are among the subjects covered. The book includes a glossary, an extensive bibliography, and an appendix which provides a list of "Reference Literature on Ecology," outlining major books, journals, and collections of reprinted articles in the field. Use the index at the back of the book to locate your subject.

Pianka, Eric R., *Evolutionary Ecology.* 3rd ed. New York: Harper & Row, 1983. This textbook includes information on genetics, species, mutation, niche, marine ecology, natural selection, symbiosis, and other topics. At the end of each chapter is a list of additional sources, arranged by topic. Each entry is abbreviated, so turn to the bibliography at the end of the book to find complete citations.

Colinvaux, Paul A. *Introduction to Ecology.* New York: Wiley, 1973. This textbook is useful if you have some background in ecology. It provides good overviews of subjects such as niche, ecosystems, pesticides, and species populations. Materials by authors cited within the text are listed in the "References" section at the back of the book. Following this section is a combined index-glossary.

Williams, Roger J., and Edwin M. Lansford, Jr. *Encyclopedia of Biochemistry.* New York: Van Nostrand Reinhold, 1967. This is a good source for those with a background in biology. It provides technical articles on hormones, viruses, antibiotics, genetics, mutation, and immunity. However, this encyclopedia is quite old, so you will need to update the information you find here.

Dictionaries

Some of the terms used in the materials you find may be unfamiliar to you. Use one of these dictionaries to find definitions.

Collocott, T. C. (Ed.). *Chambers Dictionary of Science and Technology*. New York: Barnes & Noble, 1972. This dictionary briefly defines terms used in the sciences.

Martin, E. A. *A Dictionary of Life Sciences*. London: Macmillan, 1976. The definitions are longer in this dictionary than in *Chambers* and illustrations accompany many of the definitions.

Finding Primary Sources and Recent Material

[Three paragraphs, essentially the same as those under this heading on pp. 516 – 517, have been omitted here.]

In the indexes listed here, you may find references to sources such as conference proceedings, technical reports, and government documents, as well as to journal articles. Read the descriptions of the indexes to determine which ones will be most appropriate for your topic.

Indexes and Abstracts

Biological Abstracts. Philadelphia: Biosciences Information Service, 1926 to the present. Use this to find materials on topics in biology, nutrition, behavioral sciences, and genetics.

Environment Index. New York: Environment Information Center, 1971 to the present. Natural history and ecology are covered in this specialized index. Consult the "Subject Terms: Keyword List" to select the terms that describe your topic and use the subject index to locate articles. To find an abstract of an article, copy the number at the end of the index entry, and find that number in the corresponding abstract volume. The abstract will be found in *Environment Abstracts*, listed below. It will contain a summary of the article, book, conference proceeding, technical report, or government document you have chosen.

Environment Abstracts. New York: Environment Information Center, 1971 to the present. Use this with *Environment Index* (described above).

Cumulated Index Medicus. Bethesda, MD: National Library of Medicine, 1960 to present. For subjects related to health or the body, this index will help you to locate magazine articles, review articles, and chapters in books. Begin with Volume 2 of the annual cumulation to locate review articles. If there is one for your subject you will have a good start on your research. Volume 2 indexes review articles by subject in the section "Bibliography of Medical Reviews." Use the subject index (Volumes 7 through 14) to locate other articles on your topic. Update your findings by using the monthly paperback supplements.

Science Citation Index. Philadelphia: Institute for Scientific Information, 1961 to present. There are several ways to use this index. You can look up your *subject* to find articles. Or, if you have found a particularly good article, you can use this index to locate other articles by the same *author,* or other articles that *cite* that article.

This index can also be used to *evaluate* your sources. An item that has been cited by many writers is likely to be one that has made a valuable contribution in its field, and it will probably be very useful for your research as well.

A citation index is very different from other indexes, so read the instructions located with the volumes before you begin. If you have trouble using this index, ask a librarian for help.

When you have followed the steps described in this bibliography, you should have located the materials you need for your research. If you have any problems, ask a librarian for help.

COMPUTER SCIENCES

Finding Background Information

When you are beginning research in an unfamiliar subject area, you first need an overview of the subject. There is no single source that provides a good discussion of all topics in the computer sciences, so read the descriptions carefully to determine which of the sources listed below will be most useful for your topic and your level of expertise in this field. Your own textbooks may also provide some information.

Ralston, Anthony (Ed.). *Encyclopedia of Computer Science and Engineering.* 2nd ed. New York: Van Nostrand Reinhold, 1983. Since this volume covers all aspects of the field in easily understandable terms, it is a good place to begin if you need to acquire a basic familiarity with your topic. The articles, which are usually two to three pages long, provide references to related articles in this encyclopedia and also list other sources of good introductory material. Use the index at the back to begin to locate articles on your topic.

Encyclopedia of Computer Science and Technology. New York: Marcel Dekker, 1975–81. 16 vols. This encyclopedia provides abundant information on computer systems and development. It explores such areas as hardware, software development, and artificial intelligence. The lengthy articles begin with basic information and progress to a more detailed and technical level. Start with the subject index in Volume 16.

The McGraw-Hill Computer Handbook. New York: McGraw-Hill, 1983. Like an encyclopedia, this handbook explains basic computer concepts. It includes in-depth discussions of some computer programming languages, computer graphics, voice recognition, microcomputers, and computer history. There is a glossary and an index in the back.

McGraw-Hill Yearbook of Science and Technology. New York: McGraw-Hill, current year. Keeping up in as fast-moving a field as the computer sciences is difficult. Each year new systems are introduced, fresh applications are devised, and unforeseen problems come to light. This yearbook covers current developments in all branches of the sciences, including the computer sciences. Use the index to identify articles relevant to your topic. This yearbook serves as a supplement to the *McGraw-Hill Encyclopedia of Science and Technology.*

Advances in Computers. New York: Academic Press, 1960 to the present. This is a good source for the advanced student. Each annual volume contains review articles of current interest to computer scientists. Review articles summarize significant publications on a particular subject, mention other sources to consult, and often indicate areas of the subject that require further research.

Sanders, Donald H. *Computers in Society.* 3rd ed. New York: McGraw-Hill, 1981. Use this to find information on the applications of computers to various walks of life. Use the index and the table of contents to find information on your topic.

Logsdon, Thomas S. *Computers and Social Controversy.* Potomac, MD: Computer Science Press, 1980. This book contains good background chapters on computer technology and memory systems, and describes ways in which computers affect our society. Consult the table of contents to find the chapters most relevant to your topic.

Finding Primary Sources and Recent Material

[Three paragraphs, essentially the same as those under this heading on pp. 516–517, have been omitted here.]

In the indexes listed below, you may find references to sources such as conference proceedings, technical reports, or government documents, as well as to journal articles. Read the descriptions of the indexes to determine which ones will be most appropriate for your topic.

Computer & Control Abstracts. London: Institution of Electrical Engineers and the Institute of Electrical and Electronics Engineers, 1973 to the present. For any aspect of computer sciences, this abstracting index will lead you to numerous articles. To locate recent articles, find your subject and its classification number in the Subject Guide at the end of each monthly issue of the index. Then, to find the page number of the abstracts, look for the classification number on the "Classification and Contents" page on the back cover of the issue.

Computer & Information Systems Abstracts Journal. Riverdale, MD: Cambridge Scientific Abstracts, 1965 to 1983. This is one place to find information on major aspects of the computer sciences, such as software, applications, and computer electronics. Begin with the subject index in the back. Beneath each subject heading you will find a list of other headings assigned to a particular article, along with the abstract number for that article. Use the abstract number to find the complete citation in the front of the volume.

Computer Literature Index (formerly *Quarterly Bibliography of Computers and Data Processing*). Phoenix, AZ: Applied Computer Research, 1977 to the present. Begin here if your topic is not too technical or if you want references to basic material rather than detailed articles or scholarly reports. This index will direct you to a broad selection of articles in such areas as the impact of computers on society and the uses of computers in various professions. Turn directly to your subject. Abbreviations of journal titles are spelled out in the back of each volume.

ACM Guide to Computing Literature (formerly *Bibliography of Current Computing Literature*). New York: Association for Computing Machinery, 1960 to the present. This index covers major fields in the computer sciences, with special emphasis on technical topics. Look in the Keyword Index for a term that describes your topic. Under that term, or *keyword*, titles of books and articles are listed. In each title, an asterisk (*) replaces the keyword to save space. A *number* is listed after the title of each item. Turn to the front of the volume, where sources are listed in numerical order, to find the complete information you will need to find that source.

Microcomputer Index. Santa Clara, CA: Microcomputer Information Services, 1980 to the present. Articles on the use of microcomputers in business, education, and the home are listed in this index. Find your subject in the subject index in the front of each volume, then use the abstract numbers to find relevant abstracts in the back.

Science Citation Index. Philadelphia: Institute for Scientific Information, 1961 to the present. There are several ways to use this index. You can look up your *subject* to find articles. Or, if you have found a particularly good article, you can use this index to locate other articles by the same *author*, or articles that *cite* that article.

This index can also be used to *evaluate* your sources. An item that has been cited by many writers is likely to be one that has made a valuable contribution in its field, and will probably be useful for your research as well.

This index is very different from other indexes, so read the instructions located with the volumes before you begin. If you have trouble using this index, ask a librarian for help.

Dictionaries

Some of the terms used in the materials you find may be unfamiliar to you. Use one of these dictionaries to find definitions.

Sippl, Charles J., and Roger J. Sippl. *Computer Dictionary.* 3rd ed. Indianapolis: Howard W. Sams, 1980. This dictionary briefly defines terms used in any area of computer sciences, including personal computers.

Weik, Martin H. *Standard Dictionary of Computers and Information Processing.* Rev. 2nd ed. Rochelle Park, NJ: Hayden, 1977. The definitions in this dictionary are a little more substantial than those in the dictionary listed above, but some recent terminology is not included.

When you have followed the steps described in this bibliography, you should have located the materials you need for your research. If you have any problems, ask a librarian for help.

NURSING

Finding Background Information

When you are beginning research in any unfamiliar subject area, you first need an overview of the subject. There is no single source that provides a good discussion of all topics in the health-care field. Your own textbooks may provide background information, or you may choose to use one of the sources listed below. Read the annotations carefully to determine which book will be most useful for your topic and level of expertise in this field.

Reich, Warren T. (Ed.). *Encyclopedia of Bioethics.* New York: Free Press, 1978. For information on topics concerned with the ethics or morality of health-care decisions, look here. Issues covered include mental health, death and dying, genetic counseling, human experimentation, informed consent, and psychopharmacology. Use the index in Volume four and the cross-references at the end of each article to find all of the information on your topic.

Wolman, Benjamin B. (Ed.). *International Encyclopedia of Psychiatry, Psychology, Psychoanalysis, and Neurology.* New York: Aesculapius Press, 1977. Use this to find background on the psychological aspects of health-care problems and services. Articles identify important writers on the subject and provide a bibliography. Most subjects are covered in several articles, so use the Subject Index in Volume twelve to find the articles most relevant to your topic. Beginning in 1983, this encyclopedia is being updated by "Progress Volumes."

Fishbein's Illustrated Medical and Health Encyclopedia. 4 vols. Westport, CN: H. S. Stuttman, 1977. Use these volumes for explanations of terms and concepts in the medical and nursing fields, such as drug abuse, geriatrics, genetic counseling, and hemodialysis.

Sorensen, Karen C., and Joan Luchman. *Basic Nursing: A Psychophysiologic Approach.* Philadelphia: W. B. Saunders, 1979. This textbook has good discussions of health-care delivery, the aged, patient education, stress, and dying. Use the index to find the material on your topic.

Encyclopaedia Britannica. Chicago: Encyclopaedia Britannica, Inc., 1985. This encyclopedia provides good background information on areas of health care that are of interest to the layperson, such as public health, aging, stress, drug use, death, euthanasia, and genetic counseling. The volumes labeled "Micropaedia," with the volume numbers in Roman numerals, are the *index* to the principal articles in the "Macropaedia," which has Arabic volume numbers.

Dictionaries

Some of the terms used in the materials you find may be unfamiliar to you. Use one of the dictionaries to find definitions.

Dorland's Illustrated Medical Dictionary. 26th ed. Philadelphia: W. B. Saunders, 1981. This dictionary provides short definitions and some illustrations.

Thomas, Charles L. (Ed.). *Taber's Cyclopedic Medical Dictionary.* 14th ed. Philadelphia: F. A. Davis, 1981. Use this for definitions of terms used in nursing and medicine.

McGraw-Hill Nursing Dictionary. New York: McGraw-Hill, 1979. This dictionary defines terms in nursing and related fields.

Finding Primary Sources and Recent Material

Using an index is one of the most efficient ways to find recent materials and original reports of research. Because nursing is a field that incorporates information from many disciplines, nursing students may need to use indexes that cover broad subjects such as social sciences and public affairs, as well as highly specialized indexes that cover medicine or nursing. These kinds of indexes require two different approaches.

In a less specialized index, such as *Social Sciences Index,* described below, you can look directly under a subject, such as "Health planning," and find either a list of sources or a suggestion of another term to try.

However, the more specialized indexes use a more detailed system of terminology. Some of these indexes require that you find the appropriate subject heading in a separate volume called *Medical Subject Headings;* others may have their own subject indexes. Read the description of the index you will be using to find out how it handles subject headings. Ask a librarian for assistance if you are uncertain about how to use one of these indexes.

Indexes and Abstracts

Cumulative Index to Nursing and Allied Health Literature. Glendale, CA: Seventh Day Adventist Hospital Association, 1977 to the present. To find material on topics in nursing or allied health disciplines, use this index. The list of subject headings used is at the back of each annual volume. Cross-references lead you to related information. Articles published before 1977 are indexed in *Cumulative Index to Nursing Literature.*

International Nursing Index. Philadelphia: American Journal of Nursing, 1966 to the present. This index covers nursing journals internationally. Begin with the *Nursing Thesaurus,* a special section of the index that leads you from commonly used nursing terms to the *Medical Subject Headings* used in this index. Non-English-language articles will have brackets around the titles and will have journal names in other languages.

Hospital Literature Index. Chicago: American Hospital Association, 1957 to the present. Use the subject section of this index to find recent articles on aspects of the delivery of health care, with emphasis on the management, planning, and financing of hospitals and related health-care facilities. To choose your subject headings, use *Medical Subject Headings.*

Abstracts of Health Care Management Studies. Ann Arbor, MI: Health Administration Press, 1977 to the present. This abstracting index is useful for topics in health-care planning, administration, and delivery. Use the subject index at the back of each volume to find the *number* and *letter* code for materials on your topic. Use the letter code to find the appropriate *section* of the volume. The sections are arranged alphabetically by letter code. In each section, the abstracts are in numerical order. Each abstract provides a summary of the journal article, book, or report.

Social Sciences Index. Bronx, NY: H. W. Wilson Co., 1974 to the present. Many journals in the social sciences include articles related to nursing and health care. This index will help you find such articles. Topics such as nursing homes, training and status of nurses, gerontology, patient satisfaction, stress, euthanasia, and human experimentation are covered in this index.

Public Affairs Information Service Bulletin. (PAIS) New York: Public Affairs Information Service, Inc., 1915 to the present. For information on public policy related to health care, use this index. Topics covered include nursing homes, mental health, terminal care, old age, and hospitals. *PAIS* indexes books, journal articles, government documents, and the reports of public and private organizations.

Cumulated Index Medicus. Bethesda, MD: National Library of Medicine, 1960 to the present. This is the primary clinical medicine index in the United States, and its emphasis is on diseases. However, many other areas of concern are also covered, such as health-care planning and delivery, medical ethics, and nursing topics. To find articles on your topic, begin with the *Medical Subject Headings,* volume 1 of each year, to determine the correct headings under which to look.

When you have followed the steps described in this bibliography, you should have located the materials you need for your research. If you have any problems, ask a librarian for help.

BUSINESS

Finding Background Information

When you are beginning research on a company, you first need an overview of the company's operations and its place within the industry of which it is a part. Listed in this section are a variety of sources for this kind of background information. The descriptions of each source will help you to decide which particular sources to use. In your reading, if you find terms you do not understand, use the business dictionaries listed below.

Corporate Information

Use these sources to find information about your *company.*

Standard and Poor's Corporation Records. 7 vols. New York: Standard and Poor's Corporation, current year. This is a good place to find basic financial and corporate data. For each corporation listed, you can find corporate background, including a brief history of the company, its principal subsidiaries, its capitalization, bond descriptions, stock data, and financial data such as annual reports and a consolidated balance sheet.
Use the index (on yellow pages) at the front of the volume to locate a particular company. The page number given in bold print leads you to the basic financial and corporate analysis. The other pages listed are updates, such as interim reports of earnings, acquisitions, etc.

Moody's Industrial Manual. New York: Moody's Investors Service, current year. As a source of background information, this manual is similar to *Standard and Poor's Corporation Records.* In some areas, however, the amount of information varies. You may have to use both sources to get the details you need.
Look up the company's name in the "Alphabetical Index" (on blue pages) at the front of the volume.

Industry Information

Use these sources for information about the *industry* within which your company operates.

Standard and Poor's Industry Surveys. New York: Standard and Poor's Corporation, current issues. This provides detailed coverage of sixty-nine major industries as well as information on individual companies within each major industry group.
Use the "Subject Guide" at the front of the volume to locate specific products or industries, or use the "Index to Companies" immediately following the "Subject Guide" to locate the correct industrial group for your specific company.
For each industrial group, you will find information divided into two sections. The *Current Analysis* (about eight pages) provides information about the latest developments and statistics on the industry. The *Basic Analysis* (about forty pages) provides an overview of the industry, with background on each segment of that particular industry.

Analysis of trends and problems, with many statistical charts and tables, is provided. At the end of the *Basic Analysis* is another important feature: tables of composite industry data and tables of comparable financial statistics for each of the major companies in the industry.

U.S. Industrial Outlook. Washington, DC: U.S. Government Printing Office, current year. This source provides information on trends and gives five-year projections for 200 industries, but not for individual companies. The information provided here is less detailed than that found in *Standard and Poor's Industry Surveys.* Use the index at the back of the volume to find the specific industry on which you need information. References to sources of additional information are listed at the end of the chapter.

Everybody's Business. San Francisco: Harper & Row, 1980. This provides information on 317 large corporations in an entertaining and readable narrative. In addition to the usual descriptions of corporate history and activities, there are also many short essays and background facts which give an insider's view of the companies.

Additional information on recent developments is provided in *Everybody's Business: 1982 Update.*

Corporate Annual Reports

Use an annual report for a detailed summary of your company's recent activities.

Corporate Annual Reports: prepared annually by each stock-issuing corporation, the annual report supplies an overview of the company's recent activities as well as a detailed financial statement. Information on products and subsidiaries and a discussion of the company's future are also generally included.

Dictionaries

Some of the terms used in the materials you find may be unfamiliar to you. Use one of these dictionaries to find definitions.

Rosenberg, Jerry M. *Dictionary of Business and Management.* New York: Wiley, 1983. Use this for brief definitions of terms from all phases of business.

McGraw-Hill Dictionary of Modern Economics. 2nd ed. New York: McGraw-Hill, 1973. This is a good source for general business and economics definitions.

Finding Periodical and Journal Articles

The best way to find recent materials on the particular aspect of the company you have chosen is by using *indexes* to periodicals and journals. There are many indexes that cover business-related topics exclusively.

In most indexes, you can find information by looking under the company's name and under the industry. Depending on your topic, you may also find it helpful to look under subject headings for particular products or product categories, or other terms related to your topic.

Almost all indexes abbreviate the titles of sources, but each provides a list of abbreviations. The list is usually in the front of each volume. Be sure to use the list of abbreviations to determine the correct titles of the sources you plan to use. You will need the full title to locate the item in the library.

The indexes listed below are those that are *most likely* to lead you to articles on your topic. If you have trouble finding information, ask a librarian at the Reference Desk for help.

Indexes

Business Index. Los Altos, CA: Information Access Corporation, 1979 to the present. This new index is a very good source for information on all aspects of companies. It indexes some newspapers as well as magazines. Since it covers only the period from 1979 to the present, you will need to use the other indexes listed below to find material from earlier years.

Because this is a computer-generated index, you must look under all possible variants of the company's name (for example, IBM *and* International Business Machines).

Specific subdivisions ("-advertising," "-organization and management," "-mergers and consolidation," "-prices," etc.) are used when there are many articles listed for a single company. When there are few articles on a company, the articles are given in one list beginning with the most recent.

Business Periodicals Index. New York: H. W. Wilson Co., 1958 to the present. Use this index to find information on all phases of business. All articles on a company are listed under the company's name.

F & S Index of Corporations and Industries. Cleveland: Predicasts, 1962 to the present. This is a good source for information about all aspects of a company's operations. It is divided into two sections. The first is an index of products which is arranged by Standard Industrial Classification (SIC) numbers. For an explanation of SIC numbers, see below. The second section is an index of companies which lists them alphabetically. This section uses specific subdivisions under company names. The subdivisions are somewhat different from those used in the other business indexes, and you may need to scan carefully to determine which ones list the articles on your topic. For example, advertising and marketing are part of the "business procedures" subdivision under company names.

The Wall Street Journal. Index. New York: Dow Jones and Co., 1957 to the present. This index is useful for current information on companies. In the *Corporate News* section, articles are listed alphabetically by the name of the company, and then in chronological order. Many brief news items are included in the index. The second section of the index, *General News,* is arranged alphabetically by subject.

Finding Statistical Information

When you are using statistical sources, you need to know exactly what the figures given represent and how they were obtained or derived. Knowing who is responsible for the statistics is often a good clue to their reliability. When you locate statistics that pertain to your topic, you need to record, *in addition to the statistics themselves:* (1) the units of measurement used (millions of dollars, thousands of population, etc.), (2) the original source of the data, (3) the time period covered, and (4) the way in which the figures were determined (an actual count, a survey, or projections, for example).

Much published statistical information in the field of business is arranged by Standard Industrial Classification (SIC) numbers. The Standard Industrial Classification was developed by the federal government to classify business establishments by the type of activity in which they are engaged. The SIC number for your company will, therefore, be based upon its major product or activity.

The statistical sources which use the SIC often contain an index to the numbers. You can also find the SIC number for your company by using the *Standard Industrial Classification Manual* listed below.

U.S. Office of Management and Budget. *Standard Industrial Classification Manual,* Washington, DC: U.S. Government Printing Office, 1972. A detailed description of the basic activities and types of establishments within all U.S. industrial groups is provided by this source. Use the "Alphabetical Index, Manufacturing Industries" (beginning on page 443) to find the SIC number for your company's product or industry.

Statistical Information

Predicasts. Cleveland: Predicasts, current year. This source provides projections and forecasts for specific *products* and *industries.* It is arranged by SIC numbers. Use the "Alphabetic Guide to the SIC's" (yellow pages) at the back of the volume.
For an interpretation or explanation of the data you find, look at the section "How to Use the Predicasts Forecasts Abstract Sections" following the yellow pages at the front of the volume.

Basebook. Cleveland: Predicasts, current year. For each industry and product, a "time series" of statistical data is given for a period of approximately fifteen years. The information given includes such measures as production, profits, sales, payroll, plant and equipment expenditures, and expenditures for research. It is arranged by modified SIC numbers. There is an "Alphabetical Guide to the SIC's" at the back of the volume.

U.S. Bureau of the Census. *Statistical Abstract of the United States.* Washington, DC: U.S. Government Printing Office, current year. This yearly summary of statistics collected by the U.S. government includes social, political, and economic characteristics of the nation, each of the states, and the major cities. Statistics on population, transportation, wages, and prices can be found here. Use the index at the back of the volume to find your subject. Introductory sections at the beginning of each chapter explain terminology, describe data collection methods, and provide background on the subject.

U.S. Bureau of the Census. *1977 Census of Manufactures.* Washington, DC: U.S. Government Printing Office, 1981. Use this set for statistics on many manufacturing industries. The series of paper volumes containing industry statistics arranged by SIC numbers may be the most useful for this assignment. Other volumes cover area statistics, locations of manufacturing plants, and production indexes for both manufacturing and mining industries. If you need to supplement the data found here with more recent figures, a reference librarian can suggest additional sources.

When you have followed the steps described in this bibliography, you should have located the materials you need for your research. If you have any problems, ask a librarian for help.

SELECTED "FORTUNE 500" COMPANIES*

Exxon (New York)
General Motors (Detroit)
Mobil (New York)
Ford Motor (Dearborn, Mich.)
Texaco (Harrison, N.Y.)
Standard Oil of California (San Francisco)
Gulf Oil (Pittsburgh)
International Business Machines (Armonk, N.Y.)
General Electric (Fairfield, Conn.)
Standard Oil (Ind.) (Chicago)
International Telephone & Telegraph (New York)
Atlantic Richfield (Los Angeles)
Shell Oil (Houston)
U.S. Steel (Pittsburgh)
Conoco (Stamford, Conn.)
Chrysler (Highland Park, Mich.)
Tenneco (Houston)
Sun (Radnor, Pa.)
Occidental Petroleum (Los Angeles)
Phillips Petroleum (Bartlesville, Okla.)
Procter & Gamble (Cincinnati)
Dow Chemical (Midland, Mich.)
Union Carbide (New York)
United Technologies (Hartford)
International Harvester (Chicago)
Goodyear Tire & Rubber (Akron, Ohio)
Boeing (Seattle)
Eastman Kodak (Rochester, N.Y.)
LTV (Dallas)
Standard Oil (Ohio) (Cleveland)
Caterpillar Tractor (Peoria, Ill.)
Union Oil of California (Los Angeles)
Beatrice Foods (Chicago)
RCA (New York)
Westinghouse Electric (Pittsburgh)
Bethlehem Steel (Bethlehem, Pa.)
R.J. Reynolds Industries (Winston-Salem, N.C.)
Amerada Hess (New York)
Esmark (Chicago)
Marathon Oil (Findlay, Ohio)
Ashland Oil (Russel, Ky)
Rockwell International (Pittsburgh)
Kraft (Glenview, Ill.)
Cities Service (Tulsa)
Monsanto (St. Louis)
Philip Morris (New York)
General Foods (White Plains, N.Y.)
Minnesota Mining & Manufacturing (St. Paul)
Gulf & Western Industries (New York)
Firestone Tire & Rubber (Akron, Ohio)
McDonnell Douglas (St. Louis)
W.R. Grace (New York)
Georgia-Pacific (Portland, Ore.)
PepsiCo (Purchase, N.Y.)
Armco (Middletown, Ohio)
Coca-Cola (Atlanta)
Deere (Moline, Ill.)
Colgate-Palmolive (New York)
Getty Oil (Los Angeles)
Aluminum Co. of America (Pittsburgh)
Consolidated Foods (Chicago)
International Paper (New York)
Ralston Purina (St. Louis)
TRW (Cleveland)
Allied Chemical (Morristown, N.J.)
American Can (Greenwich, Conn.)
Weyerhaeuser (Tacoma, Wash.)

Continental Group (Stamford, Conn.)
Borden (New York)
Charter (Jacksonville, Fla.)
Signal Companies (Beverly Hills, Calif.)
National Steel (Pittsburgh)
Iowa Beef Processors (Dakota City, Neb.)
Johnson & Johnson (New Brunswick, N.J.)
Honeywell (Minneapolis)
Sperry (New York)
Litton Industries (Beverly Hills, Calif.)
Lockheed (Burbank, Calif.)
General Dynamics (St. Louis)
Republic Steel (Cleveland)
Champion International (Stamford, Conn.)
Bendix (Southfield, Mich.)
American Brands (New York)
General Mills (Minneapolis)
IC Industries (Chicago)
Raytheon (Lexington, Mass.)
CPC International (Englewood Cliffs, N.J.)
CBS (New York)
Inland Steel (Chicago)
Owens-Illinois (Toledo)
United Brands (New York)
Dresser Industries (Dallas)
American Home Products (New York)
Textron (Providence)
Eaton (Cleveland)
FMC (Chicago)
Reynolds Metals (Richmond, Va.)
Texas Instruments (Dallas)
Warner-Lambert (Morris Plains, N.J.)
American Cyanamid (Wayne, N.J.)
Celanese (New York)
J. Ray McDermott (New Orleans)
American Motors (Southfield, Mich.)
PPG Industries (Pittsburgh)
NCR (Dayton, Ohio)
B.F. Goodrich (Akron, Ohio)
Kaiser Aluminum & Chemical (Oakland, Calif.)
Boise Cascade (Boise, Idaho)
AMAX (Greenwich, Conn.)
Carnation (Los Angeles)
Crown Zellerbach (San Francisco)
Burroughs (Detroit)
Anheuser-Busch (St. Louis)
Dana (Toledo)
Combustion Engineering (Stamford, Conn.)
Bristol-Myers (New York)
Pfizer (New York)
Borg-Warner (Chicago)
Motorola (Schaumburg, Ill.)
Teledyne (Los Angeles)
Norton Simon (New York)
Kerr-McGee (Oklahoma City)
Burlington Industries (Greensboro, N.C.)
Emerson Electric (St. Louis)
Standard Brands (New York)
Singer (Stamford, Conn.)
Northwest Industries (Chicago)
Uniroyal (Middlebury, Conn.)
Mead (Dayton, Ohio)
Ingersoll-Rand (Woodcliff Lake, N.J.)
Time Inc. (New York)
St. Regis Paper (New York)
H.J. Heinz (Pittsburgh)
Fruehauf (Detroit)

* These companies have been selected from the list of "The 500 Largest Industrial Corporations (ranked by sales)," published in *Fortune* 101, no. 9 (May 5, 1980): 274–295.

Central Soya (Fort Wayne, Ind.)
Kennecott Copper (Stamford, Conn.)
American Standard (New York)
North American Philips (New York)
Dart Industries (Los Angeles)
Merck (Rahway, N.J.)
Avon Products (New York)
Nabisco (East Hanover, N.J.)
Hewlett-Packard (Palo Alto, Calif.)
Diamond Shamrock (Dallas)
Hercules (Wilmington, Del.)
Archer-Daniels-Midland (Decatur, Ill.)
General Tire & Rubber (Akron, Ohio)
Walter Kidde (Clifton, N.J.)
Johns-Manville (Denver)
Whirlpool (Benton Harbor, Mich.)
Campbell Soup (Camden, N.J.)
Control Data (Minneapolis)
Owens-Corning Fiberglas (Toledo)
Ogden (New York)
Kimberly-Clark (Neenah, Wis.)
Eli Lilly (Indianapolis)
Pillsbury (Minneapolis)
Colt Industries (New York)
NL Industries (New York)
Levi Strauss (San Francisco)
Martin Marietta (Bethesda, Md.)
American Broadcasting (New York)
Pennzoil (Houston)
Gould (Rolling Meadows, Ill.)
White Consolidated Industries (Cleveland)
Gillette (Boston)
Allis-Chalmers (West Allis, Wis.)
Quaker Oats (Chicago)
Tosco (Los Angeles)
Scott Paper (Philadelphia)
INTERCO (St. Louis)
Williams Companies (Tulsa)
Kellogg (Battle Creek, Mich.)
J.P. Stevens (New York)
Marmon Group (Chicago)
Koppers (Pittsburgh)
Digital Equipment (Maynard, Mass.)
Squibb (New York)
McGraw-Edison (Elgin, Ill.)
National Distillers & Chemical (New York)
Cummins Engine (Columbus, Ind.)
SCM (New York)
Clark Equipment (Buchanan, Mich.)
Asarco (New York)
Revlon (New York)
Abbott Laboratories (North Chicago, Ill.)
Ethyl (Richmond, Va.)
Warner Communications (New York)
Times Mirror (Los Angeles)
Rohm & Haas (Philadelphia)
American Petrofina (Dallas)
Northrop (Los Angeles)
Emhart (Farmington, Conn.)
Crane (New York)
Murphy Oil (El Dorado, Ark.)
Allegheny Ludlum Industries (Pittsburgh)
Chromalloy American (St. Louis)
Stauffer Chemical (Westport, Conn.)
U.S. Gypsum (Chicago)
Upjohn (Kalamazoo, Mich.)
Sterling Drug (New York)
Anderson, Clayton (Houston)
Evans Products (Portland, Ore.)
Grumman (Bethpage, N.Y.)
Int'l Minerals & Chemicals (Northbrook, Ill.)
AMF (White Plains, N.Y.)
A.E. Staley Manufacturing (Decatur, Ill.)
Schering-Plough (Kenilworth, N.J.)
Corning Glass Works (Corning, N.Y.)
Geo. A. Hormel (Austin, Minn.)
Crown Cork & Seal (Philadelphia)
Oscar Mayer (Madison, Wis.)
Cooper Industries (Houston)
Union Camp (Wayne, N.J.)

Joseph E. Seagram & Sons (New York)
Polaroid (Cambridge, Mass.)
SmithKline (Philadelphia)
General Signal (Stamford, Conn.)
Lear Siegler (Santa Monica, Calif.)
Sunbeam (Chicago)
Heublein (Farmington, Conn.)
Louisiana-Pacific (Portland, Ore.)
Diamond International (New York)
Timken (Canton, Ohio)
Phelps Dodge (New York)
MCA (Universal City, Calif.)
Brunswick (Skokie, Ill.)
Wheeling-Pittsburgh Steel (Pittsburgh)
Air Products &Chemicals (Allentown, Pa.)
Westvaco (New York)
Commonwealth Oil Refining (San Antonio)
GAF (New York)
White Motor (Farmington Hills, Mich.)
Libbey-Owens-Ford (Toledo)
Black & Decker Manufacturing (Towson, Md.)
Sherwin-Williams (Cleveland)
Baxter Travenol Laboratories (Deerfield, Ill.)
Chesebrough-Pond's (Greenwich, Conn.)
Clark Oil & Refining (Milwaukee)
Baker International (Orange, Calif.)
Hershey Foods (Hershey, Pa.)
Great Northern Nekoosa (Stamford, Conn.)
St. Joe Minerals (New York)
Brown Group (St. Louis)
GK Technologies (Greenwich, Conn.)
Norton (Worcester, Mass.)
Cabot (Boston)
National Can (Chicago)
Interlake (Oak Brook, Ill.)
Richardson-Merrell (Wilton, Conn.)
Superior Oil (Houston)
Pennwalt (Philadelphia)
Hammermill Paper (Erie, Pa.)
Zenith Radio (Glenview, Ill.)
Whittaker (Los Angeles)
Gannett (Rochester, N.Y.)
Amstar (New York)
Crown Central Petroleum (Baltimore)
Blue Bell (Greensboro, N.C.)
Avnet (New York)
Pitney-Bowes (Stamford, Conn.)
Johnson Controls (Milwaukee)
National Gypsum (Dallas)
AMP (Harrisburg, Pa.)
West Point-Pepperell (West Point, Ga.)
Akzona (Asheville, N.C.)
Rexnord (Milwaukee)
Campbell Taggart (Dallas)
Liggett Group (Montvale, N.J.)
Lone Star Industries (Greenwich, Conn.)
G.D. Searle (Skokie, Ill.)
ACF Industries (New York)
Harris (Melbourne, Fla.)
Knight-Ridder Newspapers (Miami)
Universal Leaf Tobacco (Richmond, Va.)
Kaiser Steel (Oakland, Calif.)
Witco Chemical (New York)
Ex-Cell-O (Troy, Mich.)
R.R. Donnelley & Sons (Chicago)
Lever Brothers (New York)
Wheelabrator-Frye (Hampton, N.H.)
Harsco (Wormleysburg, Pa.)
Scovill (Waterbury, Conn.)
Cessna Aircraft (Wichita)
International Multifoods (Minneapolis)
Certain-teed (Valley Forge, Pa.)
Cyclops (Pittsburgh)
Jos. Schlitz Brewing (Milwaukee)
McGraw-Hill (New York)
Reichhold Chemicals (White Plains, N.Y.)
Alumax (San Mateo, Calif.)
Stanley Works (New Britain, Conn.)
Newmont Mining (New York)
MAPCO (Tulsa)

Amsted Industries (Chicago)
Federal Co. (Memphis)
Parker-Hannifin (Cleveland)
Sundstrand (Rockford, Ill.)
A.O. Smith (Milwaukee)
Springs Mills (Fort Mills, S.C.)
Square D (Palatine, Ill.)
Becton, Dickinson (Paramus, N.J.)
Sperry & Hutchinson (New York)
Champion Spark Plug (Toledo)
Hughes Tool (Houston)
Midland-Ross (Cleveland)
Fleetwood Enterprises (Riverside, Calif.)
Texasgulf (Stamford, Conn.)
Revere Copper & Brass (New York)
Louisiana Land & Exploration (New Orleans)
Tektronix (Beaverton, Ore.)
Bangor Punta (Greenwich, Conn.)
Anchor Hocking (Lancaster, Ohio)
Joy Manufacturing (Pittsburgh)
Southwest Forest Industries (Phoenix)
General Host (Stamford, Conn.)
AM International (Los Angeles)
Cincinnati Milacron (Cincinnati)
Vulcan Materials (Birmingham, Ala.)
Mohasco (Amsterdam, N.Y.)
Outboard Marine (Waukegan, Ill.)
Dayco (Dayton, Ohio)
Memorex (Santa Clara, Calif.)
Perkin-Elmer (Norwalk, Conn.)
Morton-Norwich Products (Chicago)
Masco (Taylor, Mich.)
McLouth Steel (Detroit)
Lubrizol (Wickliffe, Ohio)
Sybron (Rochester, N.Y.)
National Semiconductor (Santa Clara, Calif.)
Fairchild Industries (Germantown, Md.)
National Service Industries (Atlanta)
United Merchants & Manufacturers (New York)
Scott & Fetzer (Lakewood, Ohio)
Signode (Glenview, Ill.)
Quaker State Oil Refining (Oil City, Pa.)
A-T-O (Willoughby, Ohio)
Kane-Miller (Tarrytown, N.Y.)
Cone Mills (Greensboro, N.C.)
Cluett, Peabody (New York)
Norin (North Miami, Fla.)
Dover (New York)
Federal-Mogul (Southfield, Mich.)
Norris Industries (Long Beach, Calif.)
Trane (La Crosse, Wis.)
Twentieth Century-Fox Film (Los Angeles)
Sheller-Globe (Toledo)
General Cinema (Chestnut Hill, Mass.)
New York Times (New York)
Saxon Industries (New York)
Bemis (Minneapolis)
NVF (Yorklyn, Del.)
ConAgra (Omaha)
M. Lowenstein (New York)
H.K. Porter (Pittsburgh)
Belco Petroleum (New York)
CBI Industries (Oak Brook, Ill.)
Hobart (Troy, Ohio)
Handy & Harman (New York)
Hart Schaffner & Marx (Chicago)
Purex Industries (Lakewood, Calif.)
Thiokol (Newtown, Pa.)
Columbia Pictures Industries (New York)
U.S. Filter (New York)
Dow Corning (Midland, Mich.)
Macmillan (New York)
Cannon Mills (Kannapolis, N.C.)
Nashua (Nashua, N.H.)
Hoover Universal (Saline, Mich.)
Wallace Murray (New York)
Miles Laboratories (Elkhart, Ind.)
Peabody International (Stamford, Conn.)
Washington Post (Washington)

Ferro (Cleveland)
Briggs & Stratton (Wauwatosa, Wis.)
Eagle-Picher Industries (Cincinnati)
Bell & Howell (Chicago)
Insilco (Meriden, Conn.)
Brockway Glass (Brockway, Pa.)
Arcata (Menlo Park, Calif.)
Dan River (Greenville, S.C.)
Nalco Chemical (Oak Brook, Ill.)
DPF (Hartsdale, N.Y.)
Collins & Aikman (New York)
Avery International (San Marino, Calif.)
Faimont Foods (Houston)
Harnischfeger (Brookfield, Wis.)
Ball (Muncie, Ind.)
Stokely-Van Camp (Indianapolis)
Bucyrus-Erie (South Milwaukee, Wis.)
Pacific Resources (Honolulu)
Envirotech (Menlo Park, Calif.)
General Instrument (New York)
Smith International (Newport Beach, Calif.)
VF (Wyomissing, Pa.)
Masonite (Chicago)
American Bakeries (Chicago)
EG&G (Wellesley, Mass.)
Kellwood (St. Louis)
Tyler (Dallas)
Fieldcrest Mills (Eden, N.C.)
Big Three Industries (Houston)
Coca-Cola Bottling Co. of N.Y. (Hackensack, N.J.)
American Hoist & Derrick (St. Paul)
Data General (Westboro, Mass.)
Dean Foods (Franklin Park, Ill.)
Wm. Wrigley Jr. (Chicago)
Bausch & Lomb (Rochester, N.Y.)
Gerber Products (Fremont, Mich.)
United Refining (Warren, Pa.)
Gulf Resources & Chemical (Houston)
Mattel (Hawthorne, Calif.)
Copperweld (Pittsburgh)
Arvin Industries (Columbus, Ind.)
Varian Associates (Palo Alto, Calif.)
General Refractories (Bala Cynwyd, Pa.)
Maryland Cup (Owings Mills, Md.)
Freeport Minerals (New York)
Storage Technology (Louisville, Colo.)
Northwestern Steel & Wire (Sterling, Ill.)
Koehring (Brookfield, Wis.)
H.H. Robertson (Pittsburgh)
Foxboro (Foxboro, Mass.)
Sun Chemical (New York)
Carpenter Technology (Reading, Pa.)
Questor (Toledo)
McCormick (Hunt Valley, Md.)
Dexter (Windsor Locks, Conn.)
Harcourt Brace Jovanovich (New York)
Chicago Pneumatic Tool (New York)
Dennison Manufacturing (Framingham, Mass.)
Warnaco (Bridgeport, Conn.)
Dow Jones (New York)
Ideal Basic Industries (Denver)
Talley Industries (Mesa, Ariz.)
Barnes Group (Bristol, Conn.)
Nucor (Charlotte, N.C.)
Skyline (Elkhart, Ind.)
Beckman Instruments (Fullerton, Calif.)
Bunker Ramo (Oak Brook, Ill.)
Jonathan Logan (Secaucus, N.J.)
Westmoreland Coal (Philadelphia)
Royal Crown Companies (Atlanta)
Roper (Kankakee, Ill.)
Dorsey (Chattanooga, Tenn.)
Ceco (Chicago)
Federal Paper Board (Montvale, N.J.)
McDonough (Parkersburg, W. Va.)
Metromedia (Secaucus, N.J.)
Capital Cities Communications (New York)
Keystone Consolidated Industries (Peoria, Ill.)
Magic Chef (Cleveland, Tenn.)

Metric Conversion Tables

The metric system of measurement is used in virtually every country outside the United States and is increasingly used, especially in technical contexts, in the United States. In occupational research and writing it is often necessary, therefore, to be able to convert readily from the metric system to the U.S. system and vice versa. The following tables provide multipliers for converting both ways; the multipliers have been rounded to the third decimal place and thus yield an approximate equivalent.

Metric to U.S.			U.S. to Metric		
To Convert from:	To:	Multiply the Metric Unit by:	To Convert from:	To:	Multiply the U.S. Unit by:
Length					
kilometers	miles	.621	miles	kilometers	1.609
meters	yards	1.093	yards	meters	.914
meters	feet	3.280	feet	meters	.305
meters	inches	39.370	inches	meters	.025
centimeters	inches	.394	inches	centimeters	2.540
millimeters	inches	.039	inches	millimeters	25.400
Area and Volume					
square meters	square yards	1.196	square yards	square meters	.836
square meters	square feet	10.764	square feet	square meters	.093
square centimeters	square inches	.155	square inches	square centimeters	6.451
cubic centimeters	cubic inches	.061	cubic inches	cubic centimeters	16.387

Metric to U.S.   U.S. to Metric

To Convert from:	To:	Multiply the Metric Unit by:	To Convert from:	To:	Multiply the U.S. Unit by:
Liquid Measure					
liters	cubic inches	61.020	cubic inches	liters	.016
liters	cubic feet	.035	cubic feet	liters	28.339
liters	U.S. gallons*	.264	U.S. gallons*	liters	3.785
liters	U.S. quarts*	1.057	U.S. quarts*	liters	.946
milliliters	fluid ounces	.034	fluid ounces	milliliters	29.573
Weight and Mass					
kilograms	pounds	2.205	pounds	kilograms	.453
grams	ounces	.035	ounces	grams	28.349
grams	grains	15.430	grains	grams	.065

* The British imperial gallon equals approximately 1.2 U.S. gallons or 4.54 liters. Similarly, the British imperial quart equals 1.2 U.S. quarts, and so on.

Appendix F

Approved Abbreviations of Scientific and Engineering Terms

The approved abbreviations in this appendix are those for the more commonly used technical terms only. For a complete list of approved abbreviations, you should write to the Society for Technical Communication, 815 Fifteenth Street, N.W., Washington, D.C. 20005, and ask for *Abbreviations and Symbols for Terms Used in Electronics.* STC has been kind enough to permit us to reproduce an abridged version of this publication. (If you need an approved list of symbols, write to The American Society of Mechanical Engineers, 345 East 47th Street, New York, N.Y. 10017, for a copy of ANSI/IEEE Std 260-1978 *IEEE Standard Letter Symbols for Units of Measurement.*)

General Rules for Usage

1. Abbreviations and symbols are the short forms for words. Abbreviations are used in the U.S. customary measurement system; in the modernized metric system (called International System of Units or SI), short forms are known as symbols.
2. The abbreviation or symbol is the same for the singular and plural forms of a term (e.g., the abbreviation for "pounds" is "lb," not "lbs").
3. The abbreviation is the same for the nominal and adjectival forms of a term (i.e., hyphens are *not* inserted in abbreviations for compound adjectives).
4. Abbreviations are followed by periods only when the abbreviation forms a word

and could be misinterpreted. Symbols never have periods unless they are the last word in a sentence.

5. In general, any all-lowercase abbreviation may be written in capital letters in a headline, panel engraving, illustration, or other instance where the use of lower-case letters is difficult. However, short forms that are partly capital and partly lowercase (e.g., "pF," "mA," etc.) should always appear in that form.

6. Decimal prefixes should not be combined where a single prefix is available (e.g., use "nF" instead of "mμF").

7. Guides for usage are given in the "Class" column, according to the following legend:

A — Use freely in text, illustrations, and tables.

B — Use only when accompanied by numerals.

C — Use freely in illustrations and tables, but use sparingly in text.

D — Limit use to context where meaning has been firmly and explicitly established.

These guidelines are meant to apply to material addressed to a general electronics readership; in material intended for specialists, abbreviations that are part of the jargon of that specialty may be freely used. In any case, the guidelines are offered as a supplement to, and not as a substitute for, good editorial judgment.

Prefixes for Decimal Multiples and Submultiples of Units

Multiple	Prefix	Symbol
10^{12}	tera	T
10^9	giga	G
10^6	mega	M
10^3	kilo	k
10^2	hecto	h
10	deka	da
10^{-1}	deci	d
10^{-2}	centi	c
10^{-3}	milli	m
10^{-6}	micro	μ
10^{-9}	nano	n
10^{-12}	pico	p
10^{-15}	femto	f
10^{-18}	atto	a

Standard Abbreviations and Symbols for Terms Used in Electronics

Term	Abbreviation/Symbol	Class
absolute	abs	C
alternating current	ac	A
ambient	amb	C
ampere	A	B
ampere-hour	Ah	B
amplitude modulation	a-m	A
angstrom	Å	B
antilogarithm	antilog	A
approximate, -ly	approx	C
atmosphere	atm	B
audio frequency	af	C
automatic frequency control	afc	A
automatic gain control	agc	A
automatic volume control	avc	A
average	avg	C
bits per second	b/s	D
British thermal unit	Btu	A
calorie	cal	B
cathode-ray oscilloscope	CRO	A
cathode-ray tube	CRT	A
centi (10^{-2})	c	
centigram	cg	B
centimeter	cm	B
centimeter-gram-second	cgs	D
circular mil	cmil	B
clockwise	cw	C
continuous wave	cw	C
cosecant	csc	B
cosine	cos	B
cotangent	cot	B
coulomb	C	B
counterclockwise	ccw	C
cubic centimeter	cm^3	B
cubic foot	ft^3	B
cubic foot per minute	ft^3/min	B
cubic foot per second	ft^3/s	B
cubic inch	$in.^3$	B
cubic meter	m^3	B
cubic meter per second	m^3/s	B
cycle per second (see hertz)		
day	d	B
deka- (10)	da	
deci- (10^{-1})	d	
decibel	dB	B
degree Celsius	°C	B
degree Fahrenheit	°F	B
degree Kelvin (see kelvin)		
diameter	dia	C
direct current	dc	A

(Continued)

(Continued)

Term	Abbreviation/Symbol	Class
double-pole, double-throw	dpdt⁻	C
double-pole, single-throw	dpst	C
double sideband	dsb	A
dyne	dyn	B
electromotive force	emf	A
electrovolt	eV	B
equation	Eq	B
external	ext	C
extremely high frequency	ehf	C
extremely low frequency	elf	C
farad	F	B
Figure	Fig.	B
filament	fil	C
foot	ft	B
footlambert	fL	B
foot per minute	ft/min	B
foot per second	ft/s	B
foot poundal	ft-pdl	B
foot-pound force	ft-lbf	B
frequency	freq	C
frequency modulation	fm	A
gauss	G	B
giga- (10^9)	G	
gigaelectron volt	GeV	B
gigahertz	GHz	B
gilbert	Gb	B
gram	g	B
gravity	g	B
ground	gnd	C
hecto- (10^2)	h	
henry	H	B
hertz	Hz	B
high frequency	hf	C
horsepower	hp	B
hour	h	B
inch	in.	B
inch per second	in./s	B
infrared	IR	D
input/output	I/O	D
inside diameter	ID	B
integrated circuit	IC	A
intermediate frequency	i-f	A
internal	int	C
joule	J	B
joule per kelvin	J/K	B
kelvin	K	B
kilo- (10^3)	k	
kilogram	kg	B
kilogram per cubic meter	kg/m^3	B

(Continued)

(Continued)

Term	Abbreviation/Symbol	Class
kilohertz	kHz	B
kilohm	kΩ	B
kilometer	km	B
kilovolt	kV	B
kilovoltampere	kVA	B
kilowatt	kW	B
kilowatt-hour	kWh	B
lambert	L	B
light-emitting diode	LED	D
lines per minute	lines/min	D
liquid crystal display	LCD	D
liter	l	B
logarithm	log	A
logarithm, Napierian (base$_e$)	ln	B
low frequency	lf	C
lumen	lm	B
lux	lx	B
magnetomotive force	mmf	A
maximum	max	A
mean time to repair	MTTR	D
medium frequency	mf	C
mega- (10^6)	M	
megahertz	MHz	B
megavolt	MV	B
megawatt	MW	B
megohm	MΩ	B
meter	m	B
meter per second	m/s	B
meter per second squared	m/s^2	B
metric ton	t	B
micro- (10^{-6})	μ	
microampere	μA	B
microfarad	μF	B
microhenry	μH	B
microhm	μΩ	B
micrometer	μm	B
micromho	μmho	B
microsecond	μs	B
microvolt	μV	B
microwatt	μW	B
milli- (10^{-3})	m	
milliampere	mA	B
millibar	mbar	B
milligram	mg	B
millihenry	mH	B
millimeter	mm	B
millimho	mmho	B
millimicron (see nanometer)		
milliohm	mΩ	B

(Continued)

(Continued)

Term	Abbreviation/Symbol	Class
millisecond	ms	B
millivolt	mV	B
milliwatt	mW	B
minimum	min	D
minute	min	B
mole	mol	
nano- (10^{-9})	n	
nanoampere	nA	B
nanofarad	nF	B
nanohenry	nH	B
nanometer	nm	B
nanosecond	ns	B
nanowatt	nW	B
negative	neg	C
newton	N	B
number	No.	B
oersted	Oe	B
ohm	Ω	B
original equipment manufacturer	OEM	D
ounce	oz	B
outside diameter	OD	B
pascal	Pa	B
peak	pk	B
peak inverse voltage	PIV	C
pico- (10^{-12})	p	
picoampere	pA	B
picofarad	pF	B
picosecond	ps	B
picowatt	pW	B
point-of-sale	POS	D
poise	P	B
positive	pos	C
positive-negative-positive (semiconductor)	pnp	A
potentiometer	pot	C
pound	lb	B
poundal	pdl	B
pound per square foot	lb/ft^2	B
pound per square inch	lb/in.^{2}	B
power factor	PF	D
printed circuit	PC	D
private branch exchange	PBX	D
pulse-amplitude modulation	PAM	D
pulse-code modulation	PCM	D
pulse-duration modulation	PDM	D
pulse per second	p/s	B
pulse-position modulation	PPM	D
pulse repetition frequency	PRF	A
pulse repetition rate	PRR	D
pulse-width modulation	PWM	D
radian	rad	B

(Continued)

(Continued)

Term	Abbreviation/Symbol	Class
radian per second	rad/s	B
radian per second squared	rad/s^2	B
radio frequency	rf	A
receiver	rcvr	C
reference	ref	C
roentgen	R	B
root-mean-square	rms	A
secant	sec	B
second	s	B
sensitivity	sens	C
siemens	S	B
signal to noise ratio	SNR	D
sine	sin	B
single-pole double-throw	spdt	C
single-pole single-throw	spst	C
single sideband	ssb	D
small-scale integration	SSI	D
square	sq	C
square foot	ft^2	B
square inch	in.^{2}	B
square meter	m^2	B
square meter per second	m^2/s	B
standard	std	C
steradian	sr	D
super high frequency	shf	C
synchronous, synchronizing	sync	A
tangent	tan	B
teletypewriter	TTY	D
teletypewriter exchange	TWX	D
television	TV	A
television interference	TVI	D
temperature	temp	C
temperature coefficient	TC	D
tera- (10^{12})	T	
terahertz	THz	B
tesla	T	B
transistor voltmeter	TVM	D
transmit-receive	T-R	D
ultra high frequency	uhf	A
ultraviolet	UV	D
vacuum tube voltmeter	VTVM	C
variable-frequency oscillator	VFO	C
very high frequency	vhf	A
very low frequency	vlf	A
volt	V	B
volt ohmeter	VOM	D
voltampere	VA	B
watt	W	B
watthour	WH	B
words per minute	words/min	B

Index

(Note: Page numbers in italics refer to illustrations and examples.)